Understanding Communications Systems Principles— A Tutorial Approach

RIVER PUBLISHERS SERIES IN COMMUNICATIONS

Indexing: All books published in this series are submitted to the Web of Science Book Citation Index (BkCI), to SCOPUS, to CrossRef and to Google Scholar for evaluation and indexing.

The "River Publishers Series in Communications" is a series of comprehensive academic and professional books which focus on communication and network systems. Topics range from the theory and use of systems involving all terminals, computers, and information processors to wired and wireless networks and network layouts, protocols, architectures, and implementations. Also covered are developments stemming from new market demands in systems, products, and technologies such as personal communications services, multimedia systems, enterprise networks, and optical communications.

The series includes research monographs, edited volumes, handbooks and textbooks, providing professionals, researchers, educators, and advanced students in the field with an invaluable insight into the latest research and developments.

For a list of other books in this series, visit www.riverpublishers.com

Understanding Communications Systems Principles—A Tutorial Approach

Héctor J. De Los Santos

NanoMEMS Research, LLC
Irvine, CA, USA

River Publishers

Published, sold and distributed by:
River Publishers
Alsbjergvej 10
9260 Gistrup
Denmark

www.riverpublishers.com

ISBN: 978-87-7022-375-1 (Hardback)
978-87-7022-374-4 (Ebook)

Este libro lo dedico a mis queridos padres
y a mis queridos Violeta, Mara, Hectorcito,
y Joseph.

Héctor J. De Los Santos

"Y sabemos que a los que aman a Dios todas las cosas les ayudan a bien, esto es, a los que conforme a su propósito son llamados."

"And we know that all things work together for good to them that love God, to them who are the called according to his purpose."

Romanos 8:28

Contents

Preface xv

Acknowledgements xvii

List of Figures xix

List of Tables xxxv

List of Abbreviations xxxvii

1 Introduction to Wireless Communications and Sensing Systems **1**
1.1 Scientific Beginnings: Electromagnetic Waves 1
1.1.1 Generation and Detection of EM Waves 1
1.1.1.1 Ruhmkorff Coil and Spark Gap 1
1.1.1.2 Hertz's Transmitter 2
1.1.1.3 Hertz's Receiver 5
1.1.1.4 Hertz's Experiment 6
1.1.1.5 Hertz's Analysis of the Interference Pattern 7
1.2 Engineering Beginnings: Communications and RADAR . . . 9
1.2.1 Communications 9
1.2.1.1 Communications Systems 10
1.2.1.1.1 Simplified Transmitter Building Block 11
1.2.1.1.2 Simplified Receiver Building Block 12
1.2.2 RADAR 13
1.2.2.1 RADAR Systems 13
1.2.2.2 Simplified RADAR System Building Block 14
1.3 Fundamentals of Signal Processing 15
1.3.1 Mathematical Description of Carrier Modulation . . 15
1.3.1.1 Amplitude Modulation 16
1.3.1.2 Frequency Modulation 16

1.3.1.3 Phase Modulation 16
1.3.2 Spectral Properties of Basic Modulation Approaches 16
1.3.2.1 AM Spectrum 17
1.3.2.2 FM Spectrum 18
1.3.2.3 Comparing AM and FM Spectra 21
1.3.2.4 Wideband FM 22
1.3.3 Phase Modulation Spectrum 23
1.4 Fundamentals of Information Theory 24
1.5 Summary 26
1.6 Problems 27

2 Wireless Systems Building Blocks 29
2.1 System Components and Their Performance Parameters . . . 29
2.1.1 Transmission Lines 29
2.1.2 Amplifiers 30
2.1.2.1 Gain Compression and Desensitization . . 31
2.1.2.2 Cross-Modulation 33
2.1.2.3 Intermodulation 33
2.1.2.4 Memoryless Bandpass Nonlinearities . . . 34
2.1.3 Mixers 35
2.1.4 Filters 40
2.1.5 Oscillators 40
2.1.5.1 Phase Noise of a Local Oscillator 42
2.1.5.2 Amplitude Noise 44
2.1.6 Frequency Multipliers 44
2.2 Antennas 45
2.2.1 Description of Antennas and Their Parameters . . . 45
2.2.2 Antenna Arrays [23] 51
2.2.2.1 Array Factor 52
2.2.2.2 Antenna Array Directivity 55
2.2.2.3 Antenna Array Factor 55
2.2.2.4 Prototypical Phased Array Antenna 58
2.3 Free Space Propagation Model 58
2.4 Summary 63
2.5 Problems 64

3 Communication Systems Performance Parameters 65
3.1 Introduction 65
3.2 Transmitter Performance Parameters 65

3.2.1 Modulation Accuracy 66
3.2.2 Adjacent and Alternate Channel Power 67
3.3 Receiver 67
3.3.1 Sensitivity 68
3.3.2 Noise Figure 68
3.3.3 Selectivity 68
3.3.4 Receiver Image Rejection 69
3.3.5 Receiver Dynamic Range 69
3.3.6 Receiver Spurious-Free Dynamic Range 69
3.4 Sensitivity and Dynamic Range Parameters 70
3.4.1 Definition of Receiver Sensitivity 72
3.4.2 Definition of Minimum Detectable Signal 72
3.4.3 Illustration of Signal-to-Noise Ratio 73
3.4.4 Definition of 1-dB Compression Point 75
3.4.5 Definition of Intermodulation Distortion 77
3.4.6 IP3 for Cascade of Networks 82
3.5 Definition of Dynamic Range 83
3.5.1 Noise Figure of Blocks in Cascade 84
3.5.2 Spur-Free Dynamic Range 85
3.6 Circuit Signal-to-Noise Ratio 88
3.6.1 Definition of Available Noise Power 89
3.6.2 Network Noise Figure 89
3.6.3 Single-Frequency (Spot) Noise Figure 90
3.6.4 Equivalent Noise Temperature 91
3.6.5 Effective Noise Temperature of a Network 91
3.6.6 Computing the Overall NF of Cascaded Circuits . . 93
3.6.7 Noise Figure of a Mixer 95
3.7 Summary 98
3.8 Problems 99

4 Circuit Topologies for Signal Modulation and Detection 101
4.1 Introduction 101
4.2 AM Modulation Approaches 101
4.2.1 Generation of Single-Sideband AM Signals 102
4.3 AM Demodulation Approaches 103
4.3.1 Envelope Detector 103
4.4 FM Approaches 104
4.4.1 Direct FM Modulator 104
4.5 FM Demodulation Approaches 105

4.5.1 FM Demodulation by Phase-Locked Loop 106
4.6 The Digital Modulation Technique 106
4.6.1 Amplitude-Shift Keying Modulation 107
4.6.2 Frequency-Shift Keying Modulation 107
4.6.3 Phase-Shift Keying Modulation 108
4.7 Modulation Signal Representation by Complex Envelope Form 108
4.7.1 M-ary Modulation—MPSK 109
4.7.2 Binary Phase Shift Keying Modulation—BPSK . . . 110
4.7.3 Quadrature Phase Shift Keying Modulation—QPSK 112
4.7.3.1 Modulator Circuit for QPSK 113
4.7.3.2 Circuit for QPSK Demodulation 114
4.7.4 Binary Frequency-Shift Keying Modulation Circuit . 115
4.7.4.1 Circuit for BFSK Modulation 116
4.7.4.2 BFSK Demodulation via a Coherent Detector 116
4.7.4.3 BFSK Demodulation via a Noncoherent Detector 116
4.7.5 M-ary Quadrature Amplitude Modulation Approach 117
4.7.6 Orthogonal Frequency Division Multiplexing 117
4.7.7 Direct Sequence Spread Spectrum Modulation Approach 118
4.7.7.1 Modulation and Demodulation Circuits for Direct Sequence Spread Spectrum (DS/SS) 118
4.7.8 Frequency Hopping Spread Spectrum Modulation/ Demodulation 120
4.8 Summary 121
4.9 Problems 121

5 Transmitter and Receiver Architectures **123**
5.1 Introduction 123
5.2 The Transmitter 124
5.2.1 Heterodyne Transmitter Architecture 124
5.2.2 The Homodyne Transmitter Architecture 126
5.2.2.1 Drawbacks of Homodyne transmitter architecture 127
5.2.2.1.1 LO disturbance and its corrections 127
5.3 The Heterodyne Receiver Architecture 130

5.4 The Homodyne (Zero IF/Direct-Conversion) Receiver 131
5.5 Receiver Architectures in Light of 5G [37] 133
5.5.1 Super-Heterodyne Receiver 135
5.5.2 Homodyne Receiver 137
5.5.3 The Low-IF Receiver 138
5.5.4 The Software-Defined Receiver 140
5.6 Summary 141
5.7 Problems 141

6 5G **143**
6.1 Introduction 143
6.2 5G Systems Technologies 146
6.2.1 5G Systems: mmWaves [44] 147
6.2.1.1 Propagation issues 147
6.2.1.2 Blocking 147
6.2.1.3 Atmospheric and rain absorption 148
6.2.1.4 Large arrays, narrow beams 148
6.2.1.5 Link acquisition 148
6.3 5G: Internet of Things [46, 47] 148
6.3.1 Device-to-Device Communications 149
6.3.2 Simultaneous Transmission/Reception (STR) 149
6.4 Non-Orthogonal Multiple Access [45, 49] 149
6.4.1 NOMA Approaches 150
6.5 5G Evolution 151
6.6 Summary 152
6.7 Problems 152

7 MIMO **153**
7.1 Introduction 153
7.2 The SISO Channel 153
7.2.1 The SISO Channel Model 153
7.2.2 The SISO Channel Capacity 155
7.3 The MIMO Channel Model 159
7.3.1 MIMO Channel Propagation Models [13, 60, 61] . . 161
7.3.1.1 The rayleigh distribution model 161
7.3.1.2 The Ricean distribution model 161
7.3.1.3 The Nakagami-m distribution model . . . 162
7.3.2 The Singular Value Decomposition Approach [62, 63] 162
7.3.2.1 The mechanics of the SVD approach . . . 165

7.3.2.2 MIMO interpretation of SVD example . . 170
7.4 MIMO Transmit Antenna Input Power Optimization 171
7.5 MIMO Receive Antenna Signal Processing 174
7.5.1 MIMO Array Gain 174
7.5.2 MIMO Diversity Gain 174
7.6 Massive MIMO Detection and Transmission 174
7.6.1 Massive MIMO Detection: MRC, ZFBF, and MMSE 175
7.6.2 Massive MIMO Transmission: Precoding 179
7.7 Massive MIMO Systems Architectures 180
7.8 Massive MIMO Limiting Factors 184
7.8.1 Pilot Contamination 184
7.8.2 Radio Propagation 184
7.9 Summary 185
7.10 Problems 185

8 Aerospace/Electronic Warfare RADAR 187
8.1 Introduction 187
8.2 Principles of RADARs [92–95] 188
8.2.1 Types of RADAR 188
8.2.2 Radio Detection and Ranging [93] 189
8.2.3 RADAR-Target Geometry/Coordinate System 192
8.2.4 RADAR Pulses 193
8.2.5 Range Ambiguities 194
8.2.6 Range Resolution 194
8.2.7 Range Gates 195
8.2.8 RADAR Sensitivity 197
8.2.9 Doppler Shift 200
8.2.10 Track Versus Search 202
8.2.11 RADAR Cross Section 203
8.3 RADAR Architectures 204
8.3.1 CW Doppler RADAR Architecture 205
8.3.2 FM-CW RADAR Architecture 205
8.3.3 Pulse Doppler RADAR Architecture 205
8.4 ECM Capabilities of an EW RADAR 207
8.4.1 Searching for Signal Sources 207
8.4.2 ECM Techniques: Jamming 208
8.4.2.1 Noise jamming 208
8.4.2.2 Deception jamming 209
8.4.3 ECCM Techniques 211

8.4.3.1 Pulse compression 211
8.4.3.2 Frequency hopping 212
8.4.3.3 Side lobe blanking 212
8.4.3.4 Polarization 212
8.4.3.5 Artificial-Intelligence-Based Jammer-Nulling 213
8.5 Summary 214
8.6 Problems 214

9 Tutorials **215**
9.1 Introduction 215
9.2 Tutorial 1: Introduction to SystemVue 216
9.2.1 Preliminaries 216
9.2.2 Getting Started and Schematic Window 218
9.2.2.1 Implementation of Basic Phased Array (Beamforming) Antenna Model 218
9.2.2.2 Running the Workspace file 5G_MIMO_Beamforming_ULA_1 x 4.wsv. 224
9.2.2.3 Effect of Number of Elements on ULA Directivity 230
9.2.2.4 Element Antenna Radiation Pattern 232
9.3 Tutorial 2: Codebook Design for 28GHz 5G/MIMO Antenna Array Transmission 234
9.3.1 Preliminaries 234
9.3.2 Determination of Codebook for 12 × 12 MIMO URA 238
9.4 Tutorial 3: Electronic/Warfare RADAR Performance 245
9.4.1 Preliminaries: Transmitter–Receiver Simulation . . 246
9.4.2 FM-CW RADAR Model and Simulations 247
9.4.3 Exercises 253

Bibliography **255**

Index **263**

About the Author **267**

Preface

This book addresses the principles of communications and sensing systems. In particular, the development of the field from its earliest scientific beginnings, and then its engineering beginnings, is given. The last three chapters of the book deal with current topics under development, namely, the fifth generation of cell phone wireless communications or "5G," the topic of multiple-input multiple-output communications systems or MIMO architectures, and aerospace/electronic warfare RADAR.

Understanding Communications Systems Principles contains nine chapters. Chapter 1 provides an introduction to wireless communications and sensing, in particular, how curiosity-driven scientific research led to the foundation of the field, including the properties of electromagnetic waves and their mathematical theory, and the vision that led to their utilization for wireless communications and the use of their reflecting property for effecting long-distance sensing. The fundamentals of "information theory," the technical discipline that deals with the topic of maximizing the rate of transmitting information in the presence of noise, is also introduced. Chapter 2 presents a brief introduction to the building blocks that make up wireless systems, i.e., antennas, including beamforming arrays, interconnect elements, and circuits, and the fundamentals of wave propagation in free space. Chapter 3 focuses on developing an understanding of the performance parameters that characterize a wireless system so as to enable one to gauge their quality, when comparing among systems, and how to determine them. Chapter 4 deals briefly with circuit topologies for modulation and detection. That is, how to configure circuits to effect signal modulation, in preparation for transmission, and its detection or demodulation, i.e., how to reverse the modulation process so that information may be extracted. Chapter 5 deals with the fundamental transmitter and receiver system architectures that enable the transmission of information at precise frequencies and their reception from among a rather large multitude of other signals present in space. Chapter 6 introduces 5G, its motivation, and its development and adoption challenges for providing unprecedented levels of highest speed wireless connectivity. Chapter 7 takes on the topic of MIMO, its justification, its various architectures, and some approaches to dealing with the complexity introduced by transmitting and receiving platforms containing potentially thousands of antennas for maximizing information transmission capacity. Chapter 8 addresses the topic of aerospace/electronic warfare RADAR, in particular, the types of RADAR in existence, their elements pertaining to their engineering to measure distance and velocity, and the various "jamming" techniques to impair their performance by the enemy. Chapter 9, finally, presents three tutorials utilizing the SystemVue simulation tool. Tutorial 1 introduces SystemVue and demonstrates its use for modeling the fundamental uniform linear array (ULA) beamforming antenna and the uniform rectangular array (URA) antenna. Tutorial 2 deals with the concept of codebooks and how to determine them using SystemVue. And Tutorial 3 presents a demonstration, via modeling, of the impact of various types of jamming on the ability of a RADAR assumed to be placed in a fighting aircraft and tasked with measuring the distance and velocity of an enemy fighter.

The book assumes a background in electrical engineering and an interest in communications and sensing systems. It is amenable for use by practicing communication systems engineers interested in emerging 5G, MIMO, and aerospace/electronic warfare RADAR. The book may also be used by experienced electrical engineers in other areas, who wish to transition into communications and sensing systems.

Acknowledgements

The author hereby thanks for, and gratefully acknowledges, the opportunity given to him by Mr. Rajeev Prasad, of River Publishers, to write this book. The author also gratefully acknowledges the smooth interactions with Ms. Junko Nakajima during the various production stages of the book. He further thanks Ms. Stacy K. Johnson, former University Program Manager, Pathwave Design Software and Dr. Ian Rippke, ESL Segment Product Manager, DES, EEsof, both of Keysight Technlogies, for their sustained interest in the project, and Mr. Casey Latham, University Program Manager, PathWave Software, also of Keysight, who joined the project in its last stages, for his interest and for his excellent proofreading of the manuscript. At Keysight's Technical Support, the author acknowledges the excellent assistance received from Mr. Bruce Fisher, on operating/running the SystemVue software tool, and Mr. Bing Lin, on implementing the Tutorials. The author also thanks Prof. Giovanni Miano, Università degli Studi di Napoli Federico II, Italy, and Prof. David Jenn, Naval Postgraduate School, Monterey, CA, for granting permission to use their lecture materials. Finally, the author gratefully acknowledges his wife, Violeta, for her excellent assistance preparing many of the figures along the course of the project.

Héctor J. De Los Santos

List of Figures

Figure 1.1	Details of the Ruhmkorff induction coil. *Source:* Ref. [5].	2
Figure 1.2	(a) A schematic for Hertz's transmitter showing the details for the Ruhmkorff induction coil. (b) A circuit diagram for Hertz's transmitter. *Source:* Ref. [4].	3
Figure 1.3	The waveforms for the Ruhmkorff induction coil. (a) Current I_p flowing in the primary circuit and (b) voltage Vg across the spark gap in the secondary circuit. Note that the vertical dashed line separates the times when the interrupter is closed ($t_1 < t < t_2$) from the times when it is open ($t > t_2$) and that the spark gap breaks down at time t_3. *Source:* Ref. [34.	3
Figure 1.4	(a) The current in the dipole antenna after the spark gap breaks down $t > t_3$ for a single cycle of the interrupter. The inset at the top shows the charge oscillating between the end plates creating the current. (b) An expanded scale showing the current for two cycles of the interrupter spaced by the time T_i. *Source:* Ref. [4].	4
Figure 1.5	The circular loop antenna Hertz used as his receiver. The inset shows the details for the spark gap in the loop. The length of the gap, l_g, can be adjusted using the screw, and the capacitance C_L across the gap can be varied by changing the size and the spacing of the two metal plates. The electric field applied to the loop is approximately uniform and parallel to the plane of the loop ($x-y$ plane). For Hertz's loop, the circumference is $0.37\lambda_d$. *Source:* Ref. [4]. . . .	5

Figure 1.6 (a) A schematic for the basic elements of the experiment. (b) The interference pattern (standing wave) for the rms value of the total electric field. The incident and reflected fields are assumed to be plane waves within the measurement volume. Note that the nodes are spaced by $\lambda/2$ starting at the reflector. *Source:* Ref. [4]. 7
Figure 1.7 A schematic for the apparatus used by Hertz for his experiment on the interference of electromagnetic waves in air. *Source:* Ref. [4]. 8
Figure 1.8 (a) A figure from Hertz's 1888 article. The figure has been redrawn to show only the measured electric field. The spatial zones are divided by internals named A–D. The curves are for time instants separated by one-half period of the oscillation. (b) The interference pattern obtained from (a). *Source:* Ref. [4]. 9
Figure 1.9 Communication system. *Source:* Ref. [13]. 11
Figure 1.10 Simplified block diagram of transmitter. 12
Figure 1.11 Simplified block diagram of receiver. 13
Figure 1.12 Simplified scenario of RADAR application. *Source:* Ref. [14]. 14
Figure 1.13 Simplified RADAR architecture. 15
Figure 1.14 Spectrum of single-frequency AM signal. The delta function $\pi\delta(\omega - p_1)$ represents the spectrum of the baseband signal. $\pi a_0 m\delta(\omega - \omega_0)$, together with $\frac{\pi}{2}a_0 m\delta(\omega - \omega_0 + p_1)$ and $\frac{\pi}{2}a_0 m\delta(\omega - \omega_0 - p_1)$, represents the modulated, the lower, and the upper sidebands, respectively. *Source:* Ref. [15]. 17
Figure 1.15 Spectrum of amplitude modulated baseband signal. *Source:* Ref. [15]. 19
Figure 1.16 Amplitude of FM sidebands (Bessel coefficients) as function of modulation index, M_p. p is the order of the Bessel function (0–7). *Source:* Ref. [33]. . . . 23
Figure 1.17 Communication system. *Source:* Ref. [13]. 25
Figure 1.18 Quantization of a signal. (a) Original signal. (b) Quantized signal. *Source:* Ref. [16]. 26
Figure 2.1 Transmission line parameters. (a) Ideal TL model. (b) Physical TL model. 30

Figure 2.2 Depiction of 1-dB compression point. *Source:* Ref. [33]. 32
Figure 2.3 Characterization of Class AB power amplifier. *Source:* Ref. [18]. 35
Figure 2.4 AM/PM conversion. *Source:* Ref. [18]. 35
Figure 2.5 Parameters describing amplifier model. 36
Figure 2.6 Transfer characteristic of mixer's IF output power vs. input power. The bold numbers indicate the following properties: **1**. At 0 dBm input, the output is 6 dBm, indicating a 6-dB conversion gain. **2**. At this input level, the third-order two-tone intermodulation product is 30 dB below the desired output. **3**. At a higher input value, the 3-dB compression point is indicated. **4**. At still higher input level, the intercept point is shown where the projected curves of desired output and third-order intermodulation product intersect. *Source:* Ref. [33]. 38
Figure 2.7 Spurious frequency products in mixers. The bold numbers indicate the following: **1**. The image frequency $f_{im}=f_{LO}+f_{IF}$. If a signal at this frequency is picked up by the antenna, and if it reaches the mixer input, it will beat with f_{LO} to produce a difference-frequency component equal to f_{IF}. **2**. An input signal at f_{IF} will appear in the output due to normal amplifier action. **3**. An input frequency equal to $f_{RF}/2$ may be doubled to f_{RF} by the square-law mixer term and then combine with f_{LO} to produce output at f_{IF}. **4**. An input at $f_{IF}/2$ may be doubled by the mixer and appear in the output. **5**. If the LO output includes a second harmonic at $2f_{LO}$, or if the mixer generates $2f_{LO}$, this component can beat with received inputs at $(2f_{LO} \pm f_{IF})$ to produce output at f_{IF}. *Source:* Ref. [19]. 38
Figure 2.8 (a) Theoretical and (b) practical filter responses. *Source:* Ref. [33]. 41
Figure 2.9 Model of typical bandpass filter. 41
Figure 2.10 Impact of local oscillator noise on the noise of a down-converted RF signal. *Source:* Ref. [18]. . . . 42

Figure 2.11 Manifestations of oscillator phase noise. (a) Oscillator amplitude in the time domain (jitter). (b) Oscillator amplitude in the frequency domain. (c) Impact of timing jitter on position of in-phase (I) and quadrature components (Q) of binary phase-shift-keyed (BPSK) constellation. (d) Impact of timing jitter on quadrature phase-shift-keyed (QPSK) constellation. *Source:* Ref. [18]. 43
Figure 2.11(e) Oscillator model. 43
Figure 2.12 Phase noise enhancement by frequency multipliers. *Source:* Ref. [18]. 45
Figure 2.13 Frequency multiplier model. 45
Figure 2.14 Antenna as a transition region between guided wave and free-space wave. (a) Transmitter antenna. (b) Receiver antenna. *Source:* Ref. [20]. 46
Figure 2.15 Typical antennas used in wireless communications. *Source:* Ref. [20]. 46
Figure 2.16 Normalized antenna power pattern P_n with maximum aligned with the $\theta = 0$ direction (zenith). *Source:* Ref. [21]. 47
Figure 2.17 Radiation from aperture A_e with uniform field E_a. *Source:* Ref. [21]. 50
Figure 2.18 Visualization of gain, G, and directivity, D, relationships of an antenna. *Source:* Ref. [20]. . . . 51
Figure 2.19 (a) A simple two-element antenna array. (b) A collinear array of current elements. (c) A linear array of current elements. *Source:* Ref. [23]. 52
Figure 2.20 Array factor as a function of ψ, illustrating the relation between ψ and ϕ where $\pi \leq \phi \leq 0$. *Source:* Ref. [23]. 57
Figure 2.21 (a) Array factor for a uniform five-element array. (b) A factor as a polar plot of ϕ. Note grating lobe at approximately ϕ= 120° brought in by large spacing. *Source:* Ref. [23]. 57
Figure 2.22 Array factors for five-element end-fire array. In the end-fire array, $\alpha > \beta d$ has the peculiarity that it results in a sharper main beam at the expense of higher relative side lobe level. *Source:* Ref. [23]. . 59
Figure 2.23 Prototypical architecture for uniform linear antenna. 59

Figure 2.24 Linear electric field radiator. *Source:* Ref. [33]. . . 62
Figure 2.25 Sphere depicting radiating source and unit area on a sphere of radius d, receiving the power density. *Source:* Ref. [33]. 62
Figure 2.26 Equivalent circuit of receiving antenna. *Source:* Ref. [33]. 63
Figure 3.1 Constellation distortion due to noise. Example of how ideal orthogonal quadrature phase-shift keying (OQPSK) constellation may be distorted en-route from transmitter to receiver. 66
Figure 3.2 Description of adjacent channel power. *Source:* Ref. [33]. 67
Figure 3.3 Filter response used in defining the shape factor. For receivers, the passband is taken to be the bandwidth between the points where the response has fallen by 6 dB, i.e., where it is 6 dB down or −6 dB.*Source:* Ref. [33]. 68
Figure 3.4 Illustration of image problem. *Source:* Ref. [33]. . 69
Figure 3.5 Power transfer and power spectral relationships illustrating SFDR concept. *Source:* Ref. [28]. . . . 70
Figure 3.6 Homodyne (zero-IF) receiver architecture. *Source:* Ref. [28]. 71
Figure 3.7 Heterodyne receiver. *Source:* Ref. [28]. 71
Figure 3.8 System-level receiver parameters. F is the system's noise factor. kT_0 x (1 Hz) is the thermal noise power in a 1-Hz bandwidth at T_0, B is the noise equivalent bandwidth, and G is the two-port's numeric available gain. 72
Figure 3.9 Illustration of minimum detectable signal. *Source:* Ref. [28]. 74
Figure 3.10 Illustration of SNR/CNR. *Source:* Ref. [28]. 74
Figure 3.11 Probability of error versus carrier-to-noise ratio for various modulation schemes. *Source:* Ref. [28]. . . 75
Figure 3.12 General output power versus input power characteristic. *Source:* Ref. [28]. 76
Figure 3.13 Gain versus input power. *Source:* Ref. [28]. 76
Figure 3.14 Processing of signals at frequencies f_1 and f_2 by an amplifying linear network. *Source:* Ref. [28]. . . . 77

Figure 3.15 Processing of signals at frequencies f_1 and f_2 by a nonlinear network. *Source:* Ref. [28]. 78
Figure 3.16 Output power P_{out} for one fundamental frequency signal versus input power P_{in}. *Source:* Ref. [28]. . 79
Figure 3.17 (a) Fundamental P_o and third-order distortion P_d output powers versus input power P_{in}. (b) Output frequency spectrum at a fixed input power level. *Source:* Ref. [28]. 80
Figure 3.18 Typical output spectrum of (a) linear receiver and (b) nonlinear receiver. *Source:* Ref. [28]. 82
Figure 3.19 Receiver consisting of cascaded networks. *Source:* Ref. [28]. 82
Figure 3.20 Dynamic range definition. *Source:* Ref. [28]. . . . 83
Figure 3.21 Cascaded noise figure. *Source:* Ref. [28]. 84
Figure 3.22 Receiver blocking. *Source:* Ref. [28]. 85
Figure 3.23 Sketch of spur-free dynamic range. *Source:* Ref. [28]. 86
Figure 3.24 (a). Receiver block diagram. (b) Six-stage network combined into three-stage network. *Source:* Ref. [28]. 87
Figure 3.25 Each stage of a system adds noise to the signal passing through. 89
Figure 3.26 Model for addition of network's own power to input noise P_{ns}. *Source:* Ref. [19]. 92
Figure 3.27 Noisy circuit modeled by its noise referred to the input plus noiseless circuit. *Source:* Ref. [19]. . . . 92
Figure 3.28 Cascaded building blocks. *Source:* Ref. [19]. . . . 93
Figure 3.29 (a) Mixer symbol. (b) Mixing spectrum including image. *Source:* Ref. [19]. 95
Figure 3.30 Noise scenarios in mixers. (a) Single-sideband. (b) Double-sideband. *Source:* Ref. [29]. 95
Figure P3.1 Receiver sensitivity calculation. 99
Figure P3.2 Overall receiver performance calculation. 99
Figure P3.3 Simplified receiver front end. 100
Figure 4.1 AM signal spectra. The baseband signal is centered at zero frequency and contains the maximum frequency ω_m. *Source:* Ref. [33]. 102
Figure 4.2 Illustration of SSB AM by the filtering approach. *Source:* Ref. [30]. 103

Figure 4.3 Illustration of the balanced modulator approach to SSB AM signal. *g(t)* is the modulation signal. *Source:* Ref. [30]. 103

Figure 4.4 Envelope detector circuit plus low-pass filter. *Source:* Ref. [32]. 104

Figure 4.5 Voltage-controlled oscillator as FM modulator. *Source:* Ref. [30]. 105

Figure 4.6 FM demodulation by phase-locked loop. *Source:* Ref. [30]. 106

Figure 4.7 Amplitude-shift keying. (a) Baseband signal represented by stream of bits. (b) Corresponding modulation of carrier amplitude with (a). *Source:* Ref. [31]. 107

Figure 4.8 Frequency-shift-keying. (a) Baseband signal represented by stream of bits. (b) Corresponding modulation of carrier frequency. *Source:* Ref. [31]. 107

Figure 4.9 Phase-shift keying. (a) Binary representation of modulating baseband signal. (b) Corresponding modulation of carrier frequency. *Source:* Ref. [31]. 108

Figure 4.10 Constellation diagram for BPSK. *I: in-phase*; *Q: quadrature*. 111

Figure 4.11 Power spectral density for BPSK. The bandwidth, $BW=2R_b=2/T_b$ *Source:* Ref. [25]. 112

Figure 4.12 Constellations for QPSK. 113

Figure 4.13 Power spectral density for QBPSK. The bandwidth, $\mathrm{BW}=\mathrm{R}_b$. *Source:* Ref. [25]. 114

Figure 4.14 Modulator schematic for QPSK. *Source:* Ref. [25]. 114

Figure 4.15 Circuit for QPSK demodulation. *Source:* Ref. [25]. 115

Figure 4.16 BFSK demodulator implemented by coherent detector. *Source:* Ref. [25]. 116

Figure 4.17 BFSK demodulation via a noncoherent detector. *Source:* Ref. [25]. 117

Figure 4.18 Constellation of M-ary QAM (M=16). 118

Figure 4.19 (a) Schematic of OFDM modulator. (b) Carriers modulated by rectangularly shaped data pulses are densely spaced and orthogonal in frequency. *Source:* Ref. [30]. 118

Figure 4.20 Concept of direct sequence spread spectrum modulation. *Source:* Ref. [33]. 119

Figure 4.21 Direct sequence spread spectrum. (a) Modulator. (b) Demodulator. (c) PN sequence generator diagram. *Source:* Ref. [33]. 119

Figure 4.22 Frequency hopping spread spectrum approach. (a) Modulator schematic. (b) Demodulator schematic. (c) Frequency hopping versus time. T_c is the duration of time in a given frequency band. *Source:* Ref. [33]. 120

Figure 5.1 Schematic of RADAR system. 123

Figure 5.2 Communications system. (a) Transmitter. (b) Receiver. 124

Figure 5.3 Block diagram of super-heterodyne transmitter. *Source:* Ref. [33]. 125

Figure 5.4 PA/antenna interface. (a) Employing duplexer-PA isolator. (b) Employing duplexer-PA matching network. *Source:* Ref. [33]. 126

Figure 5.5 Schematic of a homodyne (zero IF/direct-conversion) transmitter. 127

Figure 5.6 (a) Direct-conversion transmitter. (b) The coupling from the power amplifier output to the local oscillator causes LO frequency pulling. (c) Phenomena of frequency pulling by signal coupled from PA back to the LO. *Source:* Ref. [34]. 128

Figure 5.7 Direct-conversion transmitter with *offset* LO for diminishing frequency pulling. 128

Figure 5.8 Heterodyne transmitter utilizing a two-step up-conversion. The first BPF suppresses the harmonics of the IF signal while the second removes the unwanted sideband centered around $f_{LO\_1} - \mathrm{f}_{LO\_2}$. 129

Figure 5.9 Block diagram of super-heterodyne receiver. 130

Figure 5.10 Homodyne receiver architecture. 131

Figure 5.11 Simplified homodyne receiver. 132

Figure 5.12 Homodyne receiver with quadrature down-converter. 132

Figure 5.13 Proposed framework for discussing the most commonly used receiver architectures. *Source:* Ref. [37]. 133

Figure 5.14 Heterodyne receiver block diagrams. (a) Double-conversion quadrature super-heterodyne receiver. (b) Single-conversion digital-IF receiver: Here, the down-converted received signal is converted into digital form at the IF stage. *Source:* Ref. [37]. . . . 135
Figure 5.15 Block diagram of zero-IF/direct-down-conversion/homodyne receiver. *Source:* Ref. [37]. 137
Figure 5.16 Block diagram of (a) low-IF receiver which utilizes a complex BPF and (b) low-IF receiver which employs real filters. *Source:* Ref. [37]. 139
Figure 5.17 Block diagram of the "ideal" SDR receiver. *Source:* Ref. [38]. 140
Figure 6.1 Evolution of networks and mobile technologies. *Source:* Ref. [40]. 143
Figure 6.2 Historical evolution of mobile technology. *Source:* Ref. [45]. 144
Figure 6.3 Evolution of networks and mobile technologies. *Source:* Ref. [41]. 144
Figure 6.4 5G requirements. *Source:* Ref. [45]. 146
Figure 6.5 Functional architecture of 5G wireless technologies. *Source:* Ref. [40]. 147
Figure 6.6 mmWave-enabled 5G network. *Source:* Ref. [40]. . 149
Figure 6.7 A pictorial comparison of OMA and NOMA. *Source:* Ref. [49]. 150
Figure 6.8 mmWave-enabled 5G network. *Source:* Ref. [45]. . 151
Figure 6.9 Evolution paths to 5G. *Source*: Ref. [45]. 152
Figure 7.1 (a) Model of single-input single-output flat fading wireless channel; *h* is the channel gain. (b) System-level representation. 154
Figure 7.2 (a) Model of MIMO flat fading wireless channel; h_{ij} are the channel gains between transmitter *i* and receiver *j*. (b) Example of MIMO system with $n_T = 9$ and $n_R = 12$. *Source:* Ref. [51]. 159
Figure 7.3 Decomposition of a MIMO channel into independent sub-channels by decomposition into singular values. (a) Classical correlated channels. (b) Independent sub-channels representation. *Source:* Ref. [60]. . . 165

Figure 7.4 (a) SISO channel capacity with various distributions. (b) MIMO channel capacity.. (c) MIMO channel capacity. *Source:* Ref. [67]. 166

Figure 7.5 Result of power allocation per channel algorithm via the water-filling algorithm. The noise derives from the channel matrix, i.e., $\sigma_1^2 = N_1$, $\sigma_2^2 = N_2$, and so on. *Source:* Ref. [57]. 173

Figure 7.6 Possible RF architectures for mmWave systems, each embodying various tradeoffs in terms of hardware implementation, power, and complexity. (a) Analogue architecture. *Source:* Ref. [73]. . . . 176

Figure 7.7 Maximum ratio combining. *Source:* Ref. [75]. . . . 177

Figure 7.8 Prototypical MIMO model for ZF beamforming. *Source:* Ref. [79]. 178

Figure 7.9 ZFBF with one interferer. *Source:* Ref. [76]. 178

Figure 7.10 Conceptual illustration of the MIMO detection problem. *Source:* Ref. [80]. 181

Figure 7.11 Massive MIMO architecture. *Source:* Ref. [81]. . . 182

Figure 7.12 Block diagram of a limited feedback unitary precoding MIMO system. *Source:* Ref. [83]. 182

Figure 7.13 Schematic block diagram of the rotating codebook precoding system where the codebook used at feedback time *t* is $F^L_{\rho(t)}$. *Source*: Ref. [83]. 183

Figure 7.14 MIMO system architecture with antenna selection. *Source:* Ref. [86]. 183

Figure 8.1 (a) Sketch of bistatic RADAR. (b) Example of bistatic RADAR situation: Fighter plane transmits signal and launches missile, and missile receives return echo from target enemy airplane. *Source:* Refs. [92, 93]. 188

Figure 8.2 Frequencies and respective wavelengths at which RADARs operate. *Source:* Ref. [92]. 189

Figure 8.3 RADAR round trip time T_R for calculating range. *Source:* Ref. [93]. 191

Figure 8.4 Sketch of quasi-monostatic RADAR. P_T is the transmit power in W. P_R is the received power in W. G_T is the transmit antenna gain. G_R is the receive antenna gain. R is the range. σ is the target RCS (m^2). *Source:* Ref. [93]. 191
Figure 8.5 Sketch of variation in received power versus range. *Source:* Ref. [93]. 192
Figure 8.6 Coordinate system with respect to which a number of RADAR parameters is defined. *Source:* Ref. [93]. 193
Figure 8.7 RADAR pulse characteristics. *Source:* Ref. [93]. . 193
Figure 8.8 RADAR pulse ambiguity. *Source:* Ref. [96]. 194
Figure 8.9 RADAR range resolution. *Source:* Ref. [93]. 195
Figure 8.10 (a) Range gates. (b) Analog implementation. *Source:* Ref. [93]. 196
Figure 8.11 (a), (b), (c) Clutter, interference, and shadowing scenarios. *Source:* Refs. [92, 93]. 196
Figure 8.12 Typical RADAR return display. *Source:* Refs. [92, 94]. 197
Figure 8.13 System noise components. *Source:* Ref. [94]. . . . 198
Figure 8.14 (a) Schematic of coherent RADAR system. (b) Schematic of detector. *Source:* Ref. [93]. (b) Schematic of envelope detector. *Source:* Ref. [94]. 199
Figure 8.15 Schematic of envelope detector. *Source:* Ref. [93]. 199
Figure 8.16 (a) Schematic illustration of Doppler shift. (b) Relation between velocities. (c) Two-way Doppler frequency shift. *Source:* Refs. [92, 94]. 200
Figure 8.17 (a) Block circuit diagram of radar with IF. (b) Typical Doppler signal of CW radar. f_c+f_D denotes R decreasing and f_c-f_D denotes R increasing. *Source:* Ref. [94]. 201
Figure 8.18 Doppler filter banks. *Source:* Ref. [93]. 201
Figure 8.19 (a) Spectrum of Doppler shifted CW signal. (b) Coherent pulse train spectrum (fixed target – no Doppler). (c) Expanded central lobe region with target Doppler shift. *Source:* Ref. [93]. 202

Figure 8.20 (a) Monopulse RADAR technique. In the monopulse radar technique, the pulse transmitted is especially encoded so as to facilitate extracting accurate range and direction information from a single signal pulse. (b) Conical scan. This antenna, used in automatic-tracking radar systems, consists of a parabolic reflector combined with a radiating element which is caused to move in a small circular orbit about the focus of the antenna with or without change of polarization. The radiation pattern is in the form of a beam that traces out a cone centered on the reflector axis. The process is also known as nutating conical scanning. *Source*: Refs. [92, 93, 95]. 203

Figure 8.21 Typical values of RCS. *Source:* Ref. [93]. 204

Figure 8.22 Types of RADAR systems. *Source:* Ref. [93]. . . . 204

Figure 8.23 Schemes for measuring range. (a) **CW RADAR**: A continuous wave of frequency f_c is transmitted, and the returned echo signal, modified by the Doppler shift, $f_c \pm f_p$ is detected and related to the target velocity via $f_D = -d\frac{V_r}{\lambda}$. (b) **Pulse RADAR**: A periodic pulse of period T_p and duration τ is transmitted, and the time delay from transmission to echo reception, $\triangle$t, is related to the range. (c) **FM-CW RADAR**: A signal, whose frequency is linearly modulated periodically over a frequency bandwidth ***B*** is transmitted, and the time delay $\triangle$t and corresponding frequency shift $\triangle$f between the transmitted sawtooth and received sawtooth signals, namely, $\triangle t_t = \triangle f \frac{T}{B}$, is related to the range by, $R = \triangle f \frac{T}{2B} c$. *Source*: Refs. [94, 95]. 205

Figure 8.24 (a) Sketch of CW Doppler RADAR. (b) Block diagram of CW Doppler RADAR with nonzero IF receiver, sometimes called *sideband super-heterodyne. Source:* Ref. [95]. 206

Figure 8.25 (a) Sketch of FM-CW RADAR. (b) Block diagram of FM-CW RADAR using sideband super-heterodyne receiver. *Source:* Ref. [95]. 206

Figure 8.26 Sketch of pulse radar using Doppler information. *Source:* Ref. [95]. 207
Figure 8.27 RADAR jamming visualized. *Source*: Ref. [92]. . . 208
Figure 8.28 (a) Spot jamming, the ECM system targets one frequency. (b) Sweep jamming, one frequency at a time is sequentially targeted. (c) Barrage jamming, a band of frequencies are targeted simultaneously. *Source:* Ref. [91]. 209
Figure 8.29 Block diagram of DRFM. *Source:* Ref. [91]. 209
Figure 8.30 Deception jamming, multiple false targets appearing on the PPI of the radar. (b) *Source:* Ref. [91]. . . . 210
Figure 8.31 Simulation comparing the signal to noise plus jamming ratio (S/(J+N) versus detection range for a RADAR operating at 5.6 GHz. *Source:* Ref. [90]. 211
Figure 8.32 Comparison of jammer and radar's peak power vs. cross over range. *Source:* Ref. [91]. 211
Figure 8.33 Block diagram of signal model and cognitive radar platform for target recognition using jammer-nulling adaptive matched waveform design. *Source:* Ref. [99]. 213
Figure 8.34 Block diagram of signal model and cognitive radar platform for target recognition using jammer-nulling adaptive matched waveform design. The received signal plus noise jammer interference and AWGN vector is y = s+j+w where the received target echo *s* is the convolution of the transmit waveform and target response (s= x*h) j_{KB} represents the noise jammer. *Source:* Ref. [100]. . . 213
Figure 9.1 New empty workspace in SystemVue. 218
Figure 9.2 Phased uniform linear array template in SystemVue. 219
Figure 9.3 Definition of Theta (θ) and Phi (ϕ) in a spherical coordinate system. $\theta = 90^o$ and $\phi = 0^o$ is the direction along the *y*-axis. 222
Figure 9.4 Annotations containing tuning parameters. 223
Figure 9.5 SystemVue screen of workspace file 5G_MIMO_Beamforming_ULA_1 x 4.wsv. 223
Figure 9.6 Workspace Tree of workspace file 5G_MIMO_Beamforming_ULA_1 x 4.wsv. 224

Figure 9.7 Tune Window in workspace file 5G_MIMO_Beamforming_ULA_1 x 4.wsv. 225

Figure 9.8 Schematic and Results windows in workspace file 5G_MIMO_Beamforming_ULA_1 x 4.wsv. 225

Figure 9.9 Parts selector pallet windows in workspace file 5G_MIMO_Beamforming_ULA_1 x 4.wsv. 226

Figure 9.10 Steps to run analysis in workspace file 5G_MIMO_Beamforming_ULA_1 x 4.wsv. 227

Figure 9.11 Screen that comes up upon clicking on "Phase Array Analysis" in workspace file 5G_MIMO_Beamforming_ULA_1 x 4.wsv. 228

Figure 9.12 Screen that comes up upon clicking on "Calculate Now" in workspace file 5G_MIMO_Beamforming_ULA_1 x 4.wsv. 228

Figure 9.13 Screen showing Array Configuration in workspace file 5G_MIMO_Beamforming_ULA_1 x 4.wsv. . 229

Figure 9.14 (a) Screen shot of beam pattern of four-element ULA disposed along the *y*-axis. (b) Directivity along the $90°-270°$ axis cut (U-cut) in (a). (c) Directivity along the $0°-180°$ axis cut (V-cut) in (a). (d) Directivity pattern along direction in (b) vs. that along (c). 230

Figure 9.15 "Tune" window. 231

Figure 9.16 "Run Analysis" button. 232

Figure 9.17 (a) Screen shot for N=4 to 12 of beam pattern of four-element ULA disposed along the *y*-axis. (b) Directivity along the $90°-270°$ axis cut (U-cut) in (a). (c) Directivity along the $0°-180°$ axis cut (V-cut) in (a). (d) Directivity pattern along direction in (b) vs. that along (c). 233

Figure 9.18 Results of simulating ULA with N=12 and antenna elements modeled as "Cosine" radiators. (a) Radiated beam pattern. (b) Directivity. 234

Figure 9.19 Screen shot Workspace Tree showing "Optimization1" setup button. 235

Figure 9.20 Optimization Parameters windows. 237

Figure 9.21 Component in schematic and their optimizable variables. 238

Figure 9.22 How to access the phase shifter values stored internally. (a) Click on "PhaseArray1_TxDesign_Data." (b) Click on the PartInfo tab. (c) See an array of phase shift values, the result of the optimization. (d) See an array of attenuator values, the result of the optimization. 240
Figure 9.23 Running Optimization1. 241
Figure 9.24 (a) Beam pattern pointing along the Theta=45°, Phi=45° direction. (b) Directivity pattern along the V-cut vs. U-cut. 242
Figure 9.25 Schematic of URA for codebook design simulations. 243
Figure 9.26 Properties of (a) ArrayAttn2 and (b) ArrayPhase1 components of URA. 244
Figure 9.27 Overall system. 248
Figure 9.28 Jammer. 248
Figure 9.29 Transmitter. 249
Figure 9.30 Environment plus Target. 249
Figure 9.31 Receiver plus signal processing to extract target velocity (V) and range (D). 250
Figure 9.32 Linearly modulated frequency of RADAR signal transmitter over 50 μs. 251
Figure 9.33 Frequency spectrum of RADAR signal transmitted. 251
Figure 9.34 Frequency spectrum of jammer signal (Multi Spot Jamming) transmitted. 252

List of Tables

Table 1.1 Amplitudes of FM wave components. *After* [61]. . . . 22
Table 2.1 Behavioral mixer model. 39
Table 6.1 Features of wireless technologies. *Source:* Refs. [42–44]. 145
Table 6.2 Target 5G features. *Source:* Refs. [40, 44, 45]. 146
Table 7.1 SISO model parameters. *Source:* Ref. [51]. 154
Table 7.2 MIMO model parameters. *Source*: Refs. [51, 52]. . . . 160
Table 8.1 Letter designations of RADAR frequency bands. *Source:* Ref. [93]. 190
Table 8.2 PRF determined Doppler and range ambiguities. *Source:* Ref. [93]. 202
Table 9.1 Beam patterns whose codebooks were obtained. . . . 245
Table 9.2 RADAR parameters. 246
Table 9.3 Estimated target distance and velocity versus jammer power. 252

List of Abbreviations

1G	first generation wireless networks
2G	second generation wireless networks
3G	third generation wireless networks
4G	fourth generation wireless networks
5G	fifth generation wireless networks
A	amplitude
A	antenna aperture area
ACP	adjacent channel power
ACPR	adjacent and alternate channel power ratio
A_e	effective antenna area
ADC	analog-to-digital converter
AF	array factor
AI	airborne intercept
AM	amplitude modulation
ASK	amplitude-shift keying
AWGN	additive white Gaussian noise
B	bandwidth
BASK	binary amplitude-shift-keying
BB	baseband
BER	bit error rate
BF	beamforming
BPF	bandpass filter
BPSK	binary phase-shift-keying
BS	base station
BW	bandwidth
c	speed of light
C	capacity
CDMA	code division multiple access
CFT	carrier feed-through
CI	coherent integration
C/N	CNR carrier-to-noise ratio
CP	compression point
CRr	cognitive RADAR
CSI	channel state information
ε_0	permittivity of free space

dB	decibels
dBi	ratio of antenna gain to isotropic antenna gain expressed in dB
dBc	ratio of signal power to carrier power expressed in dB
dBm	ratio of power in Watts to 1mW expressed in dB
d_f	distance from antenna at which far field region starts
d_0	reference distance from antenna at which power is measured
D	antenna directivity
D2D	Device-ro-Device
DAC	digital-to-analog converter
DC	direct current
DR	dynamic range
DRFM	digital radio-frequency memory
DSB	double sideband
DSP	digital signal processing
DS/SS	direct sequence spread spectrum
DT	dwell time
E_b	energy per bit
ECCM	electronic counter countermeasures
ECM	electronic countermeasures
EIRP	effective isotropic radiated power
EM	electromagnetic
ESL	electronic system level
EVM	error vector magnitude
EW	Electronic/Warfare
F	noise factor
FH/SS	frequency hopping spread spectrum
f_c	carrier frequency
f_d	Doppler frequency
FM	frequency modulation
FM-CW	frequency modulation-continuous wave
FOV	field of view
FSK	frequency-shift keying
G	antenna gain
G_c	gain compression
G_i	isotropic gain
GB	gigabyte
GHz	giga Hertz
GMSK	Gaussian minimum-shift keying
G_t	transmitting antenna gain
G_r	receiving antenna gain
H	linear antenna height
Hz	Hertz

HPBW	half-power beam width
I	in-phase
IEEE	Institute of Electrical and Electronics Engineers
IF	intermediate frequency
IMD	intermodulation distortion
IoT	Internet of Things
JSR	jamming-to-signal ratio
k, k_B	Boltzmann's constant
IP	Internet Protocol
IP3	third-order intercept point
IR	image rejection
kW	kilo-Watt
LDS	low-density spreading
LNA	Low Noise Amplifier
LFM	linearly frequency modulated
LO	local oscillator frequency
LOS	line-of-sight
LPF	low-pass filter
LTE	Long-Term Evolution
μ_0	permeability of free space
MRC	maximum ratio combining
MEA	multi-element arrays
MIMO	Multiple-Input Multiple-Output
MDS	minimum detectable signal
MHz	mega Hertz
MLD	maximumlikelihood detection
MMICs	Monolithic Microwave Integrated Circuits
MMSE	minimum mean-square error
MTI	moving target indictor
MU	multiple-user
MUSA	multi-user shared access
NCI	non-coherent integration
NF	noise figure
NLOS	non-line-of-sight
N_o	noise density
NOMA	non-orthogonal multiple access
NRZ	nonreturn to zero
mmWave	millimeter wave
OFDM	orthogonal frequency division multiplexing
OFDMA	orthogonal frequency division multiple access
OMA	orthogonal multiple access
OQPSK	orthogonal quadrature phase-shift keying

P_a	available power
PA	power amplifier
PAA	phased array antenna
PG	processing gain
PL	path loss
PLL	phase-locked loop
PM	phase modulatio
P_{ne}	added noise power
P_{no}	output noise power
P_{ns}	available noise power
PN	pseudo-noise
P_r	received power
PRF	pulse repetition frequency
PSK	phase-shift keying
PRI	pulse repetition interval
PSD	power spectral density
P_t	transmitted power
PW	pulse width
Q	quality factor
Q	quadrature
Q	error function
QAM	quadrature amplitude modulation
QPSK	quadrature phase-shift-keying
RADAR	Radio Detection And Ranging
RAN	radio access network
RCS	RADAR cross section
RF	Radio-Frequency
R_r	antenna radiation resistance
RR	rejection ratio
RX	receiver
SAW	surface acoustic wave
SCMA	sparse code multiple access
SDR	software-defined receiver
SF	shape factor
SFDR	spurious-free dynamic range
SIC	successive interference cancellation
SISO	single-input single-output system
SM	spatial multiplexing
SNR	signal-to-noise ratio
SSB	single sideband
STR	simultaneous transmission/reception
SU	single-user

SVD	singular value decomposition
T	period of sinusoidal function
T_b	bit duration
TB	terabyte
TDD	time-division duplex
TDMA	time division multiple access
T_e	effective input noise temperature
T_0	standard noise temperature
T_e	equivalent temperature
TL	transmission line
T_s	sample duration
TV	television
TX	transmitter
ULA	uniform linear array
USB	universal serial bus
URA	uniform rectangular array
VCO	voltage controlled oscillator
VGA	variable-gain amplifier
WAN	wide area network
WiFi	Wireless Fidelity
WSS	wide-sense stationary
Z_A	antenna impedance
ZF	zero-force
ZFBF	zero-force beamforming
Z_L	load impedance

1

Introduction to Wireless Communications and Sensing Systems

1.1 Scientific Beginnings: Electromagnetic Waves

The first demonstration of the generation and detection of electromagnetic (EM) waves was that of Heinrich Hertz [1, 2]. Hertz's pioneering experiments demonstrated that EM waves may exist and propagate in free space and that they travel at a finite velocity, as Maxwell's equations predicted [3, 4].

1.1.1 Generation and Detection of EM Waves

In order to carry out his experiments, Hertz invented the first radio frequency (RF) transmitter and receiver of radio waves, operating in the 50–500 MHz frequency range. The initial experiment, conducted on November 13, 1886, provided wireless transmission between two open circuits located at a distance of about 1.5 m and proved that EM waves propagate in air with the velocity $v = c = 1/\sqrt{\mu_0 \varepsilon_0}$.

1.1.1.1 Ruhmkorff Coil and Spark Gap

The key element utilized by Hertz for generating high-power/high-frequency waves was the inductor coil of the Ruhmkorff type [5] and a spark gap (Figure 1.1).

This inductor combines two windings, namely, a primary winding *A*, which possesses a low number of turns (a few hundred) of coarse insulating wire and a secondary winding *B* that has a large number of turns (tens of thousands) and is wrapped around an iron core *C*, together with a battery *G*, which sets up a DC current in the primary winding. The DC current is caused to become intermittent by a switch of contacts *E* which is in series, thus producing changes in the magnetic flux of the coil, which results in inducing a high voltage on the secondary. The switch possesses a piece of

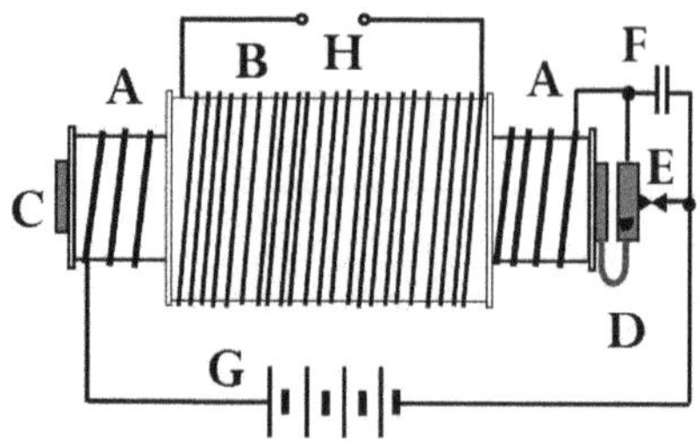

Figure 1.1 Details of the Ruhmkorff induction coil. *Source:* Ref. [5].

iron armature next to the iron core, which acts as a spring. As a result, the act of switching the current ON causes the creation of a magnetic field in the iron core which, in turn, pulls the iron arm of the switch toward the core, thus opening the contacts. When this occurs, the current in the primary is interrupted, which results in the collapse of the magnetic field and the release of the switch spring to its equilibrium position, thus making a closed circuit in the primary again. As the current in the primary increases, a magnetic field is again set up in the iron core, and the switch arm is pulled again. Every time the current in the primary is interrupted and the magnetic field turns toward zero, there is voltage induction into the secondary and, because it has a large number of turns, the resulting voltage is large enough to cause a spark to jump across the spark gap *H*.

1.1.1.2 Hertz's Transmitter

Figure 1.2(a) shows the schematic of Hertz's transmitter and Figure 1.2(b) the equivalent circuit model [4].

In the transmitter, the spark gap is connected to a dipole antenna which launches the signal into space. To explain the operation of the transmitter, we refer to its circuit model in Figure 1.2(b). Describing the circuit from left to right, we have the battery, the interrupter (represented by the parallel combination of a switch and capacitor C_i), the Ruhmkorff coil, represented by the transformer, with primary and secondary inductances and resistances L_p, R_p and L_s, R_s, respectively, and the mutual inductance M. This is followed by the resistance R_g, which represents the spark across the gap, and the elements R_d, L_d, and C_d, which represent the electrically short dipole antenna.

The sequence of events to transmit a signal pulse is as follows (Figure 1.3). At the initial time t_1 (Figure 1.3(a)) the iron piece P is making contact with the adjustable screw, and the switch in the circuit of Figure 1.3(b) closes. Then, the current I_p in the primary circuit rises with the time constant

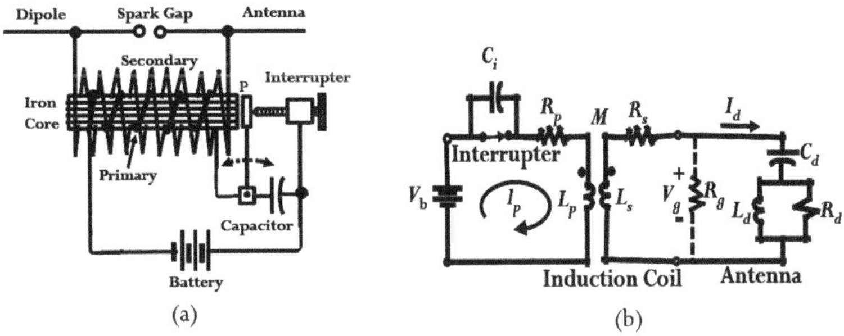

Figure 1.2 (a) A schematic for Hertz's transmitter showing the details for the Ruhmkorff induction coil. (b) A circuit diagram for Hertz's transmitter. *Source:* Ref. [4].

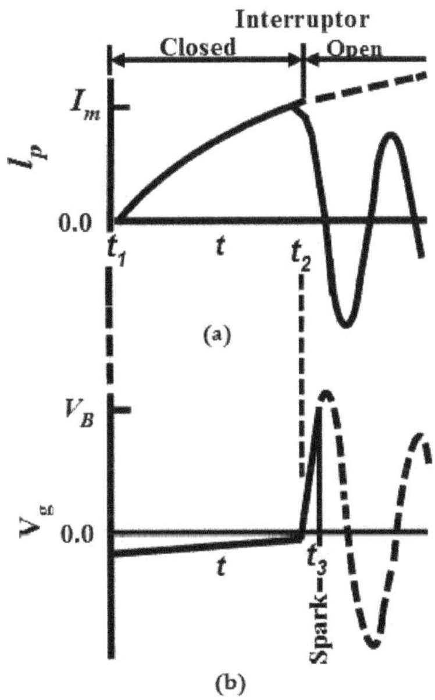

Figure 1.3 The waveforms for the Ruhmkorff induction coil. (a) Current I_p flowing in the primary circuit and (b) voltage Vg across the spark gap in the secondary circuit. Note that the vertical dashed line separates the times when the interrupter is closed ($t_1 < t < t_2$) from the times when it is open ($t > t_2$) and that the spark gap breaks down at time t_3. *Source:* Ref. [34.

$\tau_p = L_p/R_p$ (Figure 1.3(a)) and, as a result, a small voltage V_g that is developed across the spark gap in the secondary circuit. As a result of the increasing magnetic field in the iron core, when the primary current I_p arrives at its maximum value I_m at time t_2, the field strength is large enough to attract the iron piece P to the core, thus opening the switch. Upon opening of the switch, the primary circuit becomes open and, consequently, its current rapidly decreases, oscillating at the frequency of $\omega_p \approx \sqrt{L_p C_i}$ (Figure 1.3(b)).

Due to the increasing current flowing in the primary circuit, a high voltage is induced across the spark gap in the secondary circuit, thus charging the capacitance of the dipole antenna. At time t_3, this voltage across the gap attains the breakdown voltage V_{Br} of the air in the gap and a spark of low resistance R_g is produced in the gap. When this occurs, the antenna is virtually disconnected from the rest of the secondary circuit and the charge on the dipole capacitance oscillates resulting on a damped sinusoidal current I_d of frequency $\omega_d \approx 2\pi/T_d$ (Figure 1.4(a)).

I_d is given by

$$I_d(t) = I_0 e^{-\gamma_d (t/T_d)} \sin(\omega_d t) \tag{1.1}$$

where $\gamma_d = \alpha_d T_d = 2\pi\alpha_d/\omega_d$ is the logarithmic decrement, a parameter characterizing early transmitters. The physical origin of the damping is the energy lost to the radiation being launched by the antenna into space as well as the energy dissipated in the spark resistance $R_{g.}$ The radiation process occurs in pulses, that is, every time the current in the primary circuit drops, such that it is no longer possible to maintain the magnetic field keeping the switch open, the switch closes and the whole process repeats. The transmitted radiation, therefore, may be visualized as a series of damped pulses spaced

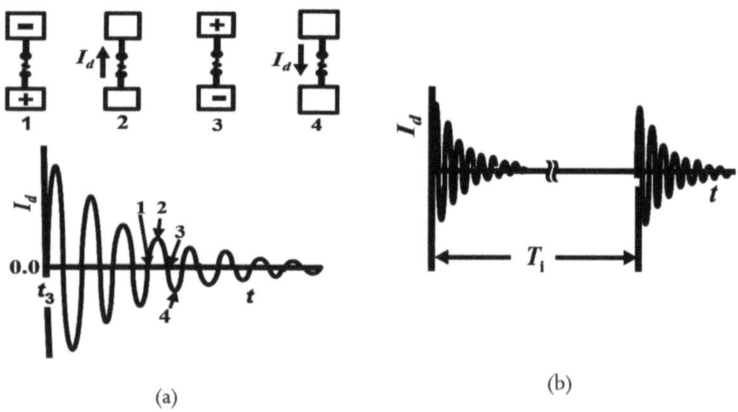

Figure 1.4 (a) The current in the dipole antenna after the spark gap breaks down $t > t_3$ for a single cycle of the interrupter. The inset at the top shows the charge oscillating between the end plates creating the current. (b) An expanded scale showing the current for two cycles of the interrupter spaced by the time T_i. *Source:* Ref. [4].

at time intervals T_i (Figure 1.4(b)). Because T_i is usually greater than a period of the damped sinusoid (Figure 1.4(b)), that is $T_i > T_d = 1/f_d$, the jitter concomitant with the commutation of the switch results in the sinusoids being incoherent from pulse to pulse. Hertz's parameters in his experiments included a frequency $f_d \approx 50.5MHz$, a logarithmic decrement in the range $0.25 \leq \gamma_d \leq 0.5$, and a dipole antenna of total length $0.23\lambda_d$. Next, we address Hertz's receiver.

1.1.1.3 Hertz's Receiver

The receiver technique utilized by Hertz was based on the circular loop antenna containing a spark gap (Figure 1.5) to detect the electric and magnetic fields of the transmitted waves. As seen in Figure 1.5, this antenna design consists of a wire ring of diameter *2a*, radius *b*, and gap of adjustable length l_g. The gap is configured as a capacitor C_L which may also be adjusted by varying the size and separation between its two metal plates. This antenna, when having a circumference small relative to the wavelength (circumference = $2rb << \lambda$), may be utilized to measure or detect the strength of an electric field, oriented parallel to the loop plane, or a received magnetic field, oriented perpendicular to the loop. The intensities of the fields are related to the intensity of the spark produced in the gap. Calibration may be effected by varying the gap length until the threshold of sparking is obtained for a known radiated/received field strength. The gap length, then, would be related to the applied field.

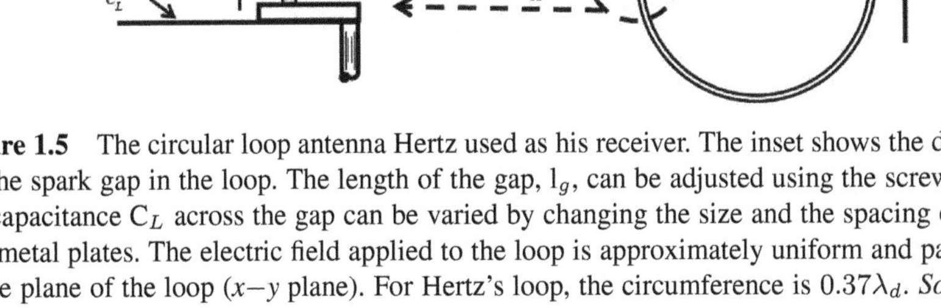

Figure 1.5 The circular loop antenna Hertz used as his receiver. The inset shows the details for the spark gap in the loop. The length of the gap, l_g, can be adjusted using the screw, and the capacitance C_L across the gap can be varied by changing the size and the spacing of the two metal plates. The electric field applied to the loop is approximately uniform and parallel to the plane of the loop (x–y plane). For Hertz's loop, the circumference is $0.37\lambda_d$. *Source:* Ref. [4].

1.1.1.4 Hertz's Experiment

To prove the existence of EM waves propagating in air, and their velocity, Hertz conceived a setup to measure the interference pattern of such waves upon reflecting from a metallic sheet [2–4]. The setup derived from considering the solution to Maxwell's equations for the EM field in free space with no sources present ($\rho = 0, \vec{J} = 0$), namely,

$$\nabla \times \vec{E} = \frac{\partial \vec{B}}{\partial t} \tag{1.2}$$

$$\nabla \times \vec{B} = \frac{1}{c^2}\frac{\partial \vec{E}}{\partial t} \tag{1.3}$$

$$\nabla \cdot \vec{E} = 0 \tag{1.4}$$

$$\nabla \cdot \vec{B} = 0 \tag{1.5}$$

where c is the speed of light and the term on the right-hand side of Equation (1.3) is the *displacement current*. In particular, from solving the wave equation of the electric field E_y of an EM *plane* wave assumed to be propagating in the z -direction, namely,

$$\frac{\partial^2 E_y}{\partial t^2} = \frac{1}{c^2}\frac{\partial^2 E_y}{\partial t^2} \tag{1.6}$$

which has a solution of the form,

$$E_y(z,t) = A\sin[\omega(t \mp z/c) + \varphi] \tag{1.7}$$

assuming an angular frequency of $\omega = 2\pi f$, the interference pattern could be used to estimate the frequency f of the EM source which, in turn, could be used to determine the wavelength λ, and the wave velocity $v = \lambda f$.

The experimental setup utilized by Hertz to create the EM interference pattern is sketched in Figure 1.6. Describing it from left to right, an electric dipole antenna oriented to radiate an electric field with time-harmonic dependence ω and polarized in the y -direction is positioned at a distance z_0 in front of a perfectly conducting planar reflector at located at $(x, y, z = 0)$. The total electric field $\vec{E}_t$, resulting from the radiated field, consists of the addition of the incident field $\vec{E}_i$ of the dipole and its reflected field $\vec{E}_r$ from the metallic sheet, namely,

$$\vec{E}_t = \vec{E}_i + \vec{E}_r. \tag{1.8}$$

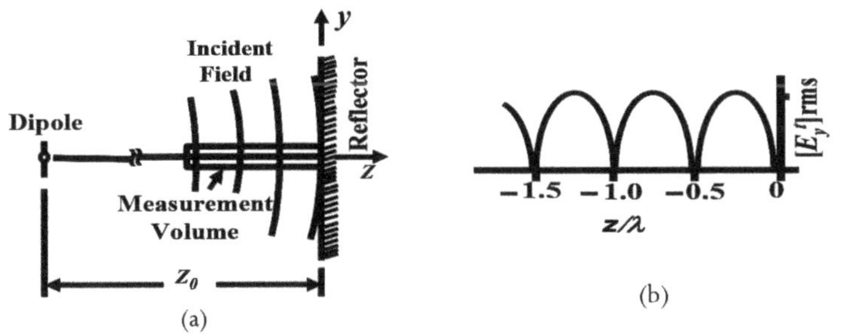

Figure 1.6 (a) A schematic for the basic elements of the experiment. (b) The interference pattern (standing wave) for the rms value of the total electric field. The incident and reflected fields are assumed to be plane waves within the measurement volume. Note that the nodes are spaced by λ/2 starting at the reflector. *Source:* Ref. [4].

In this setup (Figure 1.6), the distance z_0 was chosen large enough that the reflector would be located in the far field of the dipole radiated field and, consequently, the incident and reflected spherical wavefronts would be considered planar within the measurement volume, the space directly in from of the reflector (Figure 1.6(a)). The total field within the measurement volume is given approximately by the superposition of traveling waves propagating in opposite directions, namely,

$$E_y^t \approx A\sin[\omega(t-z/c)] - A\sin[\omega(t+z/c)] = -2A\sin(2\pi/\lambda)\cos(\omega t) \quad (1.9)$$

where the wavelength in free space is given by $\lambda = \omega/c$, and the receiver response is expressed as

$$\left[E_y^t(z)\right]_{tms} = \sqrt{2}A \mid \sin(2\pi z/\lambda) \quad (1.10)$$

which captures the measurement as the root-mean-square (rms) value of the field. Equation (1.10) gives the mathematical expression for the interference pattern shown in Figure 1.6(b), which is a standing wave having nodes spaced by $\lambda/2$, with the first node coinciding with the reflector. Next, we present Hertz's analysis of the interference pattern.

1.1.1.5 Hertz's Analysis of the Interference Pattern

Finally, we present a sketch of the apparatus utilized by Hertz in his EM wave interference pattern in air experiment (Figure 1.7).

The apparatus consists of a transmitting 60-cm-long vertical dipole antenna made out of copper wire, referred to by Hertz as the *primary conductor*, terminated in 40 cm square brass plates. The dipole antenna is

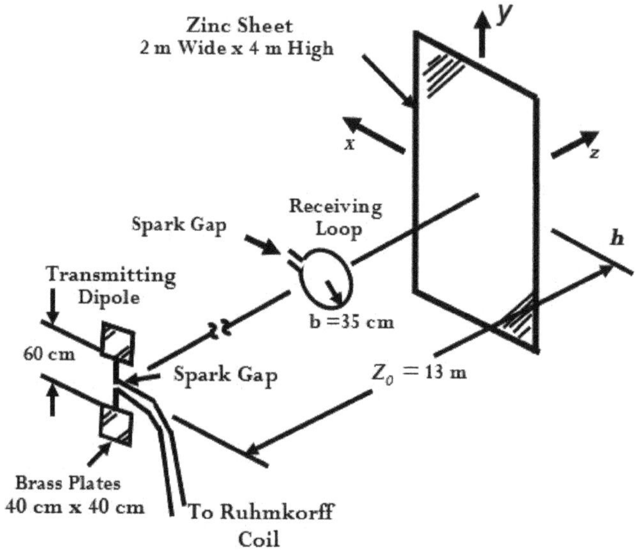

Figure 1.7 A schematic for the apparatus used by Hertz for his experiment on the interference of electromagnetic waves in air. *Source:* Ref. [4].

driven by a small spark gap connected to a Ruhmkorff coil. Then, located at a distance $z_0 = 13$m away from the dipole is a 2-m-wide by 4-m-high reflector metal sheet made out of zinc. The experiment consisted in measuring the electric field with a receiving loop antenna, referred to by Hertz as *secondary conductor*, having a radius $b = 35$ cm made out of an $a = 1$ mm diameter wire. The loop antenna in Figure 1.7 is oriented so that its plane is parallel to the ($x-y$ plane) of the mirror, and the spark gap in the loop antenna is located along the x-axis.

To effect the interference pattern measurements, Hertz recorded the voltage at the spark in the loop antenna as it was moved through along the z-direction. Figure 1.8 shows the actual interference pattern obtained by Hertz [2, 4].

Figure 1.8 shows the electric field at positions I–VII. The upward and downward arrows correspond to the measured electric field (in the $y-z$ plane) at the same position but measured at time instants that are separated by one-half the period oscillation of the transmitted wave. The wavelength of the EM waves in air was obtained by Hertz by determining the distance between the nodes of the measured interference pattern to be approximately $\lambda_m - 6.9$ m. Thus, from Hertz's estimated transmitted frequency, namely, $f_d - 50.5$ MHz,

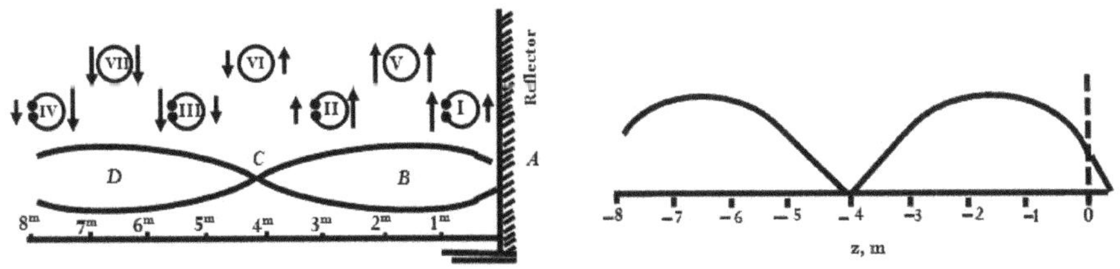

(a) (b)

Figure 1.8 (a) A figure from Hertz's 1888 article. The figure has been redrawn to show only the measured electric field. The spatial zones are divided by internals named A–D. The curves are for time instants separated by one-half period of the oscillation. (b) The interference pattern obtained from (a). *Source:* Ref. [4].

he determined the EM wave speed to be $v = f_d\lambda_m \approx 4.85 \times 10^8$ m/s. While this value was higher than the accepted value of the speed of light, Hertz surmised that the cause for the difference could be resolved by further experimentation, and, indeed, they were. In conclusion, with this experiment, Hertz established that EM waves traveled with finite velocity in air!

1.2 Engineering Beginnings: Communications and RADAR[1]

Heinrich Hertz's research was motivated by the pure pursuit of new knowledge and did not envision any practical applications. His breakthrough on establishing the propagation of EM waves at the speed of light in air, however, was the genesis of today's communications and RADAR systems. Next, we present the engineering beginnings of these systems.

1.2.1 Communications

In 1895, Alexander Popov (Russia, 1859–1906) succeeded in transmitting EM signals representing the words "Heinrich Hertz" over a distance of 250 m. He utilized his so-called "Thunderstorm Recorder," which included an antenna mounted on a balloon and a "coherer"[2] and an EM relay. The coherer

[1]RADAR: Radio Detection And Ranging.

[2]The coherer consists of a tube or capsule containing two electrodes that are spaced apart by a small distance. The space is filled with loose metal filings. Upon the application of a radio frequency signal across the electrodes, the metal particles inside the capsule cling together or "cohere." The thus-cohered metal fillings reduce the resistance of the device so that the resistance change is exploited to detect the signal [6].

was an early receiver/detector invented in 1890 by Édouard Eugène Désiré Branly (France, 1844–1940) [6]. Soon thereafter, in the same year, Guglielmo Giovanni Maria Marconi (Italy, 1874–1937) transmitted and received a coded message over a distance of 1.75 miles.

In 1896, Marconi applied for the first patent in the wireless communications field, "Improvements in Transmitting Electrical impulses and Signals, and in Apparatus therefore" [7], which disclosed the use of a transmitter and a coherer connected to a high aerial and earth.

In 1898, Karl Ferdinand Braun (Germany, 1850–1918) was granted a German patent [8] in which he disclosed the introduction of a primary tuned circuit which is inductively coupled to the secondary antenna circuitry [8]. This invention was key because it greatly enhanced the transfer of energy [9].

In the Christmas Eve of 1906, Reginald Aubrey Fessenden (Canada, 1866–1932) succeeded in transmitting the world's first wireless broadcast (radio transmission) including a speech and selected music [12].

In 1909, Marconi and Braun shared the Nobel Prize in Physics "in recognition of their contributions to the development of wireless telegraphy" [10, 11].

1.2.1.1 Communications Systems

As discussed above, Marconi and Fessenden surmised, from Hertz's original experiment and result, that the propagation of an EM wave produced by a transmitter, propagating through space at the speed of light, and being detected at a receiver located a certain distance away from the transmitter, may be exploited for transporting information. In this context, the information being transported by the EM wave may take the form of a telegraph code, as shown by Marconi, or voice and music, as shown by Fessenden. In the most general case, there is virtually no limit to the kinds of information that may be carried/transported by an EM wave, typical examples being digital data, i.e., Internet traffic, and video.

The conceptual framework for understanding communications systems was advanced in 1948 by Shannon in his monumental work, "A Mathematical Theory of Communication" [13]. The communications problem may be construed as represented in Figure 1.17. Its purpose is to transport a message from an information source to a destination, with the utmost fidelity, i.e., the message reaching the destination should ideally be identical to the one supplied by the information source. In reality, en route to the destination, the EM wave carrying the message may be distorted due to the addition of noise in the *channel*, the intervening space separating transmitter and receiver.

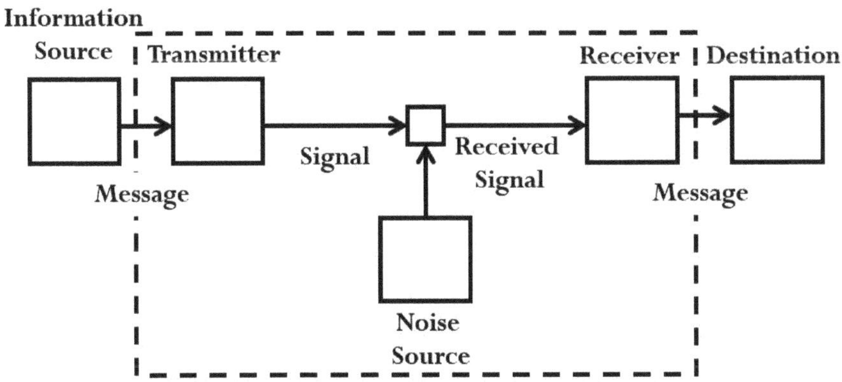

Figure 1.9 Communication system. *Source:* Ref. [13].

To describe Figure 1.9, we next focus on the part within the dashed box.

1.2.1.1.1 Simplified Transmitter Building Block

In a typical communications application, the transmitter utilizes the signal from the information source, called *baseband* signal, to modify a property of the EM wave (the *carrier*) to be transmitted, i.e., launched into space by an antenna. That the carrier usually must be a high-frequency signal derives from the need to make a practically small antenna, as the antenna dimensions are inversely proportional to its frequency of operation. This unfavorable relation between antenna size and baseband frequency is circumvented by using the baseband signal to impress a change on or, *modulate*, a high-frequency carrier.

The approach to effecting carrier modulation depends on whether the information signal is in analog or digital form. When in *analog* form and containing a single frequency, it may be represented mathematically by

$$V(t) = A_0 \cos(\omega_0 t + \varphi_0) = A_0 \cos[\varphi(t)]. \tag{1.11}$$

In Equation (1.11), factor A_0 represents the peak amplitude of $V(t), \omega_0$, represents its *radial frequency*, and T (see below) represents the *period* of the cosine function. During time T, the phase changes by 2π radians. By taking the time derivative of the phase function, $\varphi = \omega_0 t + \varphi_0$, one obtains

$$\omega_0 = \frac{d\varphi}{dt}. \tag{1.12}$$

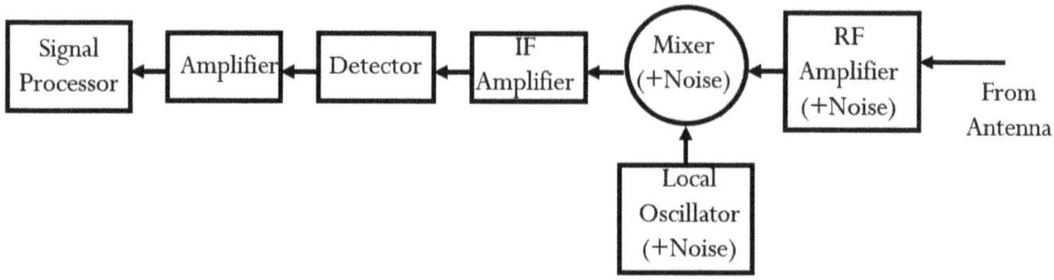

Figure 1.10 Simplified block diagram of transmitter.

Then, for a time change, T, one finds that the phase changes by 2π, i.e.,

$$\omega_0(t+T)+\varphi_0-(\omega_0 t+\varphi_0)=2\pi \tag{1.13}$$

which leads to

$$\omega_0 T=2\pi \tag{1.14}$$

or

$$T=\frac{2\pi}{\omega_0}=\frac{1}{f_0} \tag{1.15}$$

where f_0 denotes the frequency of the continuous wave (CW) measured in *Hertz.*

The cyclical and radial frequencies are related by

$$f_0=\frac{\omega_0}{2\pi}. \tag{1.16}$$

A simplified block diagram of the transmitter building, including the information source, is sketched in Figure 1.10.

It consists of an amplifier that amplifies the baseband "message" signal from the information source to a power level large enough to drive a carrier modulator. The modulator, in turn, is driven by a frequency multiplier whose function is to increase the lower frequency of a crystal oscillator to the required high carrier frequency. The modulated signal is then amplified by an RF amplifier and fed to the antenna for transmission into the *communication channel.* For wireless systems, the communications channel is typically air, water, or outer space.

1.2.1.1.2 Simplified Receiver Building Block

In a typical communications application, the receiver detects the signal transmitted at the transmitter plus the noise added to it by the channel and

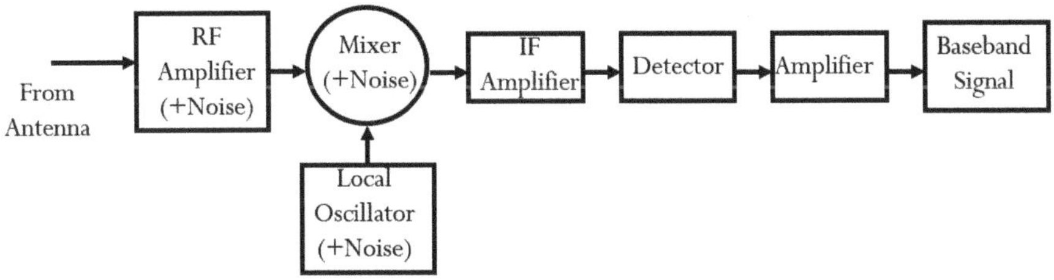

Figure 1.11 Simplified block diagram of receiver.

extracts the baseband information. With reference to Figure 1.11, from left to right, the receiver consists in the first instance of a receiving antenna and an RF amplifier which amplifies the received signal while adding minimum noise to it, i.e., a Low Noise Amplifier (LNA). Then, the output from the LNA is fed to a mixer, which is a nonlinear device that, together with the high-power signal that drives it from a local oscillator (LO), translates the frequency of the carrier to a lower intermediate frequency (IF), where it is amplified by an IF amplifier. The baseband signal contained in the IF is then extracted by a detector circuit, whose output is again amplified for further exploitation of the baseband signal.

1.2.2 RADAR

In 1904, Christian Hülsmeyer (Germany, 1881–1957) conceived the idea of applying Hertz's propagating EM waves phenomenon for preventing collisions between ships by "Seeing ships through fog and darkness by transmitting waves and detecting the echoes" [28, 42]. He applied for a German Patent titled "Means for reporting distant metallic bodies to an observer by use of electric waves" [43], which was an early implementation of a RADAR [28]. In this early RADAR system, the transmitter and receiver were located in the same platform and operated at a single frequency having a wavelength of 1 m and emitted pulses exhibiting a tunable pulse repetition frequency (PRF). The targets detected were at a maximum distance (range) of 3000 m from the platform [28].

1.2.2.1 RADAR Systems

The basic function of a RADAR is to determine the distance (range) or the speed of a target in the field of view (FOV) (Figure 1.12) [14]. This is accomplished by a system possessing a transmitter, which sends an EM wave whose line of propagation is intercepted by the target, reflects from it, and

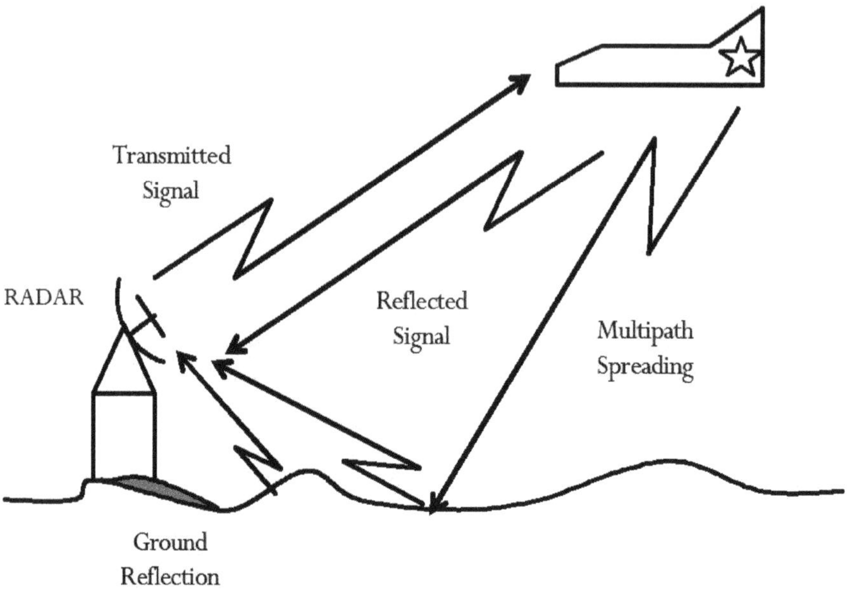

Figure 1.12 Simplified scenario of RADAR application. *Source:* Ref. [14].

travels back to the receiver where it is processed to extract the target's position and/or speed.

1.2.2.2 Simplified RADAR System Building Block

Since a RADAR radiates a signal and then examines the reflected echo, it essentially integrates a transmitter and a receiver, and its simplified building blocks may be represented as shown in Figure 1.13.

Here, we have an oscillator/frequency multiplier/modulator/power amplifier chain producing the signal to be transmitted. It should be noted that, while there is no "information source," the transmitted signal is modulated to facilitate the extraction of the target parameters, e.g., range and speed, from the returned reflected signal. The received echo signal is amplified, down-converted by a mixer, amplified, and analyzed to determine range and speed. A more detailed discussion of RADARs will be given in Chapter 8.

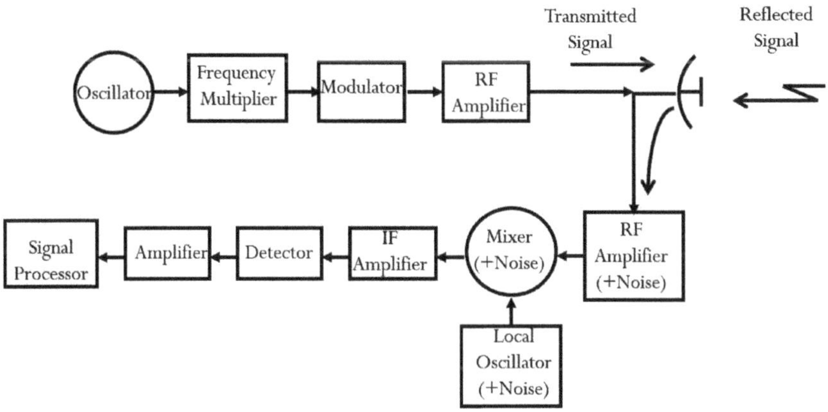

Figure 1.13 Simplified RADAR architecture.

1.3 Fundamentals of Signal Processing

As indicated in Section 1.2.1.1.1, modulation of a high frequency carrier is necessary to enable the utilization of antennas of practical size. But, when a carrier is modulated, its overall frequency spectrum and, in particular, its bandwidth, is caused by the modulating baseband (message) signal, to *increase.* Understanding the impact on carrier spectrum due to the various modulation schemes possible is facilitated by their mathematical description. This is undertaken next.

1.3.1 Mathematical Description of Carrier Modulation

A carrier, mathematically represented by

$$V(t) = A_0 \cos(\omega_0 t + \varphi_0) = A_0 \cos[\varphi(t)] \tag{1.17}$$

may be modulated by varying its amplitude, A_0, resulting in the so-called amplitude modulation (AM), varying its frequency, resulting in frequency modulation (FM), or varying its phase, resulting in phase modulation (PM). Assuming a time-varying radial frequency, namely,

$$\omega = \frac{d\varphi}{dt} \tag{1.18}$$

its integration results in the phase,

$$\varphi = \int_0^t \omega dt + \varphi_0. \tag{1.19}$$

Now, if the amplitude A_0 is also allowed to vary with time, one obtains, following substitution of Equation (1.19) into Equation (1.17),

$$V(t) = A(t)\cos\left(\int \omega dt + \varphi_0\right). \tag{1.20}$$

In what follows, the mathematical representation of these modulation approaches, namely, AM, FM, and PM, is addressed [15].

1.3.1.1 Amplitude Modulation

With respect to Equation (1.20), AM means that only the amplitude, *A(t)*, is permitted to be time-dependent, and it is assumed to adopt the form,

$$A(t) = a_0[1 + mg(t)]. \tag{1.21}$$

In Equation (1.21), *m* is the *modulation index*, *g(t)* is the *modulation function*, and a_0 is a constant.

1.3.1.2 Frequency Modulation

Again, with respect to Equation (1.20), FM means that only the radial frequency is permitted to be time-dependent, and it is assumed to adopt the form,

$$\omega(t) = \omega_0[1 + mg(t)] \tag{1.22}$$

where *m* is the modulation index, *g(t)* is the modulation function, and ω_0 is a constant.

1.3.1.3 Phase Modulation

Under PM, only the phase is permitted to be time-dependent, and it is assumed to adopt the form,

$$\varphi(t) = \omega_0\left[1 + \varphi_0 mg(t)\right] \tag{1.23}$$

where m is the modulation index, $g(t)$ is the modulation function, and ω_0 and φ_0 are constant.

1.3.2 Spectral Properties of Basic Modulation Approaches

The spectral characteristics accompanying the various modulation approaches are addressed next.

1.3.2.1 AM Spectrum

If the baseband modulation function, *g(t)*, is represented by a single cosinusoid with baseband radial frequency p_1, namely,

$$g(t) = \cos p_1 t \tag{1.24}$$

then the carrier may be written as

$$V(t) = a_0 \left[1 + m \cos p_1 t\right] \cos \omega_0 t \tag{1.25}$$

where $\omega = \omega_0, \phi_0 = 0$.

Using the trigonometric identity,

$$\cos x \cos y = \frac{1}{2}[\cos(x+y) + \cos(x-y)] \tag{1.26}$$

in Equation (1.25), we get

$$V(t) = a_0 \left[\cos \omega_0 t + \frac{m}{2} \cos\left(\omega_0 + p_1\right) t + \frac{m}{2} \cos\left(\omega_0 - p_1\right) t\right]. \tag{1.27}$$

Examination of Equation (1.27) reveals that the process of AM changes the radian frequency from the baseband, p_1, to two *sidebands*, $\omega_0 + p_1$ $\omega_0 - p_1$, centered about the carrier radial frequency, ω_0. Plotting this manifests as the spectrum shown in Figure 1.14.

When the baseband message is not a single frequency but a general signal containing a range of frequencies, it is represented by its Fourier transform,

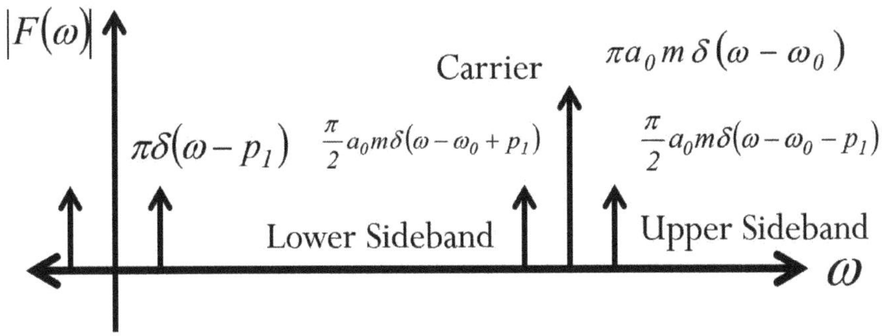

Figure 1.14 Spectrum of single-frequency AM signal. The delta function $\pi\delta\left(\omega - p_1\right)$ represents the spectrum of the baseband signal. $\pi a_0 m\delta\left(\omega - \omega_0\right)$, together with $\frac{\pi}{2} a_0 m\delta\left(\omega - \omega_0 + p_1\right)$ and $\frac{\pi}{2} a_0 m\delta\left(\omega - \omega_0 - p_1\right)$, represents the modulated, the lower, and the upper sidebands, respectively. *Source:* Ref. [15].

i.e.,

$$F_B(\omega) = \int_{-\infty}^{\infty} mg(t)e^{-j\omega t}dt. \tag{1.28}$$

Similarly, the overall modulated carrier is represented by the following Fourier transform:

$$F_M(\omega) = a_0 \int_{-\infty}^{\infty} [1 + mg(t)] \cos \omega_0 t e^{-j\omega t}dt. \tag{1.29}$$

Expanding this results in

$$F_M(\omega) = \frac{a_0}{2} \left[2\pi\delta(\omega - \omega_0) + 2\pi\delta(\omega + \omega_0) \right.$$
$$\left. m \int_{-\infty}^{\infty} g(t) \left\{ e^{-j(\omega-\omega_0)t} + e^{-j(\omega+\omega_0)t} \right\} dt \right] \tag{1.30}$$

$$F_M(\omega) = \frac{a_0}{2} \left[2\pi\delta(\omega - \omega_0) + 2\pi\delta(\omega + \omega_0) + F(\omega - \omega_0) + F(\omega + \omega_0) \right] \tag{1.31}$$

with

$$F(\omega \mp \omega_0) = m \int_{-\infty}^{\infty} g(t)e^{-j(\omega \mp \omega_0)t}dt \tag{1.32}$$

signifying a shift in frequency from baseband to a frequency near the carrier. Pictorially, this may is represented by the spectrum shown in Figure 1.15.

1.3.2.2 FM Spectrum

The FM time function has the general form,

$$V(t) = A \cos \left[\int \omega_0(1 + mg(t))dt + \varphi_0 \right] \tag{1.33}$$

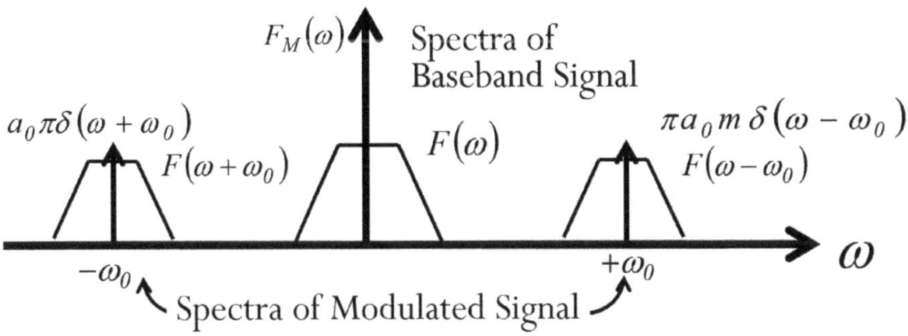

Figure 1.15 Spectrum of amplitude modulated baseband signal. *Source:* Ref. [15].

When the modulation function is a single cosinusoid, with $\varphi_0 = 0$, Equation (1.33) turns into,

$$V(t) = A\cos\left[\omega_0 t + \frac{m\omega_0}{p_1}\sin p_1 t\right]. \tag{1.34}$$

The instantaneous frequency, $\omega = \frac{d\varphi}{dt}$.

The extent of the instantaneous frequency variation is characterized by the modulation index parameter, denoted by β or M_p, which is given by

$$\beta \equiv \frac{\Delta\omega_{MAX}}{p_1} = \frac{m\omega_0}{p_1} = M_p. \tag{1.35}$$

The FM carrier, then, may be written as

$$V(t) = A\cos\left[\omega_0 t + M_p \sin p_1 t\right]. \tag{1.36}$$

Using the trigonometric identity,

$$\cos(x+y) = \cos x\cos y - \sin x\sin y, \tag{1.37}$$

we turn Equation (1.36) into

$$V(t) = A\cos\omega_0 t\cos\left(M_p\sin p_1 t\right) - A\sin\omega_0 t\sin\left(M_p\sin p_1 t\right). \tag{1.38}$$

The FM waveform may be expressed in terms of the ordinary *Bessel functions* of the first kind, which are given by

$$e^{\pm jM_p \sin p_1 t} = J_0(M_p) + 2\sum_{k=1}^{\infty} J_{2k}(M_p) \cos 2kp_1 t$$
$$\pm 2j \sum_{k=0}^{\infty} J_{2k+1}(M_p) \sin(2k+1)p_1 t \quad (1.39)$$

Utilizing this, together with the sine and cosine exponential functions, $\sin x = \frac{e^{jx}-e^{-jx}}{2j}$ and $\cos x = \frac{e^{jx}+e^{-jx}}{2}$ enable us to write, with $x = M_p \sin p_1 t$,

$$\sin(M_p \sin p_1 t) = 2\sum_{k=0}^{\infty} J_{2k+1}(M_p) \sin(2k+1)p_1 t \quad (1.40)$$

and

$$\cos(M_p \sin p_1 t) = J_0(M_p) + 2\sum_{k=1}^{\infty} J_{2k}(M_p) \cos(2k)p_1 t \quad (1.41)$$

Inserting the above equations into Equation (1.38) results in

$$V(t) = A\cos\omega_0 t \left[J_0(M_p) + 2\sum_{k=1}^{\infty} J_{2k}(M_p) \cos 2kp_1 t \right] - A\sin\omega_0 t \left[2\sum_{k=0}^{\infty} J_{2k+1}(M_p) \sin(2k+1)p_1 t \right] \quad (1.42)$$

Upon division by *A*, the first few terms of the series may be written as

$$\begin{aligned}\frac{V(t)}{A} =& J_0(M_p)\cos\omega_0 t - 2J_1(M_p)\sin p_1 t \sin\omega_0 t + \\ &+ 2J_2(M_p)\cos\omega_0 t\cos 2p_1 t - 2J_3(M_p)\sin 3p_1 t\sin\omega_0 t + \\ &+ 2J_4(M_p)\cos\omega_0 t\cos 4p_1 t - 2J_5(M_p)\sin 5p_1 t \sin\omega_0 t\end{aligned} \quad (1.43)$$

which, expressing the product terms in sum and difference, yield

$$\begin{aligned}\frac{V(t)}{A} =& J_0(M_p)\cos\omega_0 t - 2J_1(M_p)\left[\cos(\omega_0+p_1)t - \cos(\omega_0-p_1)t\right] \\ &+ J_2(M_p)\left[\cos(\omega_0+2p_1)t + \cos(\omega_0-2p_1)t\right] + \\ &+ J_3(M_p)\left[\cos(\omega_0+3p_1)t - \cos(\omega_0-3p_1)t\right] + \\ &+ J_4(M_p)\left[\cos(\omega_0+4p_1)t + \cos(\omega_0+4p_1)t\right] + \ldots\end{aligned} \quad (1.44)$$

Observing the FM carrier equation (1.44) in light of AM's carrier equation (1.27) enables making certain comparisons, which are explored next.

1.3.2.3 Comparing AM and FM Spectra

AM produces a spectrum containing a single set of sidebands (Equation (1.27)). On the other hand, FM (Equation (1.44)) is seen to produce infinitely many sidebands. In the case of FM, each sideband is located at a frequency separation kp_1 from the carrier, where k is an integer and p_1 is the modulating frequency. Furthermore, examination of the partitioning of power between the carrier and the sidebands reveals that this differs. In particular, from the total average power contained in any time function, namely,

$$\bar{P}_T = \frac{1}{T}\int_{T\to\infty} |v(t)|^2 dt \tag{1.45}$$

one finds that in the AM signal,

$$v(t) = a_0 \cos\omega_0 t + \frac{a_0 m}{2}\left[\cos\left(\omega_0 + p_1\right)t + \cos\left(\omega_0 - p_1\right)t\right] \tag{1.46}$$

making use of the orthogonality of the cosine function, which makes the integral of cross-product terms squared zero, one obtains

$$\bar{P}_T = \frac{a_0^2}{2} + \frac{a_0^2 m^2}{8} + \frac{a_0^2 m^2}{8}. \tag{1.47}$$

Here, the first term represents the power in the carrier, whereas the second and third terms represent, respectively, the power in the lower and upper sidebands. It is clear from Equation (1.47) that the carrier power is independent of the properties of the modulating function (*m*). In the case of FM, however, since the carrier envelope is constant, the total power calculation results in

$$\bar{P}_T = \frac{A^2}{2}. \tag{1.48}$$

Thus, according to Equation (1.48), in FM, the total average power in the modulated carrier (carrier plus sidebands) is constant. It turns out that squaring the FM carrier signal (1.44), inserting into (1.45), and applying orthogonality conditions, one obtains

$$J_0^2\left(M_p\right) + 2\sum_{k=1}^{\infty} J_k^2\left(M_p\right) = 1. \tag{1.49}$$

Thus, $J_0^2\left(M_p\right)$ represents the *fraction* of total power in an FM signal that is contained in the carrier. Now, because $M_p = \frac{\Delta\omega_M}{p_1}$, it follows that the

Table 1.1 Amplitudes of FM wave components. *After* [61].

$\Delta\omega_M$	p_1	M_p	$J_0(M_p)$	$J_1(M_p)$	$J_2(M_p)$	$J_3(M_p)$	$J_4(M_p)$	$J_5(M_p)$	$J_6(M_p)$	$J_7(M_p)$
1000	3000	1/3	0.9725	0.1644	0.03					
1000	1500	1/2	0.8930	0.3138	0.06					
1000	1000	1	0.7652	0.4401	0.1149	0.0196				
1000	500	2	0.2339	0.5767	0.3528	0.1289	0.034			
1000	333	3	−0.2501	0.3391	0.4861	0.3091	0.1320	0.043		
1000	250	4	−0.3971	−0.06604	0.3641	0.4302	0.4302	0.1321	0.0491	
1000	100	5	−0.1776	−0.3276	0.1697	0.2404	0.3912	0.2611	0.1311	0.0533

power in the carrier depends on the properties of the modulating function. The distribution of power in the various terms of Equation (1.49) may be appreciated from an example. Suppose we have a cosinusoidal FM wave. Let us take $\Delta\omega_M = 1000\text{sec}^{-1}$ and allow p_1 to vary between 3000 and 200 sec^{-1}. This produces the values shown in Table 1.1.

Examination of the decay of the coefficient amplitudes leads to the conclusion that all the significant sidebands fall within a bandwidth, denoted by the *FM bandwidth*, expressed as [15]

$$\omega_T \approx 2(\Delta\omega_M + p_1). \tag{1.50}$$

1.3.2.4 Wideband FM

If we free the narrowband constraint, then, from Equation (1.42), which is repeated below

$$\begin{aligned} V(t) =& A\cos\omega_0 t\left[J_0\left(M_p\right) + 2\sum_{k=1}^{\infty} J_{2k}\left(M_p\right)\cos 2kp_1 t\right] - \\ & A\sin\omega_0 t\left[2\sum_{k=0}^{\infty} J_{2k+1}\left(M_p\right)\sin(2k+1)p_1 t\right] \end{aligned} \tag{1.51}$$

and using

$$\text{Im}\{Z\} = \text{Re}\{-jZ\} \tag{1.52}$$

we have

$$\begin{aligned} V(t) = \text{Re}\Bigg\{ Ae^{j\omega_0 t} & \left[J_0\left(M_p\right) + 2\sum_{k=1}^{\infty} J_{2k}\left(M_p\right)\cos 2kp_1 t\right. \\ & \left.\left. +2\sum_{k=0}^{\infty} J_{2k+1}\left(M_p\right)\sin(2k+1)t\right]\right\}. \end{aligned} \tag{1.53}$$

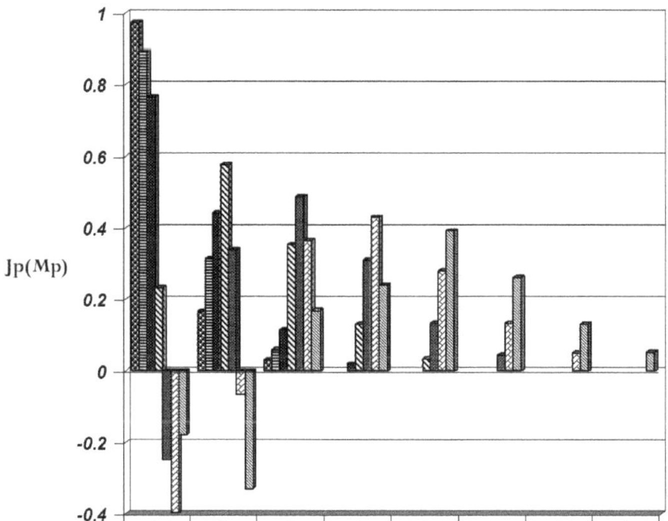

	1	2	3	4	5	6	7	8
Mp=1/3	0.9725	0.1644	0.03					
Mp=1/2	0.893	0.3138	0.06					
Mp=1	0.7652	0.4401	0.1149	0.0196				
Mp=2	0.2339	0.5767	0.3528	0.1289	0.034			
Mp=3	-0.2501	0.3391	0.4861	0.3091	0.132	0.043		
Mp=4	-0.3971	-0.06604	0.3641	0.4302	0.2811	0.1321	0.0491	
Mp=5	-0.1776	-0.3276	0.1697	0.2404	0.3912	0.2611	0.1311	0.0533
	$J_0(Mp)$	$J_1(Mp)$	$J_2(Mp)$	$J_3(Mp)$	$J_4(Mp)$	$J_5(Mp)$	$J_6(Mp)$	$J_7(Mp)$

Figure 1.16 Amplitude of FM sidebands (Bessel coefficients) as function of modulation index, M_p. p is the order of the Bessel function (0 7). *Source:* Ref. [33].

It may be appreciated from Equation (1.53) that the even order sidebands (those with coefficient $J_{2k}(M_p)\cos 2kp_1t$) are in phase with the carrier, whereas the carrier lags all the odd-order harmonics by a constant phase of 90°. Assuming the M_p=1 case, then, from Table 1.1, it is found that only the first three sidebands are significant, namely, those with amplitudes: $J_0(1) = 0.765$, $J_1(1) = 0.44$, and $J_2(1) = 0.115$.

1.3.3 Phase Modulation Spectrum

The term "angular modulation" is often used to refer simultaneously to PM and FM.

There are some differences, however, that need to be exposed, and we address these next.

Similar to FM, the envelope of a PM signal is held constant while its phase changes according to

$$\varphi = \varphi_0 m g(t) + \omega_0 t. \tag{1.54}$$

In addition, consider the carrier to be given by the expression

$$V(t) = A \cos \varphi. \tag{1.55}$$

Then the phase-modulated time function is given by

$$V(t) = A \cos \left[\omega_0 t + \varphi_0 m g(t)\right] \tag{1.56}$$

so that the instantaneous frequency of the phase-modulated signal is, differentiating Equation (1.54),

$$\omega(t) = \frac{d\varphi}{dt} = \omega_0 + \varphi_0 m \frac{dg(t)}{dt}. \tag{1.57}$$

For the case in which the modulating signal is sinusoidal,

$$g(t) = \sin p_1 t \tag{1.58}$$

and inserting it into Equation (1.56), we obtain

$$V(t) = A \cos \left[\omega_0 t + \varphi_0 m \sin p_1 t\right]. \tag{1.59}$$

Comparing Equation (1.86) with Equation (1.61) for FM, one finds that both PM and FM, generate infinitely many sidebands.

1.4 Fundamentals of Information Theory

Thus far, we have addressed the fundamental aspects of transmitters, carrier modulation, and receivers in a communications system. We will next address the fundamentals of the *information source* building blocks (Figure 1.17) by beginning with the *fundamentals* of information theory.

Information theory may be construed as the discipline that deals with how to represent the signals produced by the information source (Figure 1.17) so that their *rate of transmission* from source to destination is maximized, given the limitations imposed by hardware, transmitter power, channel bandwidth, channel noise, and receiver sensitivity.

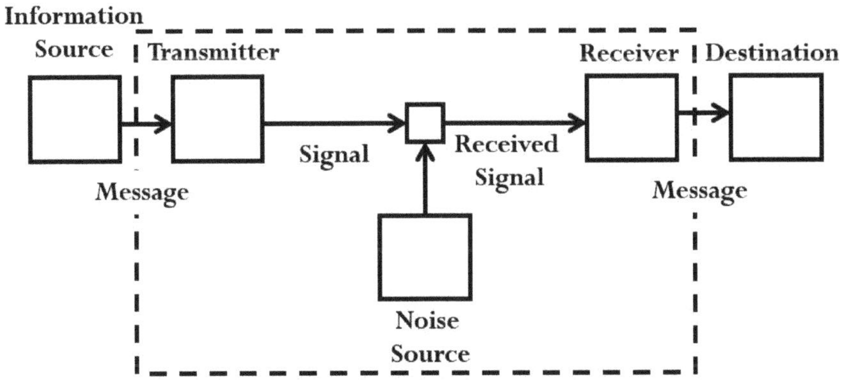

Figure 1.17 Communication system. *Source:* Ref. [13].

The amount of *information* contained in a source may, intuitively, be related to the number of possible *distinct* messages it may contain. Thus, for example, of the two information sources, namely, one being a USB thumb drive with a storage capacity of 32 GB and the other being a portable hard drive, with a storage capacity of 2 TB, clearly, the latter could store more messages and, thus, more information than the former. The number of distinct messages available in the information source, on the other hand, is related to the uncertainty of the messages since, clearly, the more messages available for transmission, the more difficult it will be to "guess" what the next received message will be.

In practice, the question arises as to how fast a certain amount of information may be transmitted from a source to a destination, i.e., what is the rate of transmission of information for a given channel? This so-called *channel capacity* was mathematically defined by Shannon [13, 15].

Quantifying channel capacity, to facilitate comparisons among different communication systems, is facilitated by measuring the amount of information present in an information source. Even if the information signal to be transmitted is analog, its information *content* may be expressed in binary digits, *bits* [13, 15]. The process by which this *digitization* is accomplished is called *quantization* and entails dividing the time function representing the signal into time intervals, choosing the most likely analog value of the signal within a given time interval and converting that value to a digital binary number (Figure 1.18).

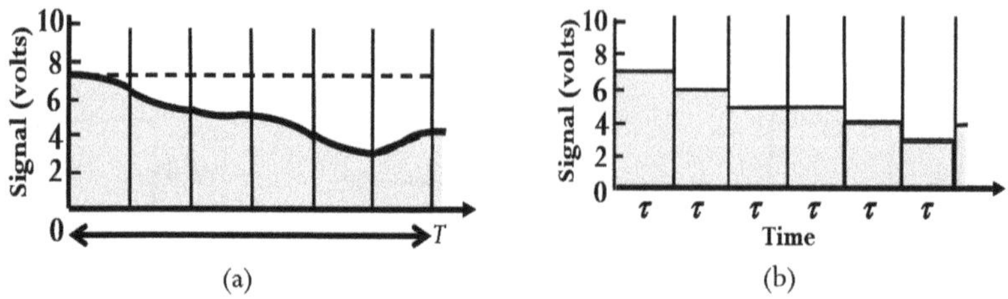

Figure 1.18 Quantization of a signal. (a) Original signal. (b) Quantized signal. *Source:* Ref. [16].

If a time interval T, over which a signal is observed, is subdivided into smaller time intervals τ, and within these subintervals, the signal amplitude is represented by one of n possible quantization levels, then it is said that in the signal possesses, an amount of information that is *defined* as being equal to [16]

$$\textit{Information} = \frac{T}{\tau}\log_2 n \text{ bits} \tag{1.60}$$

and the maximum rate of transmitting the information, the *system capacity*, C, is [16].

$$C = \frac{\textit{Information}}{T} = \frac{1}{\tau}\log_2 n \text{ bits/second.} \tag{1.61}$$

The importance of the system capacity parameter, C, is that there has to be a specific relationship between the *rate* of information to be transmitted and it. In particular, C must be greater than the rate of information one desires to transmit; otherwise, there will be *errors* in the received signal because the additional rate of information in excess of C, not being able to "fit" within the channel, will be lost and, therefore, less than the information transmitted will be received. This brings us to the topic of relating the system capacity to the *information content* of the messages available at the information source which, however, is beyond the scope of this book [13, 15].

1.5 Summary

In this chapter, we have presented an introduction to wireless communications and sensing systems. The introduction traces the development of the field from the very scientific beginnings, typified by the discoveries of electricity, magnetism, EM waves, their generation and detection, and the epoch-making Hertz's proof of EM wave propagation at the speed of light which verified Maxwell's theory. We then turned to the engineering beginnings, in particular, in what pertains to wireless communications and RADAR (sensing) applications, which concluded with an introduction of their simplified system-level building blocks. This was followed by a discussion of several signal processing aspects of communications and sensing systems, namely, AM, PM, FM, and their mathematical representation as well as spectral properties. Finally, we presented the basics of the broad topic of information theory to elucidate the fact that the properties of the information source do play a role in the overall system performance, as they may be tailored to minimize the impact of channel-induced noise, thus enhancing channel capacity.

1.6 Problems

1. Write the modern Maxwell's equations for wave propagation in vacuum.
2. Assume the wavelength of the EM waves in air obtained by Hertz by determining the distance between the nodes of the measured interference pattern was $\lambda_m = 3$ m. If the frequency transmitted by him had been $f_d =$ 100 MHz, what EM wave speed would he have determined?
3. What amount of information is contained in a signal observed over a time interval *T*, subdivided into smaller time intervals τ, and where the signal amplitude may be represented by one of *n* possible quantization levels?

2

Wireless Systems Building Blocks

2.1 System Components and Their Performance Parameters System Components

As could be appreciated from the basic discussion of communications and RADARs (sensing) systems in Chapter 1, these systems are predicated upon the configuration of the following components:

1) Transmission lines
2) Amplifiers
3) Mixers
4) Filters
5) Oscillators
6) Frequency multipliers
7) Antennas

In what follows, we present a description of these building blocks, from the perspective of their utilization in systems engineering studies.

2.1.1 Transmission Lines

The interconnection of the various building blocks that make up a system is carried out through transmission lines (TLs). TLs should preserve the amplitude and frequency spectrum of the signals they convey from one component to another. Two models of TLs are usually employed, namely, an ideal one and a physical one.

In the ideal TL model, the assumptions of no attenuation and infinite bandwidth are made (Figure 2.1).

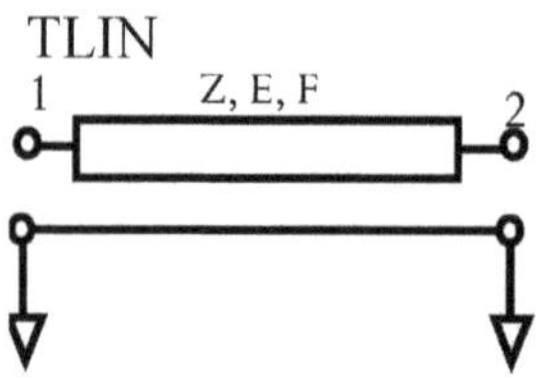

Name	Description	Units	Default
Z	Characteristic impedance	Ohm	50.0
E	Electrical length	deg	90
F	Reference frequency for electrical length	GHz	1

(a)

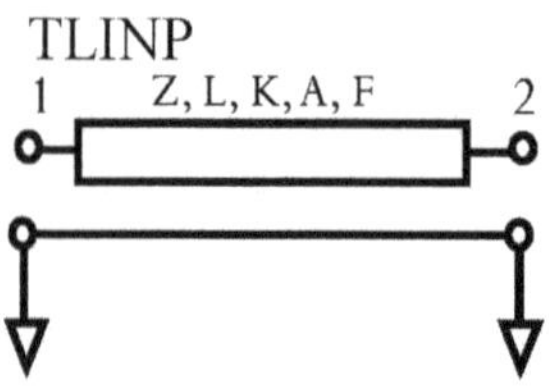

Name	Description	Units	Default
Z	Characteristic impedance	Ohm	50.0
L	Physical length	mil	1000.0
K	Effective dielectric constant	None	2.1
A	Attenuation	dB/meter	0.0001
F	Frequency for scaling attenuation	GHz	1
TanD	Dielectric loss tangent	None	0.0002
Mur	Relative permeability	None	1
TanM	Permeability	None	0
Sigma	Dielectric conductivity	None	0
Temp	Physical temperature	°C	None1

(b)

Figure 2.1 Transmission line parameters. (a) Ideal TL model. (b) Physical TL model.

2.1.2 Amplifiers

As is well known, amplifiers are required whenever there is a need to increase the power level of a signal. This function should, ideally, be effected without distorting the signal, i.e., without modifying its spectral content, independently of the signal amplitude. It is found, however, that in practical applications, the desired signal is accompanied by one or more interfering signals (i.e., interferers) and, since the amplifier gain is generally a nonlinear function of its input power, the combination of multiple signals manifests as interactions with the amplifier nonlinearities and causes its output to contain spectral components not present in the original input signal. To capture the nonlinear behavior of amplifiers, a number of parameters related to its transfer function are employed. These are addressed next.

The nonlinear transfer characteristic curve of an amplifier may be written as

$$y(t) = F(x(t)) = \sum_{n=1}^{\infty} a_n x^n(t) \cong \sum_{n=1}^{N} a_n x^n(t). \tag{2.1}$$

If we truncate the series at the third term, this may be expressed as

$$y(t) = a_0 + a_1 x(t) + a_2 x^2(t) + a_3 x^3(t) \tag{2.2}$$

where the first two terms represent the linear part, and the third and fourth terms represent the nonlinear part of the characteristic. The types of distortion generated by this type of nonlinearity are described next.

2.1.2.1 Gain Compression and Desensitization

Let us assume that the following signals are present at the input of an amplifier, namely,

$$S_d(t) = A_d \cos 2\pi f_0 t \tag{2.3}$$

$$S_{I1}(t) = A_{I1} \cos 2\pi f_1 t \tag{2.4}$$

$$S_{I2}(t) = 0 \tag{2.5}$$

where S_d, S_{I1}, and S_{I2} represent, respectively, the desired signal and two possible interferers, the latter with zero amplitude. These signals, when inserted into Equation (2.2), result in

$$\begin{aligned} y(t) = {} & a_1 \left[S_d(t) + S_{II}(t) + S_{I2}\right] + a_2 \left[S_d(t) + S_{II}(t) + S_{I2}) t)\right]^2 \\ & + a_3 \left[S_d(t) + S_{II}(t) + S_{I2}(t)\right]^3 + \ldots . \end{aligned} \tag{2.6}$$

This, in turn, upon insertion of Equations (2.3)–(2.5), yields

$$\begin{aligned} y(t) = {} & a_1 \left[A_d \cos 2\pi f_0 t + A_{I1} \cos 2\pi f_1 t\right] \\ & + a_2 \left[A_d^2 \cos^2 2\pi f_0 t + A_{I1}^2 \cos^2 2\pi f_1 t + 2A_d A_{I1} \cos 2\pi f_0 t \cos 2\pi f_1 t\right] \\ & + a_3 \left[A_d^3 \cos^3 2\pi f_0 t + A_{I1}^3 \cos^3 2\pi f_1 t + 3A_d^2 A_{I1} \cos^2 2\pi f_0 t \cos 2\pi f_1 t \right. \\ & \left. + 3A_d A_{I1}^2 \cos 2\pi f_0 t \cos^2 2\pi f_1 t\right] + \ldots \end{aligned} \tag{2.7}$$

and upon further simplification,

$$y(t) = a_1 A_d \left[1 + \frac{3a_3}{4a_1} A_d^2 + \frac{3a_3}{2a_1} A_{I1}^2\right] \cos 2\pi f_0 t + \ldots \tag{2.8}$$

Examination of Equation (2.8) reveals that, in the *absence of interference*, and with the desired signal having an amplitude $A_d << 1$, the *small signal gain* is equal to a_1. This is so because for a weakly nonlinear system, all the terms will be negligibly small, when compared to the first term $a_1 A_d cos 2\pi f_0 t$. As the amplitude of the desired signal increases, however, the gain captured by the terms within the square bracket will vary with the input signal level. The second term on Equation (2.8), $3a_3A_d^2/4a_1$, in particular, becomes significant gradually. When the condition in which the sign of a_3 and that of a_1 are opposite, the output, *y(t)*, will be smaller than that predicted by linear theory, causing the phenomenon denoted as *gain compression*. In decibels (dB), the gain under compression G_c is given by

$$G_c = 20\log\left|a_I\left(1+\frac{3a_3}{4a_1}A_d^3\right)\right|. \tag{2.9}$$

It is clear from Equation (2.9) that G_c decreases as A_d increases. The phenomenon of gain compression is characterized by the so-called *1-dB compression point* parameter. This is the signal level at which $A_d = A_{-1}$, i.e., at which the gain is 1 dB lower than the small signal gain of the amplifier. The value of A_{-1} is obtained by setting

$$G_c = 20\log\left|a_l\left(1+\frac{3a_3}{4a_1}A_{-1}^3\right)\right| = -1 \tag{2.10}$$

and solving for A_{-1}, which results in

$$A_{-1} = \sqrt{\left(1-10^{-\frac{1}{20}}\right)\frac{4}{3}\left|\frac{a_1}{a_3}\right|} = \sqrt{0.145\left|\frac{a_1}{a_3}\right|}. \tag{2.11}$$

Graphically, the concept of 1-dB compression point may be depicted as in Figure 2.2.

In the presence of interference, it is found that there will be a decrease in the output signal whenever the amplitude A_{I1} of the interferer increases, if

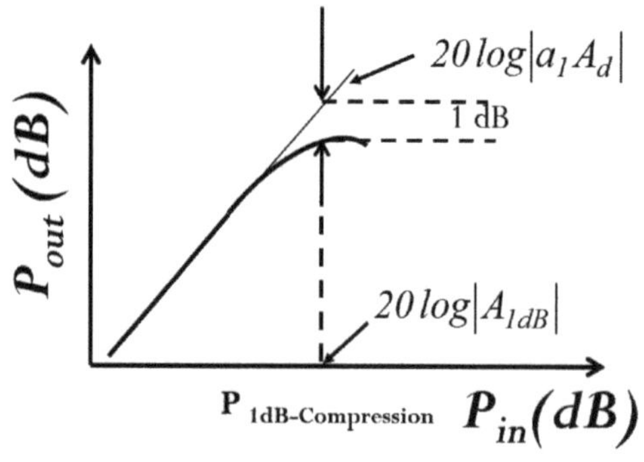

Figure 2.2 Depiction of 1-dB compression point. *Source:* Ref. [33].

$a_3<0$; see Equation (2.11). This interference is responsible for a diminution in gain by the amount $3a_3A_{I1}^2/2a_1$. Being known as *desensitization*, this nonlinear effect manifests itself in that, when it occurs, the gain afforded a weak desired signal which will be very small. In particular, the gain decrease in the presence of strong interference is *twice the rate* as compared to that in the gain compression case. Therefore, at high enough levels of interference, the gain will decrease to zero and the desired signal may even be totally *blocked.*

2.1.2.2 Cross-Modulation

If the input $x(t)$ is a combination of a weak desired signal $S_d(t)$ together with a strong interferer signal $S_{I1}(t)$, the latter undergoing an amplitude modulation (AM) $1+m(t)$, namely,

$$\left.\begin{array}{l} S_d(t) = A_d \cos 2\pi f_0 t \\ S_{I1}(t) = A_{I1}[1+m(t)]\cos 2\pi f_1 t \\ S_{I2}(t) = 0 \end{array}\right\}. \tag{2.12}$$

Under this circumstance, the output, $y(t)$, of the nonlinear system (2.2) is given:

$$y(t) = a_1 A_d \left[1 + \frac{3a_3}{4a_1} A_d^2 + \frac{3a_3}{2a_1} A_{I1}^2 \left[I + m^2(t) + 2m(t)\right]\right] \cos 2\pi f_0 t + \ldots. \tag{2.13}$$

The coefficient of the cosine function, in square brackets, contains desensitization and compression terms, previously introduced, plus two new terms, namely, $(3a_3/2a_1)\,A_{I1}^2 m^2(t)$ and $(3a_3/a_1)\,A_{I1}^2 m(t)$. These new terms capture the fact that the AM on the strong interferer is transferred unto the desired signal by way of the interaction with the nonlinearity. This phenomenon, which accompanies the nonlinearity of the transfer characteristic, is denoted as *cross-modulation.*

2.1.2.3 Intermodulation

Let us now take on the circumstance in which the input signal $x(t)$ contains more than one interferer, $S_{I1}(t)$ and $S_{I2}(t)$, together with desired signal $S_d(t)$,

$$\left.\begin{array}{l} S_d(t) = A_d \cos 2\pi f_0 t \\ S_{I1}(t) = A_{I1} \cos 2\pi f_1 t \\ S_{I2}(t) = A_{I2} \cos 2\pi f_2 t \end{array}\right\}. \tag{2.14}$$

Then, the resulting output may be expressed as

$$\begin{aligned} y(t) =& a_1 A_d \left[1 + \frac{3a_3}{4a_1} A_d^2 + \frac{3a_3}{2a_2} \left(A_{I1}^2 + A_{I2}^2 \right) \right] \cos 2\pi f_0 t \\ & + a_2 A_{I1} A_{I2} \left[\cos 2\pi \left(f_1 + f_2 \right) t + \cos 2\pi \left(f_1 - f_2 \right) t \right] \\ & + \frac{3}{4} a_3 \left[A_{I1}^2 A_{I2} \cos 2\pi \left(2f_1 \pm f_2 \right) t + A_{I1} A_{I2}^2 \cos 2\pi \left(2f_2 \pm f_1 \right) t \right] + \ldots \end{aligned} \tag{2.15}$$

In this equation, the terms with frequencies $(f_1 + f_2)$ are called *second-order intermodulation products,* whereas those with frequencies $(2f_1 \pm f_2)$ and $(2f_2 \pm f_1)$ are called *third-order intermodulation products.*

2.1.2.4 Memoryless Bandpass Nonlinearities

In wireless communications systems, the frequency response of power amplifiers (PAs) is denoted as of the bandpass type. These PAs are usually assumed to lack any storage elements, i.e., they are memoryless and, therefore, may be represented by nonlinear gain (AM−AM), phase distortion, and amplitude-to-phase conversion or AM−PM [18] characteristic.

When the signal being processed has a narrowband of frequencies, with carrier f_0, it may be represented by

$$x(t) = A(t) \cos \left[2\pi f_0 t + \phi(t) \right] \tag{2.16}$$

and the PA response given by

$$y(t) = f[A(t)] \cos \left(2\pi f_0 t + \phi(t) + g[A(t)] \right). \tag{2.17}$$

In Equation (2.17), the factor *f[A(t)]* captures the nonlinear *AM−AM conversion* gain and *g(t)* captures the amplitude-to-phase *AM−PM conversion.* A depiction of the typical magnitude and phase characteristics of a PA are given in Figure 2.3.

In Figure 2.4, the degree of AM/PM *conversion* is obtained from the slope of the phase shift vs. P_{in} or the amount of phase modulation (PM) caused by an input envelope variation of 1 dB; see Figure 2.4. AM/PM *transfer*, on the other hand, is the equivalent PM induced on an unmodulated carrier by another carrier with 1 dB of AM [18].

The typical amplifier model for system-level analysis is given in Figure 2.5.

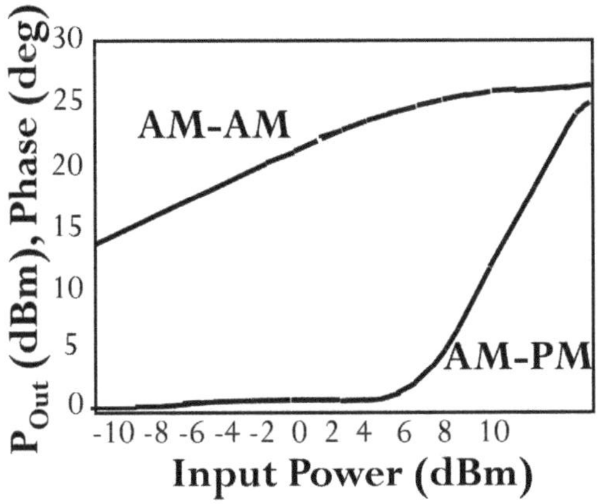

Figure 2.3 Characterization of Class AB power amplifier. *Source:* Ref. [18].

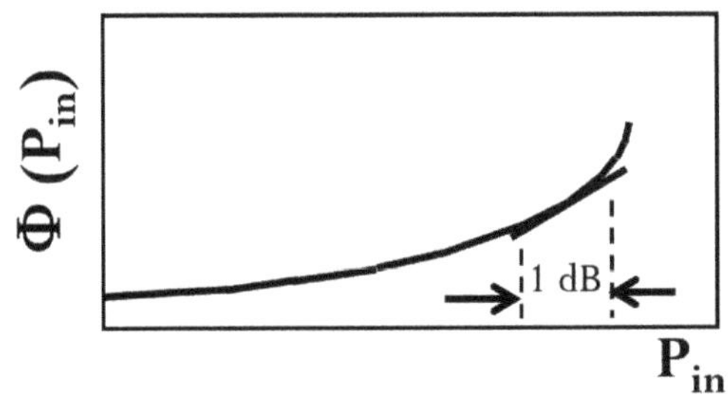

Figure 2.4 AM/PM conversion. *Source:* Ref. [18].

2.1.3 Mixers

As indicated in Chapter 1, the carrier to be transmitted must have a high frequency in order to render a relatively small and efficient antenna. The mixer is the nonlinear circuit utilized to translate a lower-frequency baseband information signal to a higher frequency; it may also be used to translate a higher frequency to a lower frequency at the receiver. The mixer ideally exhibits a square-law current–voltage transfer curve that yields an output signal containing the *sum and difference* frequencies of the two input signals. The performance of a mixer is characterized by the following parameters:

(i) Conversion gain (loss): This is the ratio, in dB, of the output (intermediate frequency (IF)) signal power to the input (radio frequency (RF)) signal power.

Name	Description	Units	Default
S21	Forward transmission coefficient, use x+jy, polar(x,y), dbpolar(x,y) for complex value	None	dbpolar(0,0)
S11	Forward reflection coefficient, use x+jy, polar(x,y), dbpolar(x,y),vswrpolar(x,y)) for complex value	None	polar(0,0)
S22	Reverse reflection coefficient, use x+jy, polar(x,y), dbpolar(x,y),vswrpolar(x,y)) for complex value	None	polar(0,180)
S12	Reverse transmission coefficient, use x+jy, polar(x,y), dbpolar(x,y) for complex value	None	0
NF	Noise figure [NF mode for NFmin = 0]	dB	None
NFmin	Minimum noise figure at Sopt[(NFmin, Sopt,Rn) mode used for NFmin>0]	dB	None
Sopt	Optimum source reflection for minimum noise figure [(NFmin, Sopt,Rn) mode used for NFmin>0]	None	None
Rn	Equivalent noise resistance[(NFmin, Sopt,Rn) mode used for NFmin>0]		
Z1	Reference impedance for port 1 (must be a real number)	None	
Z2	Reference impedance for port 2 (must be a real number)	None	
GainComp Type	Gain compression type	None	List
GainComp Freq	Frequency at which gain compression is specified	None	
ReferTo Input	Specify each of the gain compression options notified in polynomial order for various magnitude modes with respect to the input or output power of the device	None	Output
SOI	Second-order intercept	dBm	None
TOI	Third-order intercept	dBm	None
Psat	Power saturation point (always referred to input, regardless of the value of the ReferToInput parameter)	dBm	None
GainComp Sat	Gain compression at Psat	dB	5
GainCom Power	Power level at gain compression specified by GainComp	dBm	None
GainComp	Gain compression to phase modulation	dB	1
AM2PM	Amplitude modulation to phase modulation	deg/dB	None
PAM2PM	Power level at AM2PM	dBm	None

Figure 2.5 Parameters describing amplifier model.

(ii) Noise figure: This is the signal-to-noise ratio (SNR) at the input (RF) port divided by the SNR at the output (IF) port.
(iii) Isolation: This captures the amount of signal that "leaks" or "feeds through" from one mixer port to another.
(iv) Conversion compression: This is the RF input power level above which the curve of IF output power versus RF input power level deviates from linearity.
(v) Dynamic range: This is the range of amplitudes over which the mixer can operate without performance degradation. It is a function of the conversion compression and the noise figure.
(vi) Two-tone third-order intermodulation distortion: This is the amount of third-order distortion produced by the presence of a second received signal at the RF port.
(vii) Intercept point: This is the extrapolated (imaginary) point at which the fundamental response and the third-order spurious response curves intersect.
(viii) Desensitization: This is the compression at the desired signal frequency produced by the presence of a strong interfering signal on an adjacent frequency.
(ix) Harmonic intermodulation distortion: This occurs when the mixer-generated harmonics of the input signals mix. The frequencies of these distortion products are given by those where m and n represent the harmonic order.
(x) Cross-modulation distortion: This is the amount of modulation that is transferred from a modulated carrier to an unmodulated carrier when the combination of both signals is applied to the RF port.

The terminology of mixer performance is illustrated in Figure 2.6.

It is found that practical mixers produce responses to more signals than just the desired RF and local oscillator (LO) signals. The set of these unwanted responses are denoted as *spurious response*. They include outputs at the IF f_{IF} due to signals at frequencies *different from* the desired received frequency f_{RF}. Other frequency components originate from: (1) the antenna, in the absence of an RF preamplifier stage; (2) nonlinear behavior of RF amplifier; (3) the mixer itself; (4) harmonic frequencies of the oscillator from which the mixer drive is derived. A depiction of these spurious frequencies is shown in Figure 2.7 [19].

The fact that many possibilities exist, as indicated in Figure 2.7, illustrates the need for adequate selectivity in front of the mixer stage, and for good

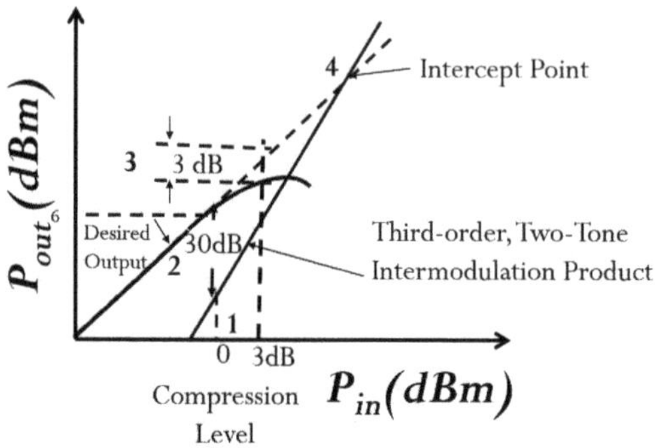

Figure 2.6 Transfer characteristic of mixer's IF output power vs. input power. The bold numbers indicate the following properties: **1**. At 0 dBm input, the output is 6 dBm, indicating a 6-dB conversion gain. **2**. At this input level, the third-order two-tone intermodulation product is 30 dB below the desired output. **3**. At a higher input value, the 3-dB compression point is indicated. **4**. At still higher input level, the intercept point is shown where the projected curves of desired output and third-order intermodulation product intersect. *Source:* Ref. [33].

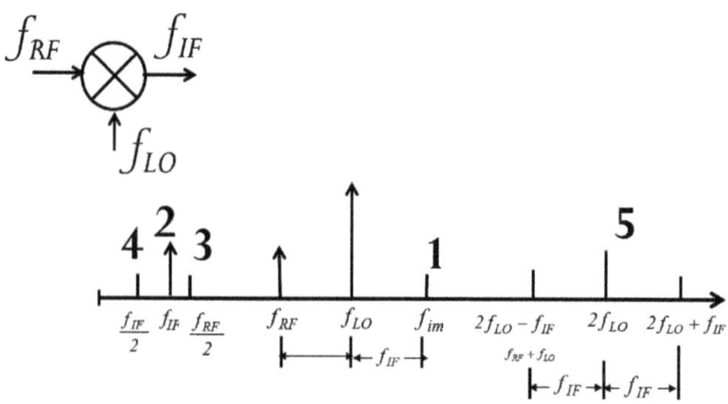

Figure 2.7 Spurious frequency products in mixers. The bold numbers indicate the following: **1**. The image frequency $f_{im}=f_{LO}+f_{IF}$. If a signal at this frequency is picked up by the antenna, and if it reaches the mixer input, it will beat with f_{LO} to produce a difference-frequency component equal to f_{IF}. **2**. An input signal at f_{IF} will appear in the output due to normal amplifier action. **3**. An input frequency equal to $f_{RF}/2$ may be doubled to f_{RF} by the square-law mixer term and then combine with f_{LO} to produce output at f_{IF}. **4**. An input at $f_{IF}/2$ may be doubled by the mixer and appear in the output. **5**. If the LO output includes a second harmonic at $2f_{LO}$, or if the mixer generates $2f_{LO}$, this component can beat with received inputs at $(2f_{LO} \pm f_{IF})$ to produce output at f_{IF}. *Source:* Ref. [19].

linearity in the RF stage, to avoid the generation of spurious frequencies at that point. A set of parameters capturing the mixer behavior is given in Table 2.1.

Table 2.1 Behavioral mixer model.

Name	Description	Units	Default
Sideband	Specify the sideband/image option for the mixer	None	Both
OutputSideSsuppression	Output sideband suppression (only relevant for Sideband=LOWER, UPPER)	dB	−200
InputImage Rejection	Input image rejection (only relevant for Sideband=LOWER IMAGE REJECTION, UPPER IMAGE REJECTION)	dB	−200
ConvGain	Conversion gain; use x+jy, polar(x,y), dbpolar(x,y) for complex	None	dbpolar(0,0)
RevConv Gain	Reverse conversion gain, use x+jy, polar(x,y), dbpolar(x,y) for complex value	None	polar(0,0)
SP11	S11, RF port reflection, polar(x,y), dbpolar(x,y), vswrpolar(x,y) for complex data	None	polar(0,0)
SP12	S12, IF port to RF port leakage, use x+jy, polar(x,y), dbpolar(x,y), vswrpolar(x,y) for complex data	None	polar(0,0)
SP13	S13, LO port to RF port leakage, use x+jy, polar(x,y), dbpolar(x,y), vswrpolar(x,y) for complex data	None	polar(0,0)
SP21	S21, RF port to IF port leakage, use x+jy, polar(x,y), dbpolar(x,y), vswrpolar(x,y) for complex data	None	polar(0,0)
SP22	S22, IF port reflection, use x+jy, polar(x,y), dbpolar(x,y), vswrpolar(x,y) for complex data	None	polar(0,0)
SP23	S23, LO to IF port leakage, use x+jy, polar(x,y), dbpolar(x,y), vswrpolar(x,y) for complex data	None	polar(0,0)
SP31	RF port to LO leakage (real or complex number) S31, RF port to LO port leakage, use x+jy, polar(x,y), dbpolar(x,y), vswrpolar(x,y) for complex value	None	polar(0,0)
SP32	IF port to LO leakage (real S32, IF port to LO leakage, use x+jy, polar(x,y), dbpolar(x,y), vswrpolar(x,y) for complex value or complex number	None	polar(0,0)
SP33	LO port reflection (real or complex num S33, LO port reflection, use x+jy, polar(x,y), dbpolar(x,y), vswrpolar(x,y) for complex value or complex number	None	polar(0,0)
PminLO	Minimum LO power before starvation	dBm	−100
DetBW	Detector bandwidth for LO limiting	Hz	1E100
NF	Double-sideband noise figure	dB	None
NFmin	Minimum double-sideband noise figure at Sopt	dB	None

Continued

Table 2.1 *Continued*

Name	Description	Units	Default
Sopt	Optimum SOurce Reflection for Minimum Noise Figure, use x+jy, polar(x,y), dbpolar(x,y) for complex value	None	None
Rn	Equivalent noise resistance		
Z1	Reference impedance for port 1 (must be a real number)	50	
Z2	Reference impedance for port 2 (must be a real number)	50	
Z3	Reference impedance for port 3 (must be a real number)	50	
GainComp Type	Gain compression type	None	List
GainComp Freq	Reference frequency for gain compression if gain compression is described as a function of frequency		
ReferTo Input	Specify gain compression with respect to input or output power of device	None	Output
SOI	Second-order intercept	dBm	None
TOI	Third-order intercept	dBm	None
Psat	Power saturation point (always referred to input, regardless of the value of the ReferToInput parameter)	dBm	None
GainComp Sat	Gain compression at Psat	dB	5.0
GainComp Power	Power level in dBm at gain compression for X dB compression point, specified by GainComp	dBm	None

2.1.4 Filters

As their name implies, filters allow the transmission of certain frequency bands while substantially blocking all others. Their performance is characterized by the smallest attenuation they introduce within the passband. Furthermore, the application may demand a linear phase (constant delay) throughout the passband. In theory, the frequency response of a filter should manifest amplitude transitions with zero width, i.e., abrupt, between passband to stopband (Figure 2.8(a)). In practice, however, this is not possible, and the actual response can only approximate theory (Figure 2.8(b)).

A typical filter model, employed in systems analyses, is shown in Figure 2.9.

2.1.5 Oscillators

The carrier frequency sent by a transmitter is generated by an oscillator circuit. Because the receiver must be tuned to this frequency, in order to distinguish it from other frequencies belonging to other transmitters, this frequency must be stable, i.e., it must be constant to within certain

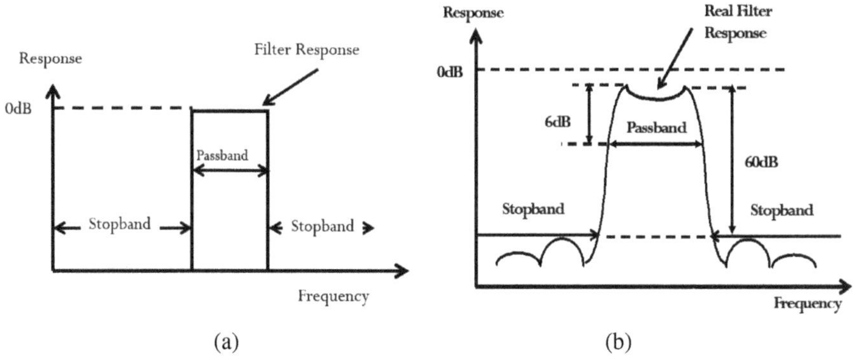

Figure 2.8 (a) Theoretical and (b) practical filter responses. *Source:* Ref. [33].

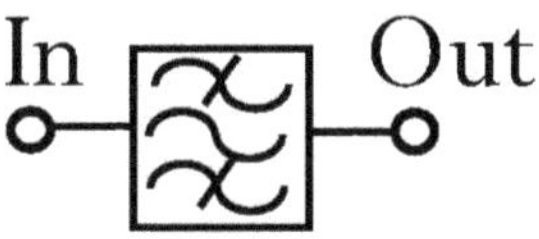

Name	Description	Units	Default
Fcenter	Passband center frequency	GHz	1.5
BWpass	Passband edge-to-edge width	GHz	1.0
Ripple	Passband ripple	dB	1
BWstop	Stopband edge-to-edge width	GHz	1.2
Astop	Attenuation at stopband edge	dB	20
StopType	Stopband input impedance type: OPEN or SHORT	None	Open
BWstop	Stopband edge-to-edge width	GHz	1.2
MaxRej	Maximum rejection level	dB	None
N	Filter Order (if N>0, it overwrites GDpass	None	0
IL	Passband insertion loss	dB	0
Qu	Unloaded quality factor for resonators, default setting is an infinite Qu and expresses a dissipationless resonant circuit	None	1e308
Z1	Input port reference impedance	Ohm	50
Z2	Output port reference impedance	Ohm	50
Temp	Temperature	°C	None
Z2	Output port reference impedance	Ohm	50

Figure 2.9 Model of typical bandpass filter.

specifications. The power of this signal is also important, as it is clearly a determinant of how far away from the transmitter it can reach. The frequency stability of the signal generated by an oscillator is denoted as its *phase noise* and the stability of its output signal amplitude as its *amplitude noise.* In general, an expression of the oscillator waveform, which includes both phase and amplitude noises, may be written as

$$A_{Osc}(t) = (A + a_n(t)) \cos (2\pi f_0 t + \phi_n(t)) . \tag{2.18}$$

In Equation (2.18), the phase and amplitude noises are, respectively, captured by $\phi_n(t)$ and $a_n(t)$.

2.1.5.1 Phase Noise of a Local Oscillator

The PM of the LO is very important when used to drive, e.g., a mixer; it is transferred onto the incoming signal (see Figure 2.10) [18]. This PM $\phi_n(t)$ may be of two types, namely: (1) deterministic, i.e., single specific frequencies (δ functions), or (2) random, i.e., a continuous range of frequencies.

Whether deterministic or random, the nature of the oscillator phase noise is observable in both the time and the frequency domains. In the time domain, phase noise manifests itself as a "jitter" in the time position of a digital clock (pulse) (Figure 2.11(a)). In the frequency domain, the phase

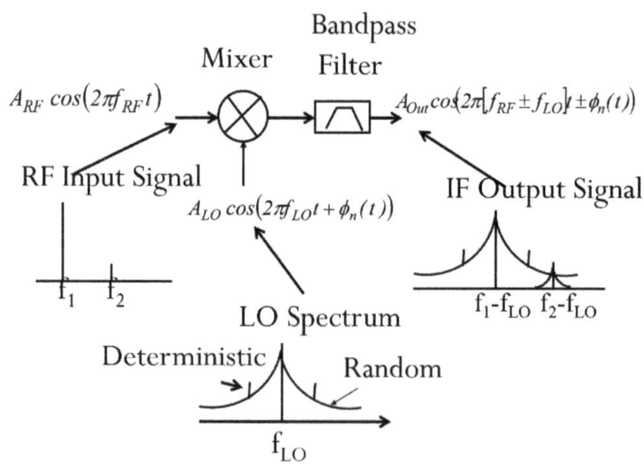

Figure 2.10 Impact of local oscillator noise on the noise of a down-converted RF signal. *Source:* Ref. [18].

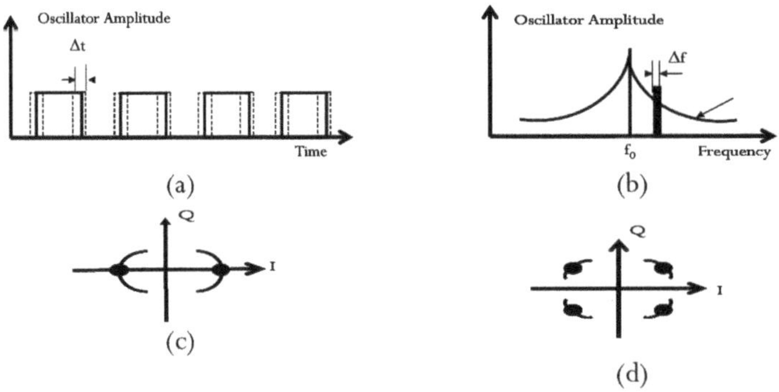

Figure 2.11 Manifestations of oscillator phase noise. (a) Oscillator amplitude in the time domain (jitter). (b) Oscillator amplitude in the frequency domain. (c) Impact of timing jitter on position of in-phase (I) and quadrature components (Q) of binary phase-shift-keyed (BPSK) constellation. (d) Impact of timing jitter on quadrature phase-shift-keyed (QPSK) constellation. *Source:* Ref. [18].

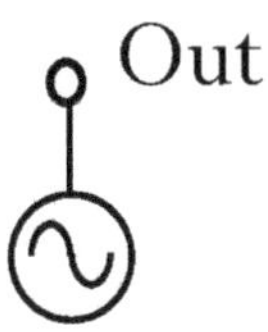

Name	Description	Units	Default
Freq	Frequency	GHz	1
P	Output power	W	dbmtow(0)
Rout	Output resistance	Ohm	50
PhaseNoise	Phase noise data	None	List(10Hz, -20dB, 100Hz, -40dB, 1KHz, -50dB)

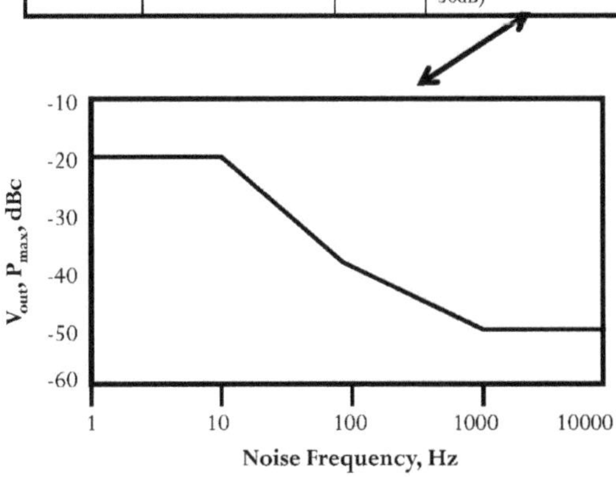

Figure 2.11(e) Oscillator model.

noise is manifested as a broadening Δf of the frequency of interest about f_0 (Figure 2.11(b)). In a digital communications system, the impact of the phase noise in the signals transmitted is captured in the so-called error vector magnitude (EVM). EVM is the displacement of the transmitted digital in-phase and quadrature amplitudes from the ideal ones, i.e., see Figure 2.11(c) and (d) [50].

2.1.5.2 Amplitude Noise

Setting $\phi_n(t)=0$, in Equation (2.18), yields the oscillator output signal as

$$A'_{Osc}(t) = (A + a_n(t)) \cos(2\pi f_0 t) \tag{2.19}$$

which represents an amplitude modulated signal. The frequency spectrum of this signal includes the main frequency f_0 and, in the narrowband case, two sidebands. If the time-dependent part of the amplitude $a_n(t)=0$, then the oscillator output signal will follow the expression

$$A''_{Osc}(t) = A \cos(2\pi f_0 t + \phi_n(t)). \tag{2.20}$$

This equation is similar to that of a frequency modulated signal, i.e., its spectrum is broadened around the desired frequency, f_0. This will be appreciated in the discussion of the LO that follows the next sub-section. Figure 2.11(e) shows the typical behavioral model of an oscillator, including its frequency, output power, and phase noise properties.

2.1.6 Frequency Multipliers

The most pure electronic frequency sources utilized in communication systems are crystal oscillators. Unfortunately, the piezoelectric crystal resonators they are based on operate at frequencies up to about 100 MHz. To increase the frequency of these oscillators, it is necessary to employ a frequency multiplier (FM) circuit, which produces as its output a harmonic of its input signal. Since FMs exploit a nonlinear transfer characteristic to generate harmonics, their properties and those of PAs are similar. In particular, the amplitude of the harmonics depends on both the level of the driving signal and the sharpness of the nonlinearity. In the case of FMs, however, the effect of multiplication is to also contribute to the increase in phase noise. For instance, random noise, characterized by the time function $\Phi_n(t)$, is added to the input FM signal, and then it would appear at the output multiplied by N, the FM multiplication factor. Therefore, if one were to treat the sidebands as phase modulating the carrier frequency with $\Phi_n(t)$, the

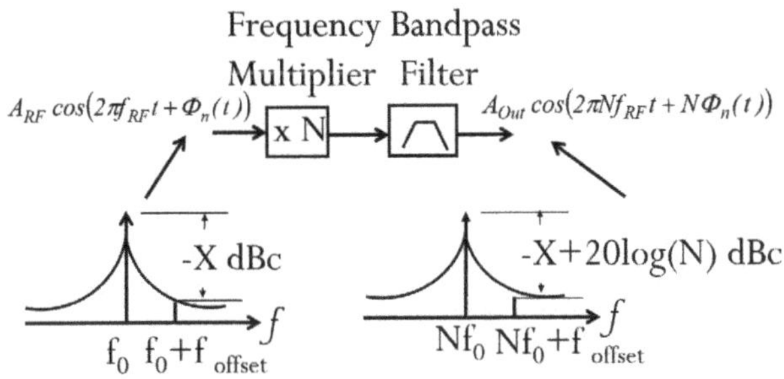

Figure 2.12 Phase noise enhancement by frequency multipliers. *Source:* Ref. [18].

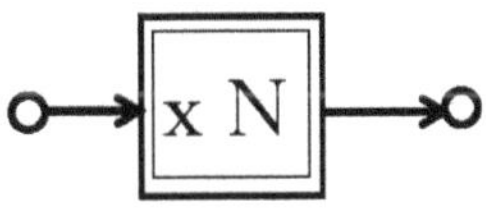

Name	Description	Units	Default
S11	Complex reflection coefficient for port 1	None	0
S22	Complex reflection coefficient for port 2	None	0
G1	Power gain of input tone	dB'	3
G2	Power gain of second harmonic relative to input tone	dB	None
G3	Power gain of third harmonic relative to input tone	dB	None
G4	Power gain of fourth harmonic relative to input tone	dB	None
G5	Power gain of fifth harmonic relative to input tone	dB	None
G6	Power gain of sixth harmonic relative to input tone	dB	None
G7	Power gain of seventh harmonic relative to input tone	dB	None
G8	Power gain of eighth harmonic relative to input tone	dB	None
G9	Power gain of ninth harmonic relative to input tone	dB	None
Pmin	Minimum input power for specified conversion	dBm	-40
Z1	Reference impedance for port 1	Ohm	50
Z2	Reference impedance for port 2	Ohm	50

Figure 2.13 Frequency multiplier model.

effective modulation index M_p would be increased by N, and the sidebands by 20log(N) dB (Figure 2.12) [18].

FMs may be represented by the behavioral model shown in Figure 2.13.

2.2 Antennas

2.2.1 Description of Antennas and Their Parameters

Once the information source signal is impressed upon the carrier, the transmitter must launch its corresponding electromagnetic (EM) wave into space via an antenna. The fundamental function of an antenna is to couple the power output of the transmitter to, or the power received by the receiver from, space. To do so, the transmitter's last amplifier drives a waveguide that

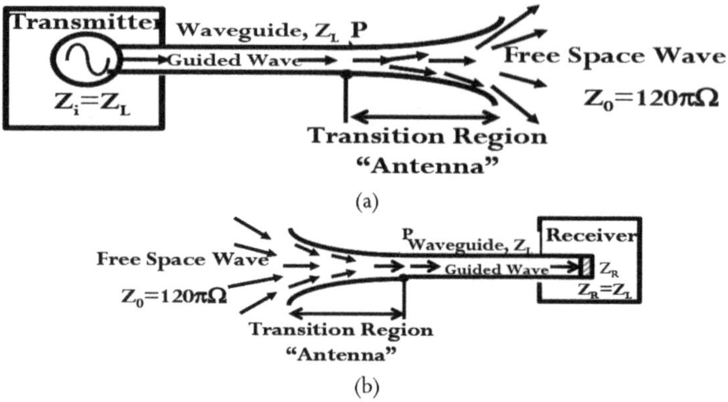

Figure 2.14 Antenna as a transition region between guided wave and free-space wave. (a) Transmitter antenna. (b) Receiver antenna. *Source:* Ref. [20].

is impedance-matched to an antenna. This waveguide embodies a region of transition, with its input matched to the amplifier's output impedance and its output matched to the antenna's input (Figure 2.14(a)), which subsequently launches the EM wave. Once the radiated EM wave reaches the receiver antenna, which is impedance-matched to free space, a waveguide transition region from the free space impedance to the input impedance of the receiver (Figure 2.14(b)) is effected.

In wireless communications, one can find a variety of antenna types operating in the 10 kHz (wavelength λ_0 = 30 km) up to about 300 GHz (wavelength λ_0 = 1 mm) range of frequencies. Some antenna types are shown in Figure 2.15 [20]. In general, the type of antenna chosen depends on various design factors, in particular, the specific application, and the desired radiation

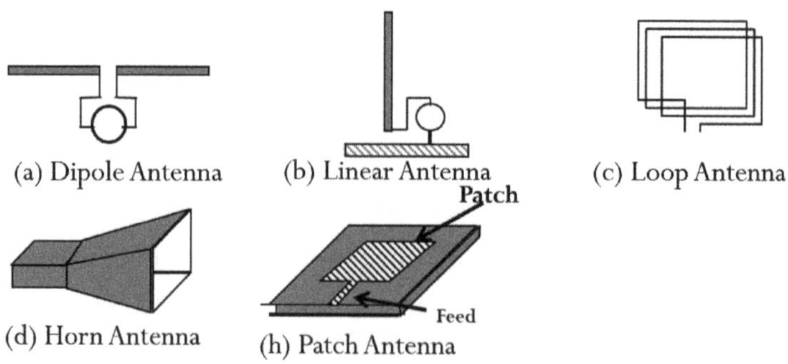

Figure 2.15 Typical antennas used in wireless communications. *Source:* Ref. [20].

properties. The weight, volume, and mechanical stability also may play a role. The role of the antenna dimensions becomes more important as the wavelength decreases [20].

The performance parameters of an antenna are its *directivity* and its *gain* [21–23]. These are related to its *effective area* and efficiency [21, 23]. The analytical formulation of these parameters, expressed in terms of a spherical coordinate system, is shown in Figure 2.16 [21].

In Figure 2.16, P_n is the normalized antenna power pattern, defined as [21]

$$P_n(\theta, \phi) = \frac{P(\theta, \phi)}{P(\theta, \phi)_{\max}} \text{ (dimensionless)} \tag{2.21}$$

where $P(\theta, \phi)$ is the radiated power, expressed as

$$P(\theta, \phi) = S_r r^2 = \frac{1}{2}\frac{E^2(\theta, \phi)}{Z} r^2 = \frac{1}{2} H^2(\theta, \phi) Z r^2 \qquad \left(\mathrm{Wsr}^{-1}\right). \tag{2.22}$$

The Poynting vector, which captures the transport of radiated power, is given by

$$S_{av} = \frac{1}{2} \int_s \mathrm{Re}\left(\vec{E} \times \vec{H}^*\right) \cdot d\vec{s} \tag{2.23}$$

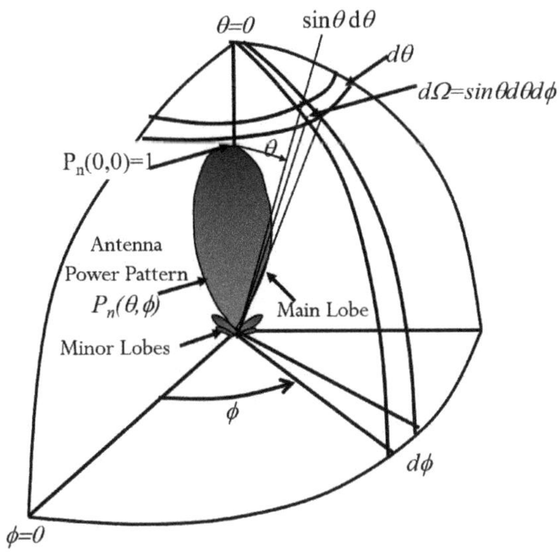

Figure 2.16 Normalized antenna power pattern P_n with maximum aligned with the $\theta = 0$ direction (zenith). *Source:* Ref. [21].

where, in Equation (2.22), r is the distance from the antenna to the measurement point, S_r is the power density radiated in the radial direction, $E(\theta, \phi)$ is the component of total transverse electric field as a function of angle, $H(\theta, \phi)$ is the component of total transverse magnetic field as a function of angle, and Z is the intrinsic impedance of the medium. Because P_n is the normalized (dimensionless) power, its integral over the solid angle (θ, ϕ) that the total radiated pattern occupies (the *beam solid angle*) Ω_{A} is given by [21]

$$\Omega_A = \int_{\phi=0}^{\phi=2\pi} \int_{\theta=0}^{\theta=\pi} P_n(\theta, \phi) \sin\theta d\theta d\phi = \iint_{4\pi} P_n(\theta, \phi) d\Omega. \tag{2.24}$$

Conceptually, all of the radiation of the transmitting antenna may be construed as being contained in an *effective solid angle* through which its radiated power would flow if this power were constant equal to the maximum power throughout this angle. Usually, this angle is taken to be approximately equal to the half-power beam width (HPBW); that is, this is the angle spanning either side of the maximum, at which the power has dropped to one-half the maximum power. In some applications, it is desirable that the energy radiated by the antenna be concentrated in a given direction. This characteristic is specified by its directivity, D [21]. D is expressed as

$$D = \frac{\text{maximum radiation intensity}}{\text{average radiation intensity}} = \frac{P(\theta, \phi)_{\max}}{P_{av}} \quad \text{(dimensionless)} \tag{2.25}$$

which is the ratio of the peak radiation intensity to the power radiated W divided by the maximum solid angle 4π sr, namely,

$$\begin{aligned} D &= \frac{P(\theta, \phi)_{\max}}{W/4\pi} - \frac{4\pi P(\theta, \phi)_{\max}}{\iint_{4\pi} P(\theta, \phi) d\Omega} - \frac{4\pi P(\theta, \phi)_{\max}}{\iint_{4\pi} \left[P(\theta, \phi)/P(\theta, \phi)_{\max}\right] d\Omega} \\ &= \frac{4\pi}{\iint_{4\pi} P_n(\theta, \phi) d\Omega} \end{aligned} \tag{2.26}$$

Applying Equation (2.24), we may write Equation (2.26) as

$$D = \frac{4\pi}{\Omega_A} \quad \text{(dimensionless)}. \tag{2.27}$$

From the definition (2.27), one can surmise that, since D is inversely proportional to the beam solid angle Ω_{A}, it is intuitively clear that the smaller the Ω_A, the greater the directivity. In other words, the smaller the solid angle in which the radiated power is concentrated, the greater the antenna directivity [21]. As a reference, an isotropic antenna, which radiates equally in all directions, $P_n(\theta,\phi)$=1 and Ω_A=4π, has a directivity of unity, $D=4\pi/\Omega_A=1$. In contrast, a dipole antenna, whose normalized radiation pattern is given by, $P_n(\theta,\phi) = \sin^2\theta$, has a *directivity*, $D = 4\pi \Big/ \int\limits_{\psi=0}^{2\pi}\int\limits_{\theta=0}^{\pi} \sin^2\theta d\theta d\psi = 1.5$. Thus, the maximum radiation intensity of a dipole antenna is 1.5 times greater than that of an isotropic antenna, which radiates equally in all directions [21].

Besides the directivity, which expresses the ability of an antenna to concentrate the radiated power in a given direction, another important characteristic compares its *radiation pattern* to that of an isotropic antenna. In other words, how sharp the beam pattern is. This parameter is denoted antenna gain and it is given by

$$G = \frac{\text{maximum radiation intensity}}{\substack{\text{maximum radiation intensity from a reference antenna}\\ \text{with the same power input}}}. \tag{2.28}$$

The gain is, thus, determined by taking the ratio of the peak radiated power intensity of the antenna in question, *P'*, to that of a reference isotropic antenna, P_0, assuming the same input power level is applied to both. It is given by [21]

$$G_i = \frac{P'}{P_0}. \tag{2.29}$$

P' refers to the peak power radiated by a 100% efficient antenna, P_{max}, multiplied by its efficiency, η, or

$$P' = \eta P_{\max} \tag{2.30}$$

where $\eta < 1$. Insertion of Equation (2.30) into Equation (2.29) gives

$$G_i = \eta\frac{P_{\max}}{P_0} = \eta \cdot D \tag{2.31}$$

Clearly, then, the gain and directivity of a 100% efficient antenna are equal. Now, since they represent relative measures referred to an isotropic antenna, it is common to express the gain and directivities in *decibels over isotropic*, or *dBi*, that is,

$$G_i(dBi) = 10\log G_i \tag{2.32}$$

and similarly for the directivity [21].

A receiving antenna is tasked with collecting energy from space. Since the radiated power is expressed in power per unit area, it is obvious that the greater the area of a receiving antenna, the greater the amount of power it can intercept from space. Because of this, the parameter *antenna effective area* is utilized to characterize a receiving antenna. Based on this concept, it is said that a receiving antenna possesses a certain *aperture area* A, which intercepts energy from a passing EM wave and is defined as

$$A = \frac{\text{received power}}{\text{power density of the incident wave}} = \frac{I^2 R_L}{S} \tag{2.33}$$

where A, the antenna aperture, has units of m^2, I is the rms terminal current, in units of *Amperes*, S is the Poynting vector of the incident wave (W/m^2), and R_L is the load resistance (Ω). The current I is given by

$$I = \frac{V}{Z_L + Z_A} \tag{2.34}$$

where V is the rms emf that the intercepted EM wave induces in the antenna, and Z_A and Z_L are, respectively, the antenna and load impedances. Maximization of the antenna aperture occurs when the emf V is highest, which has been shown to occur when the antenna is oriented in such a way that the conditions ($Im\{Z_L\}$=-$Im\{Z_A\}$) and ($Re\{Z_A\}$=R_r) are met, where R_r represents the radiation resistance of the antenna. Under these impedance matching conditions, the aperture area A is given a special name, namely, the *effective antenna area* A_e and is expressed as (Figure 2.17)

$$A_e = \frac{V^2}{4SR_r}. \tag{2.35}$$

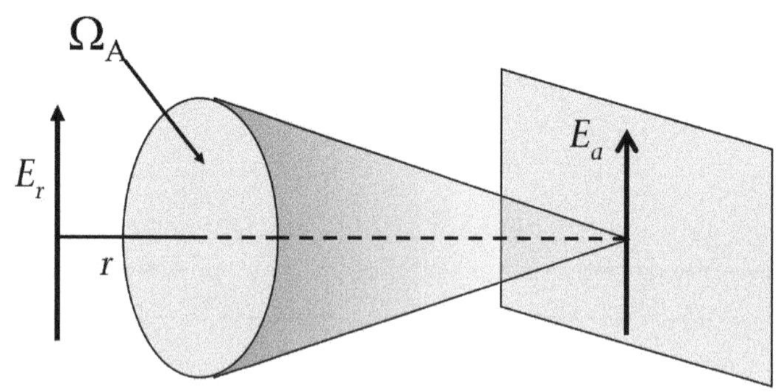

Figure 2.17 Radiation from aperture A_e with uniform field E_a. *Source:* Ref. [21].

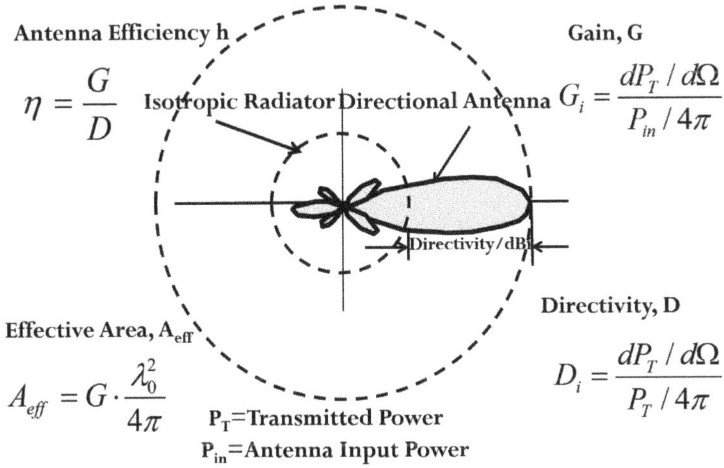

Figure 2.18 Visualization of gain, G, and directivity, D, relationships of an antenna. *Source:* Ref. [20].

A general relationship unites an antenna's effective aperture A_e, its beam solid angle Ω_A, and the wavelength of the EM wave being intercepted, namely [21],

$$\lambda^2 = A_e \Omega_A \quad \left(m^2\right). \tag{2.36}$$

This allows the directivity to be related to the effective aperture and wavelength by

$$D = \frac{4\pi}{\lambda^2} A_e. \tag{2.37}$$

We conclude that the directivity is inversely proportional to the wavelength. In other words, we develop the intuition that the higher the frequency, the greater the directivity. A summary of these relations is shown in Figure 2.18.

2.2.2 Antenna Arrays [23]

In many applications, it is found that it would be desirable to aim the direction of transmission or reception of the antenna into certain preferred directions. Because the directions of interest may change continually, and so should the antenna properties, a single antenna with fixed properties would be unable to meet this situation. Arrays of antennas, on the other hand, where the relative phase excitation of the currents driving each element in the array is made variable, may accomplish the desired goal. The simplest antenna array is the

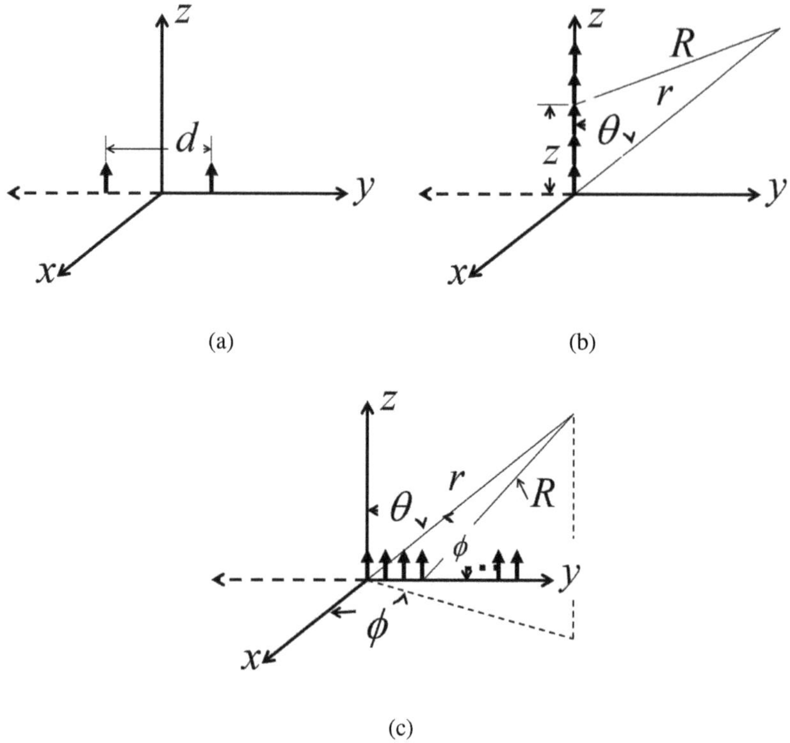

Figure 2.19 (a) A simple two-element antenna array. (b) A collinear array of current elements. (c) A linear array of current elements. *Source:* Ref. [23].

uniform linear array (ULA) (Figure 2.19), where all inter-element distances are equal.

Other popular antenna arrays exploit the nonuniformity of the distance between elements and the shape or configuration of the array of the antenna elements, which range from linear arrays to planar arrays in which the relative locations of the array may be, e.g., rectangular lattice, triangular lattice, etc. The elements in an antenna array are dipoles, patches, horns, etc., depending on the frequency of application.

2.2.2.1 Array Factor

The performance of an antenna array derives from the principle of superposition. In particular, with reference to Figure 2.19(a), suppose the two current elements are excited 180° out of phase. Then, in the plane that is perpendicular to the line bisecting the line joining the two antenna elements, the field must be zero. On the other hand, when examined along the *y*-axis, it is found that the field does not cancel but, in fact, would add if the separation *d* is close to half a wavelength. This implies that, in general, the net field is a function of phase and amplitude, the spacing between elements, and the direction of observation with respect to the elements. This type of antennas is most often referred to as phased array antennas (PAAs).

The performance of the ULA in Figure 2.19(b) may be captured by a set of *N* element currents, where they may be mathematically expressed as

$$\begin{aligned} I_z &= \delta(x)\delta(y)\left[I_0\delta(z) + I_1\delta(z-d) + I_2\delta(z-2d) + \ldots\right] \\ &= \delta(x)\delta(y)\sum_{n=0}^{N-1} I_n\delta(z-nd) \end{aligned} \quad . \tag{2.38}$$

In Equation (2.38), I_n is the current of the *n*th element, which is a complex number, thus having a magnitude and a phase, and *d* is the distance between elements in the *z*-direction. To find the vector potential of this array of elements, we invoke superposition, whereby

$$A_z = \iiint \frac{e^{-j\beta R}}{4\pi R}\delta(x)\delta(y)\sum_{n=0}^{N-1} I_n\delta(z-nd)dxdydz. \tag{2.39}$$

To evaluate this expression, from the geometry in Figure 2.19(c), one obtains that the distance *R* to the observation point may be approximated by

$$R \equiv r - z\cos\theta. \tag{2.40}$$

Then, as usual, we substitute Equation (2.40) in the exponential, but in the denominator, take it as $R \equiv r$. This gives the vector potential as

$$\begin{aligned} A_z &= \frac{e^{-j\beta r}}{4\pi r}\int e^{-j\beta R}\sum_{n=0}^{N-1} I_n\delta(z-nd)dz \\ &= \frac{e^{-j\beta r}}{4\pi r}\sum_{n=0}^{N-1} I_n e^{j\beta nd\cos\theta} \end{aligned} \tag{2.41}$$

From the vector potential, we can obtained the electric field as

$$\begin{aligned} E_\theta &= j\omega\mu A_z\sin\theta \\ &= j\omega\mu\frac{e^{-j\beta r}}{4\pi r}\sin\theta\sum_{n=0}^{N-1} I_n e^{j\beta nd\cos\theta}. \end{aligned} \tag{2.42}$$

In this expression, the summation factor is usually denoted as array factor (AF)

$$AF = \sum_{n=0}^{N-1} I_n e^{j\beta nd\cos\theta}. \tag{2.43}$$

If we express the complex nature of the *n*th-element excitation current as $I_n = A_n e^{jn\alpha}$ and the polar angle with respect to the line that contains the elements as ϕ (see Figure 2.19(c)), then Equation (2.43) would be

$$AF = \sum_{n=0}^{N-1} A_n e^{j\beta nd\cos\varphi + \alpha} \tag{2.44}$$

where α is the uniform phase shift from element to element as we run through the array. The interpretation of Equation (2.44) is as follows. If α is negative, then there is a phase lag or delay between the times at which elements are excited. If $\alpha = 0$, then all excitation currents are in phase, and all elements are excited at the same time. Equation (2.44) may also be written as

$$AF = \sum_{n=0}^{N-1} A_n e^{j\psi} \tag{2.45}$$

where $\psi = n\beta d\cos\phi + \alpha$. The magnitude or the square of AF is also referred to as space factor; it contains all the pattern directivity information and also the detailed element information. AF is a function of the number of elements, *N*, and their spacing, *d*.

A useful concept that helps in understanding how an antenna array functions is the "principle of pattern multiplication" [22]. According to Stutzman and Thiele [22], this principle states that the overall electric field pattern of an array whose elements are identical is given by multiplying the pattern of one of the individual elements and the pattern of an array of ideal isotropic point sources having the same locations, the same relative amplitudes, and the same phase as the elements of the original array. In other words, the complex normalized electric field pattern of an antenna array may be mathematically expressed as

$$F(\theta,\phi) = g_a(\theta,\phi)\cdot f(\theta,\phi) \tag{2.46}$$

where $g_a(\theta,\phi)$ is the normalized pattern of one of the antenna elements of the array, i.e., the element pattern, and $f(\theta,\phi)$ is the normalized AF.

2.2.2.2 Antenna Array Directivity

For a ULA, the mathematical form of directivity, D, can be written as [24]

$$D = \frac{4\pi U_{\max}}{P_{rad}} \tag{2.47}$$

where

$$U_{\max} = \frac{|AF|}{\max(|AF|)^2} \tag{2.48}$$

and

$$P_{rad} = \int_0^{2\pi}\int_0^{\pi} U(\theta)\sin(\theta)d\theta d\phi \tag{2.49}$$

where P_{rad} is the total radiated power and U_{max} is the maximum radiation intensity. The directivity of a ULA in the broadside array of isotropic elements is given by the approximation

$$D = 2\frac{Nd}{\lambda} \tag{2.50}$$

which is normally expressed in dB

$$D(dB) = 10\log_{10} 2\frac{Nd}{\lambda}. \tag{2.51}$$

Then, for a uniform element spacing, $d=\lambda/2$, we have

$$D(dB) = 10\log_{10} N. \tag{2.52}$$

2.2.2.3 Antenna Array Factor

One of the motivations behind antenna arrays mentioned at the beginning of this topic was their ability to steer the direction of radiation or reception continually. This is easily illustrated, again, with the ULA.

If the origin of the ULA is taken at one end of the array, i.e., $n = 0$, then the AF may be expressed by the sum

$$AF = A_0 \sum_{n=0}^{N-1} e^{jn\psi} \tag{2.53}$$

which may be recognized as a geometric series. This series, as is well known, sums to

$$\sum_{n=0}^{N-1} e^{jn\psi} = \frac{e^{jN\psi} - 1}{e^{j\psi} - 1} \tag{2.54}$$

whereby the AF may be written as

$$AF = A_0 \frac{e^{j\frac{1}{2}N\psi}\left(e^{j\frac{1}{2}N\psi} - e^{-j\frac{1}{2}N\psi}\right)}{e^{j\frac{1}{2}N\psi}\left(e^{j\frac{1}{2}\psi} - e^{-j\frac{1}{2}\psi}\right)} = A_0 e^{j\frac{1}{2}(N-1)\psi} \frac{\sin\frac{1}{2}N\psi}{\sin\frac{1}{2}\psi}. \quad (2.55)$$

If, on the other hand, the origin of the ULA is taken at the center of the array, and we call it the $n = 0$ element, and the number of elements is odd, then, the AF may also be expressed as

$$AF = A_0 \sum_{n=-\frac{1}{2}(N-1)}^{\frac{1}{2}(N-1)} e^{jn\psi} \quad (2.56)$$

which is also a geometric series but with the factor term, $A_0 e^{-\frac{1}{2}j(N-1)\psi}$, which allows us to write Equation (2.56) as

$$AF(\psi) = A_0 \frac{\sin\frac{1}{2}N\psi}{\sin\frac{1}{2}\psi}. \quad (2.57)$$

All we have done is redefine the labeling of the array elements; the two expressions for AF represent the same physical array behavior. By making the substitution $AF(\psi) \to AF(\psi + 2\pi)$, one finds that, in fact,

$$AF(\psi + 2\pi) = AF(\psi). \quad (2.58)$$

Thus, the AF is periodic with period 2π.

To visualize $AF(\psi)$, we proceed to examine a plot of the right-hand side of Equation (2.58), namely, $\sin\frac{1}{2}N\psi/\sin\frac{1}{2}\psi$ (Figure 2.20). This figure shows a plot of the AF as a function of ψ and a graphical representation of the equation $\psi = \beta d\cos\phi + \alpha$.

It is seen that, as the quantity βd rotates as a function of ϕ, only the portion of the AF for which $\beta d - \alpha \leq \psi \leq \beta d + \alpha$ corresponds to physical values of the angle ϕ. The range of $AF(\psi)$ is denoted by visible range and corresponds to a practical radiation pattern. The details of an actual five-element array are depicted in Figure 2.21.

The salient properties of a ULA may be appreciated from Figure 2.21(a). There we see that, if all the elements are excited in phase, $\alpha = 0$, then the maximum of the AF is at $\psi = 0 = \beta d\cos\phi$, which, solving for ϕ, gives $\phi = 90^o$. This is the so-called radiation in the “broadside” direction

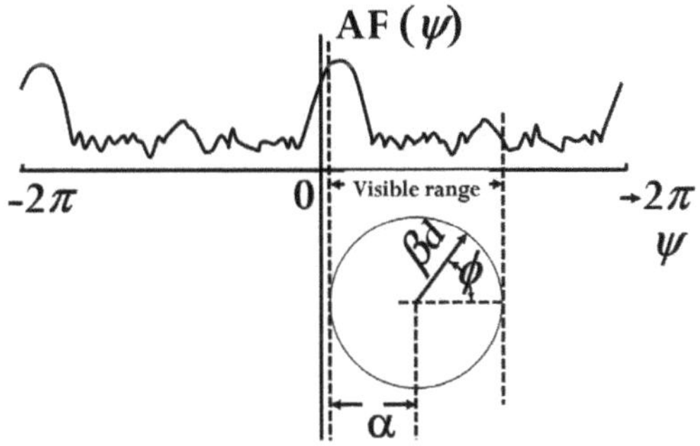

Figure 2.20 Array factor as a function of ψ, illustrating the relation between ψ and ϕ where $\pi \leq \phi \leq 0$. *Source:* Ref. [23].

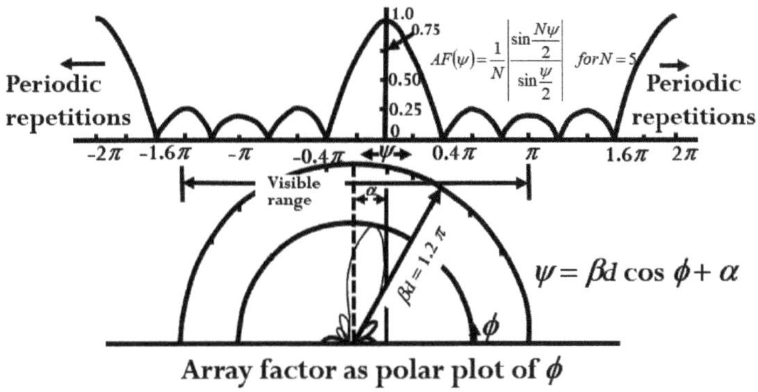

(a)

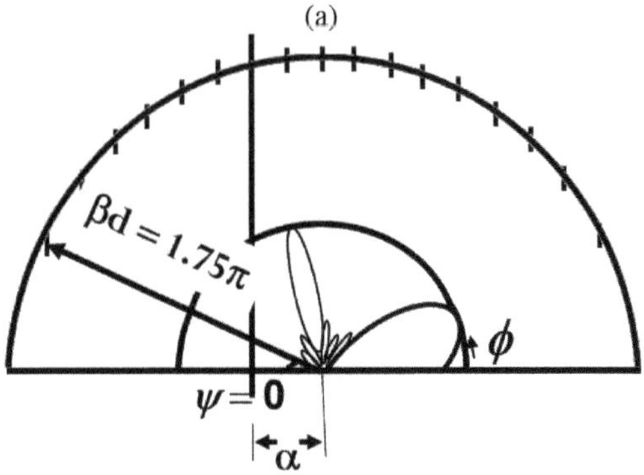

(b)

Figure 2.21 (a) Array factor for a uniform five-element array. (b) A factor as a polar plot of ϕ. Note grating lobe at approximately ϕ= 120° brought in by large spacing. *Source:* Ref. [23].

(perpendicular to the line) of the array. If the distance between elements is restricted to $\beta d < (2\pi - \pi/N)$, i.e., then, only one principal lobe is observed in the radiation pattern. On the other hand, if βd attains larger values, then many more lobes become visible in the radiation pattern (Figure 2.21(b)). These lobes are referred to as "grating" lobes. Lower amplitude lobes are called "side lobes."

When $\alpha \neq 0$, then there is a phase shift between the element-to-element excitation and, rather than the maximum of the radiation being at $\phi = 90^o$, it is at another angle ϕ_m. This other angle is obtain from

$$\psi = \beta d \cos\phi + \alpha \rightarrow \psi_m = \beta d \cos\phi_m + \alpha = 0 \tag{2.59}$$

from where

$$\alpha = -\beta d \cos\phi_m. \tag{2.60}$$

In general, then, we may express

$$\psi = \beta d \left(\cos\varphi - \cos\phi_m\right). \tag{2.61}$$

From Equation (2.61), it is seen that the position of the beam maximum may be varied by varying the element-to-element phase shift ϕ_m. This is accomplished by phase shifters and results in *scanning* the beam.

Lastly, when $\alpha = \pm\beta d$, the radiation maximum falls at the angles $\phi = 0^o$ or 180°, respectively. These are the directions in line with the array of elements; see Figure 2.22. Obtaining a single lobe is achieved by using an element-to-element spacing restricted to $\beta d < (\pi - \pi/N)$.

2.2.2.4 Prototypical Phased Array Antenna

As seen above, by judiciously controlling the amplitude and phases of the elements in an antenna array, the direction of its resulting radiation beam pattern may be adjusted. The fundamental architecture of a simple PAA, namely, a phased uniform linear antenna (ULA), is depicted in Figure 2.23. In the tutorials, such a system will be studied.

2.3 Free Space Propagation Model

The most fundamental contributor to signal attenuation during propagation is due to free space propagation. This contribution determines the strength of the signal arriving at the receiver when there is a clear line-of-sight path between it and the transmitter [25]. Typical situations where this mode of attenuation is prominent include microwave links, satellite communications,

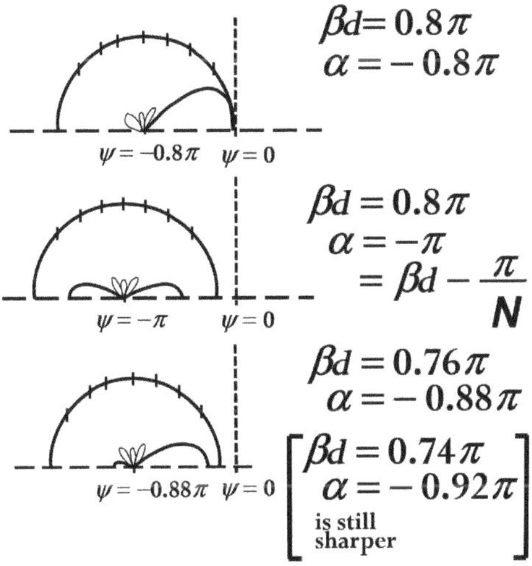

Figure 2.22 Array factors for five-element end-fire array. In the end-fire array, $\alpha > \beta d$ has the peculiarity that it results in a sharper main beam at the expense of higher relative side lobe level. *Source:* Ref. [23].

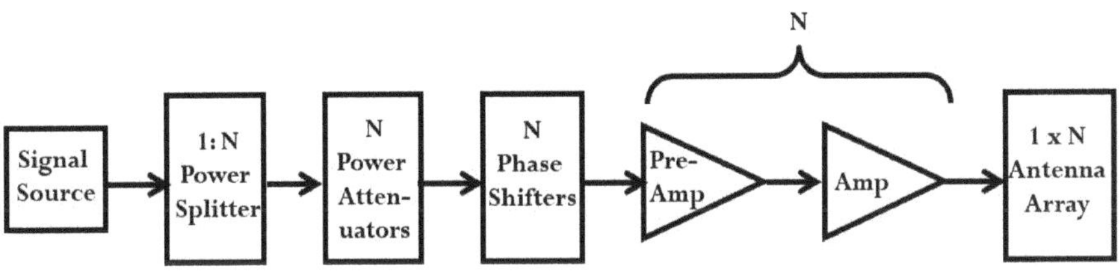

Figure 2.23 Prototypical architecture for uniform linear antenna.

and wireless sensing. Free space propagation is characterized by the so-called *Friis* formula, which expresses the fact that power arriving at a received is a decaying function of the distance of separation between transmitter and receiver raised to some power [25]. For a transmitted power P_t, a transmitter–receiver separation distance d, a transmitter antenna of gain G_t, a receiver antenna of gain G_r, a received power $P_r(d)$, a wavelength λ (m), and a system loss factor not related to propagation, *Friis* is given by [25]

$$P_r(d) = \frac{P_t G_t G_r \lambda^2}{(4\pi)^2 d^2 L}. \tag{2.62}$$

As shown previously, the antenna gain may be obtained from its effective aperture area, A_e, and the carrier frequency wavelength λ as

$$G = \frac{4\pi A_e}{\lambda^2} \tag{2.63}$$

where λ is related to the carrier frequency f in Hertz by

$$\lambda = \frac{c}{f} = \frac{2\pi c}{\omega_c} \tag{2.64}$$

where ω_c is in radians per second and c is the speed of light in meters/s. Clearly, P_t and P_r must be expressed in the same units, and G_t and G_r are dimensionless quantities. Finally, L represents miscellaneous losses, e.g., filter losses; a value of $L = 1$ indicates that there is no loss in the system hardware. The Friis formula, Equation (2.51), shows that the received power falls off as the square of the T-R separation distance. This implies that the received power decays with distance at a rate of 20 dB/decade [25].

An important parameter for characterizing the power loss between transmitter and receiver is the so-called *path loss* (PL). This parameter is essential when calculating the total loss of a "link." Mathematically, the PL is the difference, dB, between the effective transmitted and received powers, namely [25]

$$PL(dB) = 10 \log \frac{P_t}{P_r}. \tag{2.65}$$

Including both the transmitter and receiver antenna gains, the PL may be rewritten as

$$PL(dB) = -10 \log \left[\frac{G_t G_r \lambda^2}{(4\pi)^2 d^2} \right]. \tag{2.66}$$

When the antenna gains are excluded from the calculation, i.e., assuming them to be unity, then the PL may be expressed as [25]

$$PL(dB) = -10 \log \left[\frac{\lambda^2}{(4\pi)^2 d^2} \right]. \tag{2.67}$$

Certain assumptions must be met to apply Friis' formula. In particular, the receiving antenna must be at a distance d in the *far field* of the transmitting antenna. If D is the largest linear dimension of an antenna, then the far-field region may be taken as given by

$$d_f = \frac{2D^2}{\lambda} \tag{2.68}$$

where d_f >>D and $d_f \sim \lambda$. So, the receiving antenna will be in the far field of the transmitting antenna if 1) it is located at a distance of approximately *a* wavelength and 2) it is located at a distance much greater than the maximum linear antenna dimension. In propagation models, a reference distance d_0 is utilized to avoid the unpleasant case of having to calculate $P_r(d)$ for d=0. Under this circumstance, the power received at a distance d>d_0 is scaled from that at d_0 as follows [25]:

$$P_r(d) = P_r(d_0)\left(\frac{d_0}{d}\right)^2 \quad d \geq d_0 \geq d_f. \tag{2.69}$$

When dealing with mobile radio systems, it turns out that, due to the typical area of coverage being of the order of several km^2, the received power throughout the area may vary by many orders of magnitude. In this case, it is customary to express it in dBm [25]

$$P_r(d)dBm = 10\log\left[\frac{P_r(d_0)}{0.001W}\right] + 20\log\left(\frac{d_0}{d}\right) \quad d \geq d_0 \geq d_f \tag{2.70}$$

where $P_r(d_0)$ is in units of Watts, d_0 assumes the typical value of *1* m for indoor environments, and 100 m or 1 km for outdoor environments [25].

Since a transmitter antenna launches a radiated power "magnitude" into free space, but, fundamentally, what propagates is the EM field, the question comes up as to how do the antenna properties relate to the radiated EM field. For a linear radiator of height H ($H \ll \lambda$), on which a current of amplitude i_0 flows and the point P of observation of the field is at a distance d making an angle θ (see Figure 2.24), it can be shown that the electric field is given by [25]

$$E_r = \frac{i_0 H \cos\theta}{2\pi\varepsilon_0 c}\left\{\frac{1}{d^2} + \frac{c}{j\omega_c d^3}\right\} e^{j\omega_c(t-d/c)} \tag{2.71}$$

$$E_\theta = \frac{i_0 H \cos\theta}{4\pi\varepsilon_0 c^2}\left\{\frac{j\omega_c}{d} + \frac{c}{d^2} + \frac{c^2}{j\omega_c d^3}\right\} e^{-j\omega_c(t-d/c)} \tag{2.72}$$

$$H_\phi = \frac{i_0 H \cos\theta}{4\pi c}\left\{\frac{j\omega_c}{d} + \frac{c}{d^2}\right\} e^{j\omega_c(t-d/c)}. \tag{2.73}$$

Three field types are identified in the above, namely, the *radiation field* components, which decay as *1/d*, the *induction field* components, which decay as $1/d^2$, and the *electrostatic field* components, which decay as $1/d^3$. Far from the antenna, only the radiation fields with the 1/d decay survive [21, 25].

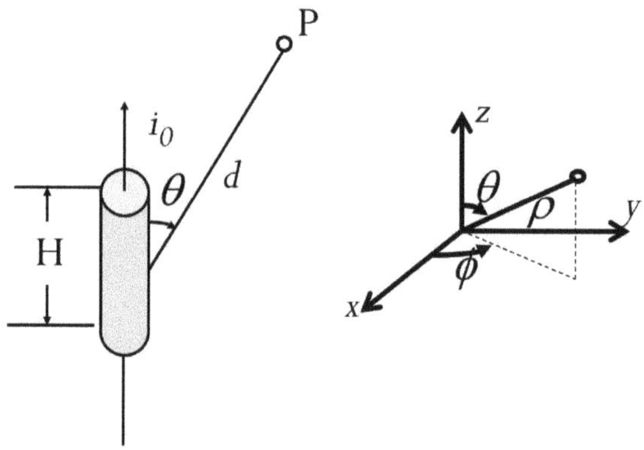

Figure 2.24 Linear electric field radiator. *Source:* Ref. [33].

One of the key parameters of an antenna is the total power it radiates. This is given by the product of the transmitted power and its corresponding antenna gain, P_tG_t, and is called the *effective isotropic radiated power* (EIRP). If the antenna is an isotropic radiator, then the amount of radiated power measured a distance d away from the antenna is obtained as the product of the radiated power *density*, namely,

$$P_d = \frac{EIRP}{4\pi d^2} = \frac{P_tG_t}{4\pi d^2} = \frac{E^2}{R_{fs}} = \frac{E^2}{\eta} W/m^2 \tag{2.74}$$

and the effective area of the receiving antenna, where R_{fs} is the intrinsic impedance of free space, $\eta = 120\pi\ \Omega(377\Omega)$ (Figure 2.25) [25].

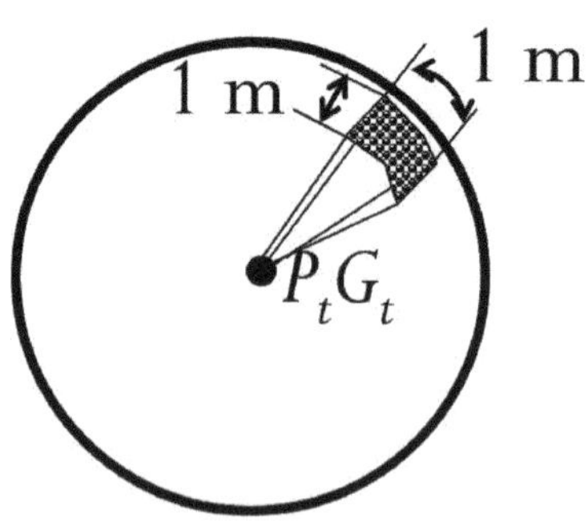

Figure 2.25 Sphere depicting radiating source and unit area on a sphere of radius d, receiving the power density. *Source:* Ref. [33].

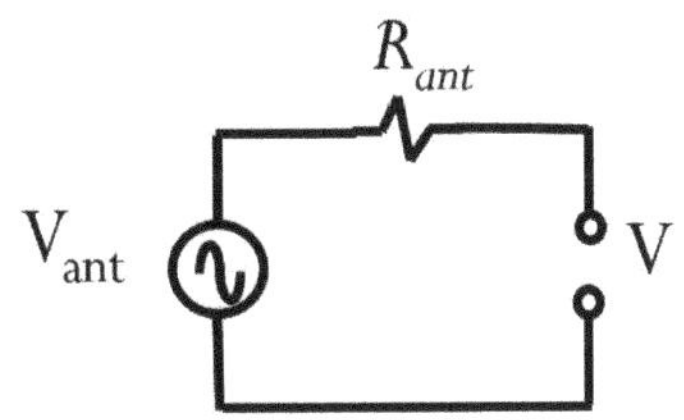

Figure 2.26 Equivalent circuit of receiving antenna. *Source:* Ref. [33].

The power density in free space R_{fs}= 377 Ω is given by

$$P_d = \frac{|E|^2}{377\Omega} W/m^2. \tag{2.75}$$

Of the receiving antenna has an effective aperture area A_e, then it receives a power

$$P_r(d) = P_d A_e = \frac{|E|^2}{120\pi} \cdot A_e = \frac{P_t G_t G_r \lambda^2}{(4\pi)^2 d^2} = \frac{|E|^2 G_r \lambda^2}{480\pi^2} W. \tag{2.76}$$

The question arises as to what voltage may be detected as a result of a power P_r being received at a distance d from the transmitter. With reference to Figure 2.26, the voltage in question is the open circuit voltage at the receiving antenna terminals

$$P_r(d) = \frac{V^2}{R_{ant}} = \frac{[V_{ant}/2]^2}{R_{ant}} = \frac{V_{ant}^2}{4R_{ant}}. \tag{2.77}$$

Equation (2.77) is obtained from the equivalent circuit of the receiving antenna (Figure 2.26), which is characterized by its open circuit voltage and its radiation resistance [25]. To maximize the detected voltage, the antenna output voltage, V, must be matched to the receiver circuit.

2.4 Summary

We began this chapter by introducing a variety of circuits/systems building blocks, in particular, their description and typical behavioral models and their parameters for systems analysis. These included, TLs, amplifiers, mixers, filters, oscillators, FMs, and antennas and, in particular, 1) efficiency, 2) effective area, 3) gain, and 4) directivity. Then, we introduced the rudiments of antenna arrays, defining the AF, the overall array electric field for identical elements, and the directivity for a ULA. We concluded with the topics of free-space propagation and PL, together with an elucidation of the relationship between power and electric field, the EIRP, and the measured received electric field and its relation to the received power.

2.5 Problems

1. Perform a product search and make a table of PA frequency, power gain, output power, 1-dB compression point, and IP3.

Frequency (GHz)	Power gain	Output power (dBm)	P1dB (dBm)	IP3 (dBm)
1				
2				
3				
...				
100				

   Do you notice any trends? Describe them.
2. How does the phase noise of a signal prior to frequency multiplication compare to that after frequency multiplication? Is it smaller or greater?
3. Why is frequency multiplication used?
4. What factors should be considered when choosing an antenna with high directivity if the pointing direction is fixed and if the pointing direction is variable?
5. What is Friis formula? What is it used for?
6. For an antenna radiating in free space, plot the PL as function of distance from the antenna for the following frequencies: 0.1, 1, 10, and 100 GHz.

3

Communication Systems Performance Parameters

3.1 Introduction

This chapter addresses the parameters describing the performance of a communications system. In the first place, we discuss the transmitter's maximum output power, modulation accuracy, and adjacent and alternate channel power. In the second place, we discuss the parameters pertaining to the receiver, in particular, its sensitivity, noise figure (NF), selectivity, image rejection (IR), dynamic range (DR), and spurious-free DR. In the third place, we deal with the topics of noise and nonlinearity. These include an introduction of a variety of noise sources, in particular, thermal noise and shot noise. Then, the topic of signal-to-noise ratio (SNR) is addressed, including available noise power, NF, noise temperature, NF of cascaded linear networks, and mixer NF. We then continue with a discussion of the calculation of the noise for a cascade of building blocks. In particular, we review effects of nonlinearity, namely, gain compression and intermodulation, and then address the topics of cascaded nonlinear stages, second-order intercept point, intermodulation distortion formulas, third-order intercept point (IP3), and cascaded third-order input intercept point (IIP3).

3.2 Transmitter Performance Parameters

The parameters describing the performance of a typical transmitter are its (1) maximum output power, (2) modulation accuracy, and (3) adjacent and alternate channel power. We begin by addressing the last two.

3.2.1 Modulation Accuracy

The modulation accuracy characterizes the fidelity of the transmitted waveform according to the *error vector magnitude* (EVM) or *waveform quality factor* ρ [27] measurement technique?

EVM captures the difference between the actual and theoretical symbol locations depicted on the modulation vector constellation diagram (Figure 3.1) [27].

EVM is quantified as follows [27]:

$$\boldsymbol{EVM} = \sqrt{\frac{\boldsymbol{E}\{|\overline{e}|^2\}}{\boldsymbol{E}\{|\overline{a}|^2\}}} \tag{3.1}$$

where $\boldsymbol{E}\{.\}$ denotes the expectation value of ensemble averages. The waveform quality factor, ρ, is the correlation coefficient between the actual waveform $Z(t)$ and the ideal waveform $R(t)$ and is expressed as follows:

$$\rho = \frac{\left|\sum_{k=1}^{M} \boldsymbol{R_k Z_k^*}\right|}{\sum_{k=1}^{M} |\boldsymbol{R_k}|^2 \cdot \sum_{k=1}^{M} |\boldsymbol{Z_k}|^2} \tag{3.2}$$

When the correlation between the ideal and the error signal is negligible, it can be shown that [59]

$$\rho \cong \frac{1}{1 + \boldsymbol{EVM}^2} \tag{3.3}$$

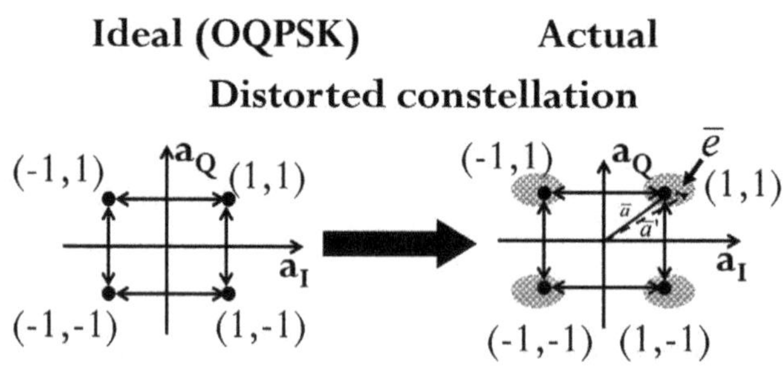

Figure 3.1 Constellation distortion due to noise. Example of how ideal orthogonal quadrature phase-shift keying (OQPSK) constellation may be distorted en-route from transmitter to receiver.

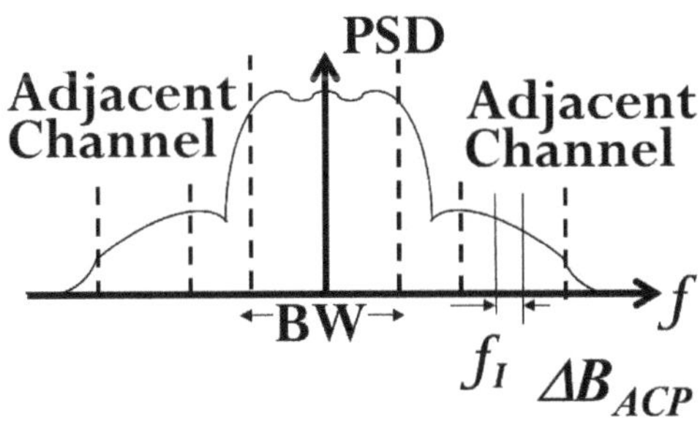

Figure 3.2 Description of adjacent channel power. *Source:* Ref. [33].

The contributions to EVM come from a variety of sources, including inter-symbol interference, close-in phase noise, carrier leakage (carrier feed-through-CFT), I & Q imbalance, power amplifier (PA) nonlinearity, in-channel bandwidth noise, etc. Therefore, the total EVM as a function of these may be expressed as [27]

$$\begin{aligned} EVM &= \sqrt{\sum_k EVM_k^2} \\ &= \sqrt{EVM_{ISI}^2 + EVM_{CFT}^2 + \sum_i EVM_{Nphasse,i}^2 + \dots} \end{aligned} \tag{3.4}$$

3.2.2 Adjacent and Alternate Channel Power

The adjacent and alternate channel power is defined as the ratio of the power integrated over an assigned bandwidth in the adjacent/alternate channel to the total transmitted power (Figure 3.2). It is expressed as [27]

$$ACPR = \frac{\int_{f_1}^{f_1+\Delta B_{ACP}} PSD(f)df}{\int_{f_0-BW/2}^{f_0+BW/2} PSD(f)df} \tag{3.5}$$

3.3 Receiver

The typical performance parameters for a receiver are its (1) sensitivity, (2) NF, (3) selectivity, (4) IR, (5) DR, and (6) spurious-free DR (SFDR). These are defined in what follows.

3.3.1 Sensitivity

The receiver sensitivity captures the minimum input carrier voltage that will produce the desired signal-to-noise power ratio (SNR) at the output of the intermediate frequency (IF) section. It embodies the ability of the receiver to detect a signal with given SNR, i.e., the lowest input signal power that guarantees the needed SNR at the RX output for a given power.

3.3.2 Noise Figure

The NF measures the degradation of the SNR as the signal progresses from the input (i.e., antenna) to the output (i.e., baseband).

It is expressed as [28]

$$NF = \frac{input\ SNR}{output\ SNR} = \frac{P_{si}/P_{ni}}{P_{so}/P_{no}} \tag{3.6}$$

where P_{si}, P_{sn}, and P_{so}, P_{no} are the power and noise at the input and the output, respectively. NF is usually given in decibels: NFdB = 10 $\log_{10}$ NF.

3.3.3 Selectivity

The ability of the receiver to discriminate a targeted signal in the presence of interferers (images) is given by the ratio of the bandwidths of the stopband to the passband; see Figure 3.3.

That is, selectivity is the parameter that quantifies the ability of the receiver to select the wanted signal out of all the other ones around it. This is determined mainly by the filters in the IF section.

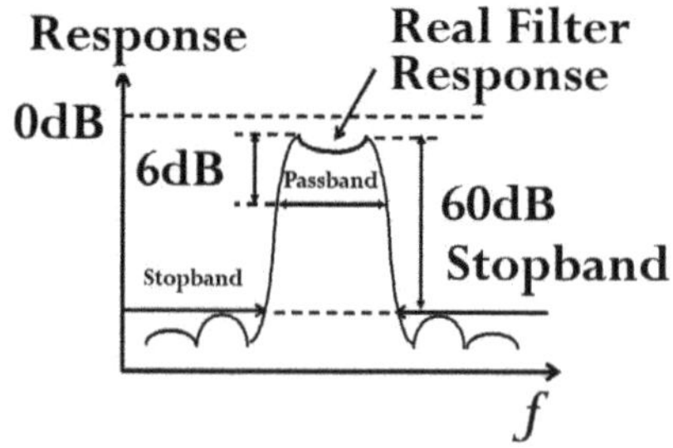

Figure 3.3 Filter response used in defining the shape factor. For receivers, the passband is taken to be the bandwidth between the points where the response has fallen by 6 dB, i.e., where it is 6 dB down or −6 dB.*Source:* Ref. [33].

The quantitative measure of selectivity is the *shape factor* (SF), defined with respect to Figure 3.3 by

$$SF = \frac{BW - 60dB}{BW - 6dB}. \tag{3.7}$$

Radio receivers employ filters with very high levels of performance and thus enable them to select individual signals even in the presence of many other nearby signals. A filter with a passband of 3 kHz at −6 dB and a passband of 6 kHz at −60 dB for the stopband, for instance, would have a SF of 2:1.

3.3.4 Receiver Image Rejection

The ratio of the receiver (RX) gain for the desired signal to that for the image, measured in dB, is denoted by IR. Figure 3.4 depicts the image problem. When signals are located at the same distance from the local oscillator (LO) than the desired radio frequency (RF) signal, they will also be simultaneously down-converted with the RF in such a way that they will both have the same IF. If it so happens that the desired RF signal is weaker than the image, then the image will corrupt it, thus making it hard to distinguish the desired RF signal.

3.3.5 Receiver Dynamic Range

The ratio of the largest input signal tolerated by the RX to its sensitivity measured in dB is denoted by DR.

3.3.6 Receiver Spurious-Free Dynamic Range

The SFDR combines a measure of the receiver linearity and noise performance (Figure 3.5). The SFDR is the sum of the output power of, say,

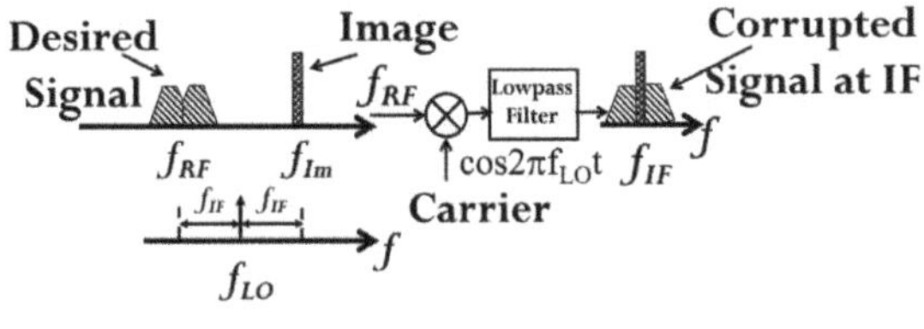

Figure 3.4 Illustration of image problem. *Source:* Ref. [33].

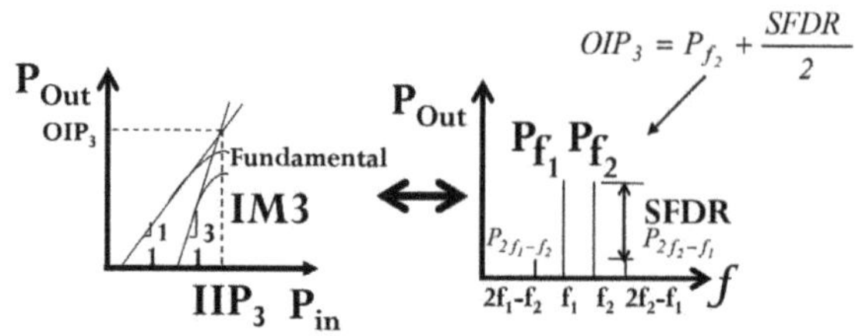

Figure 3.5 Power transfer and power spectral relationships illustrating SFDR concept. *Source:* Ref. [28].

tone f_2, and half the difference, in dB, in the output power of a tone f_2 and the output power at the third-order product.

3.4 Sensitivity and Dynamic Range Parameters

Designing a communications receiver involves engaging into an iterative process that ultimately results in achieving a compromise among conflicting performance goals [29]. The reason for this is posed by the need to both detect the smallest signals in the context of the noise introduced by the receiver as well as withstanding reception of the largest signals in the context of maximum acceptable distortions arising from circuit nonlinearities. The compromise in question manifests itself by way of the difference between the largest and the smallest signals the receiver system is capable of processing.

In achieving the lowest noise performance receiver, a designer has to develop an understanding of the fundamental causes of noise in the receiver circuits. Similarly, due to the distortion generated by large interfering signals, whose products fall within the receiver passband, it becomes difficult to receive small signals, thus reducing message reliability and occasioning suboptimal large-signal performance. Furthermore, it is found that obtaining the best large signal performance is accompanied by higher noise degradation, which, in turn, limits the weak signal reception. Getting the best small- and large-signal receiver performance, then, requires undertaking an optimization tradeoff [28].

A typical receiver adopts one of two fundamental architectures, namely, the *homodyne* (zero-IF) (Figure 3.6) and the *heterodyne* (Figure 3.7).

The homodyne receiver effects the translation of the desired incoming RF signal directly into the baseband, without an IF; because of this, it is also referred to as direct conversion or zero-IF. The IF signals typically occupy

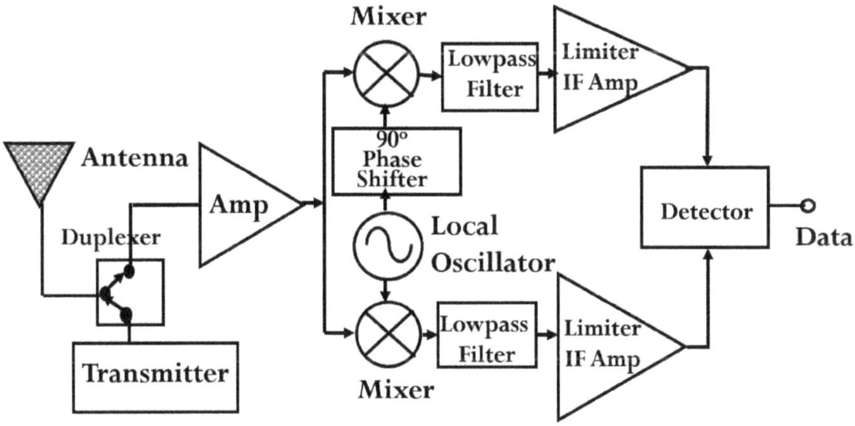

Figure 3.6 Homodyne (zero-IF) receiver architecture. *Source:* Ref. [28].

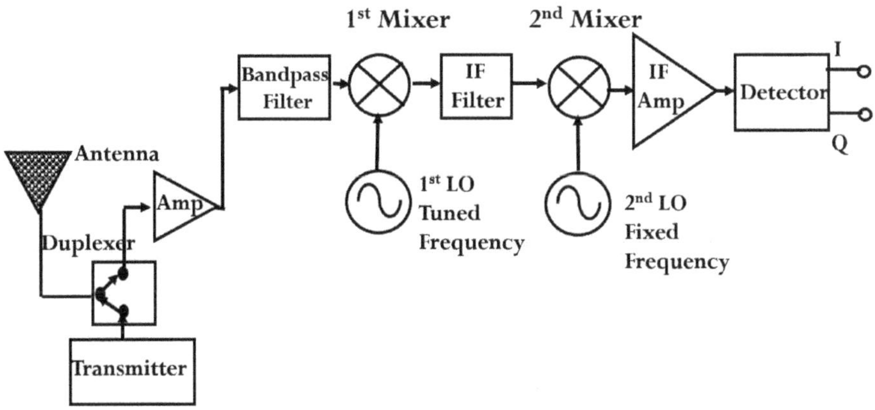

Figure 3.7 Heterodyne receiver. *Source:* Ref. [28].

frequencies that are well below the carrier frequencies, in particular, nearing DC on the low end.

The heterodyne receiver effects the translation of the desired incoming RF signal into one or more intermediate frequencies before demodulation. The last IF circuitry recovers the modulation information.

The dual-conversion super-heterodyne receiver employs successive mixers to translate the desired RF signal to two IF frequencies (Figure 3.8). This is accomplished by driving the mixer circuits with LO signals tuned at particular spacings above or below the RF signal. Mixing the RF and

LO signals produce a difference frequency known as the IF frequency. The dual down-conversion receiver employs two corresponding down-conversion mixers.

3.4.1 Definition of Receiver Sensitivity

One of the most important parameters that determines the overall performance of a communication system is its *sensitivity.* This is directly related to communication range and bit error rate (BER). The sensitivity of a receiver denotes the signal level that is required in order to achieve a given *quality* of the received information. When dealing with digital communications systems, the quality of the received information is given by the *BER* at a specific SNR. Quantitatively, the sensitivity is the absolute input power level that produces the desired SNR. It is computed from the *minimum detectable signal* (MDS) and the required carrier-to-noise ratio [28]. When referred to the input, the sensitivity is written as the sum of the MDS and the desired output SNR

$$Sensitivity\ (dBm) = \mathrm{MDS}(dBm) + \frac{C}{N} \tag{3.8}$$

where *C/N* is the output SNR that produces the desired performance [28].

3.4.2 Definition of Minimum Detectable Signal

The MDS is defined with respect to the noise floor at the IF (Figure 3.8). For a receiver with input temperature T_0, the noise floor at the IF is obtained as

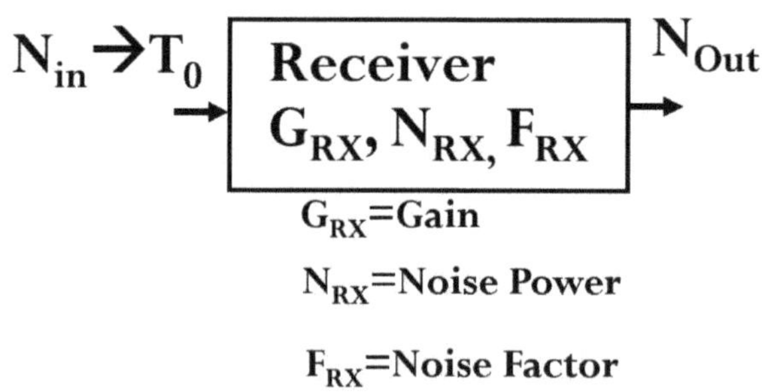

Figure 3.8 System-level receiver parameters. F is the system's noise factor. kT_0 x (1 Hz) is the thermal noise power in a 1-Hz bandwidth at T_0, B is the noise equivalent bandwidth, and G is the two-port's numeric available gain.

follows [28]. From the corresponding input noise level of

$$N_{in} = kT_0B \tag{3.9}$$

if the receiver possesses a gain G_{RX}, a noise power N_{RX}, a noise factor F_{RX}, and a bandwidth B, then the output noise is

$$N_{Out} = (N_{RX} + kT_0B)\, G_{RX} = F_{RX} kT_0 B G_{RX} \tag{3.10}$$

which encapsulates the process of the receiver taking the input noise power, adding to it the receiver noise power, and amplifying both simultaneously by the receiver gain to yield the output noise. With respect to the input noise, the noise introduced by the receiver may be expressed by

$$N_{RX} = (F_{RX} - 1)\, kT_0B \tag{3.11}$$

so that

$$N_{Out} = F_{RX} kT_0 B G_{RX} = F_{RX} \cdot kT_0(1Hz) \cdot \left(\frac{B}{1Hz}\right) \cdot G_{RX} \tag{3.12}$$

which may be expressed in decibels as

$$\begin{aligned} 10\log\left(N_{Out}\right) &= 10\log\left[kT_0(1Hz) \cdot \left(\frac{B}{1Hz}\right) \cdot F_{RX} \cdot G_{RX}\right] \\ &= -174dBm + 10\log B + N_{RX} + G_{RX}(dB) \end{aligned} \tag{3.13}$$

The noise floor is equal to the signal power N_{Out} delivered to the output, i.e., it is equal to the *output noise floor*. The MDS is thus defined as

$$MDS = -174dBm + 10\log B + N_{RX}. \tag{3.14}$$

Again the noise at the input passes through the receiver bandwidth, gets amplified by the receiver gain, and, upon addition to the receiver's own noise, appears at the output as noise floor, N_{Out}. If, from the output noise power in dBm, we subtract the receiver gain in dB, we obtain the input MDS. Thus, the MDS equation determines the input signal level required to deliver an output signal to a load equivalent to the output noise floor (Figure 3.9) [28].

3.4.3 Illustration of Signal-to-Noise Ratio

The noise floor at the output of a receiver is, as we have seen, determined by the MDS at the input which elicits that output signal power. In practice, in

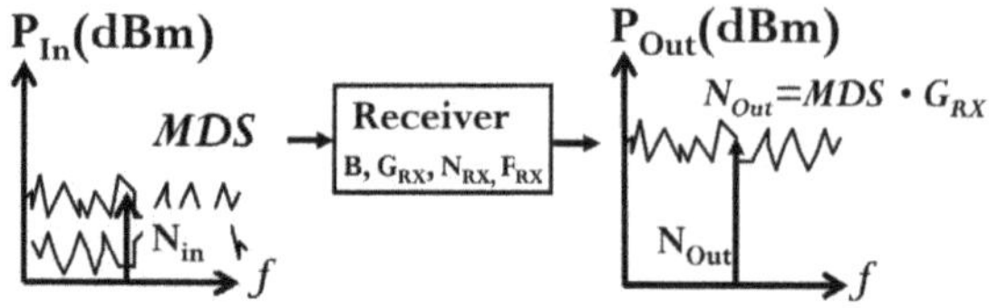

Figure 3.9 Illustration of minimum detectable signal. *Source:* Ref. [28].

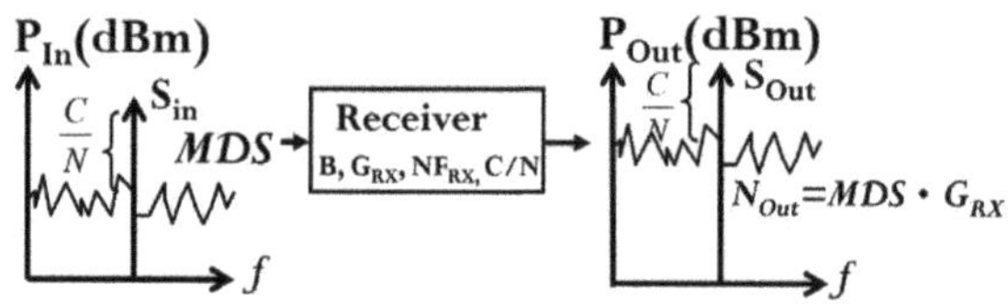

Figure 3.10 Illustration of SNR/CNR. *Source:* Ref. [28].

order to be able to extract the desired information, the signal at the receiver output must be over the noise floor. Accordingly, a specific SNR at the receiver output is necessary in order to recover the received information with a specific quality level. This specific SNR is sometimes also referred to as *carrier-to-noise ratio*, C/N. The actual input noise floor when the receiver is terminated at its input by a passive network, however, is given by kT_0B, while the MDS references the source noise and noise added by the receiver to its input terminals; thus, this is an imaginary noise floor due to noise added by the receiver (referred to the input).

As shown in Figure 3.10, when the MDS is used as the noise reference at the receiver input, the carrier-to-noise level is the same at both the input and the output. Therefore, the input signal S_{in} must be above the MDS level by a certain amount, which means that the output signal S_{Out} is over the output noise floor also by this same amount. Since the receiver adds noise, when the source noise is utilized as reference, the real SNR at the input is greater than the SNR at the output [28].

The required SNR at the receiver output is a function of the system parameters contributing to it, including the modulation scheme, the distortion of the group delay in the IF filters, the detector linearity, and the distortion introduced [28].

The carrier-to-noise ratio required is determined by the desired quality of the information received. In the case of a digital receiver, the primary measure of quality is the BER, which is the probability P(e) that any received

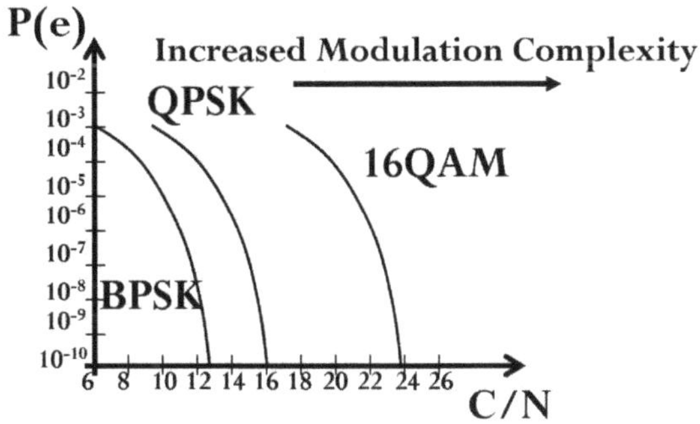

Figure 3.11 Probability of error versus carrier-to-noise ratio for various modulation schemes. *Source:* Ref. [28].

bit is in error. In particular, the BER is defined as the number of bit errors divided by the number of bits transmitted, and one of the things that degrade it is the jitter, which, in turn, results from noise in the receiver. In addition, the BER is also affected negatively by the level of detector nonlinearity, the frequency response of the filter, and the bandwidth. Figure 3.11 shows the BER as a function of the receiver's MDS. In particular, the greater the amount by which the desired signal exceeds the noise floor, the lower the BER (P(e)). It is found, then, that the probability of error increases dramatically as SNR decreases, which results in BER changing by several orders of magnitude with small increments in SNR. The BER worsens, however, with a decrease in signal level. In order to achieve an acceptable BER, one must utilize a number of modulation schemes with increased complexity that, however, requires higher SNRs [28].

3.4.4 Definition of 1-dB Compression Point

In practice, all systems are nonlinear. This nonlinearity manifests itself in the fact that its input–output relationship departs from linearity. When this occurs, the output becomes the combination of the fundamental plus the distortion products. This is depicted in Figure 3.12.

As a result of the nonlinearity, the power gain given by the ratio $G = P_{Out}/P_{in}$ stops being constant (Figure 3.13).

The departure from linearity in the power transfer plot is captured by the concept of *1-dB compression point* ($P_{1-\text{dB}}$ CP). This is defined as the input signal level at which the corresponding output power is 1 dB below what it

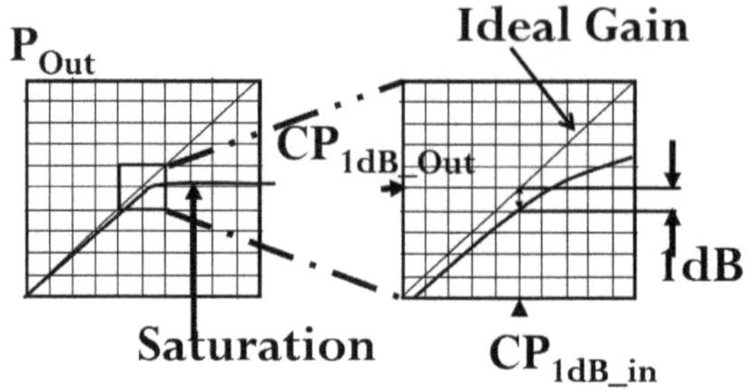

Figure 3.12 General output power versus input power characteristic. *Source:* Ref. [28].

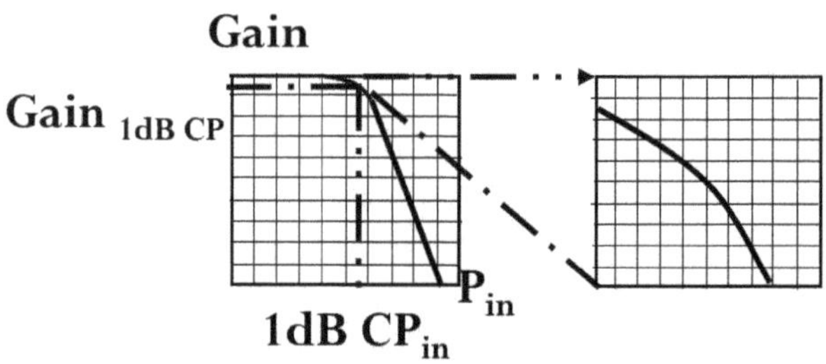

Figure 3.13 Gain versus input power. *Source:* Ref. [28].

would have been in a strictly linear circuit. As the amplitude of the input signal increases, so does the number of higher-order harmonics contributing to the output and, as a result, the output power versus input power continues to deviate from the ideal linearity. This deviation from linearity continues until the input signal reaches a point at which the output power corresponding to the fundamental frequency attains a maximum value; this point is the *saturation point*. Beyond this point, further increases in input signal cause the gain to drop (Figure 3.13). Once the circuit is "compressed," the gain experienced by all signals applied to it decreases. Any strong input signal into the receiver, whether desired or undesired, can cause compression.

The phenomenon of compression is important for the following reasons. When a weak signal is input into a receiver together with a strong undesired signal, it is observed that the strong undesired signal causes compression, as a result of which, if the smaller desired signal is close to the sensitivity level, the reduction in gain reduces the carrier-to-noise level which, in turn, increases the BER (Figure 3.11).

The amount of input power required to cause a gain drop of 1 dB is defined as the 1-dB compression point. In some occasions, the output power corresponding to the 1-dB compression, rather than the input, is given. The former is denoted as CP1dBin, indicating the amount of power incident at the input of the system that reduces the small signal gain of equation $G(dB) = 10\log(P_{Out}/P_{in})$ by 1 dB. The relationship between the output and input 1-dB compression points, CP_{1dBout} and CP_{1dBin}, is given as follows [28]:

$$CP_{1\,dBout} = CP_{1dBin} + Gain - 1dB \tag{3.15}$$

or

$$CP_{1dBin} = CP_{1dBout} - Gain + 1dB. \tag{3.16}$$

3.4.5 Definition of Intermodulation Distortion

Besides causing the phenomenon of compression, the nonlinearity in receiver systems results in *distortion products*. This refers to frequency components that may be identified in the output signal spectrum of the system. For instance, when applying two RF signals, f_1 and f_2, at the input of a strictly linear amplifier (Figure 3.14), the output will show both signals amplified and delivered to the output load.

As suggested by Figure 3.14, the output spectrum contains the same frequencies present at the input, except that these have been amplified. This situation reflects the fact that increasing the input power by 1 dB results in a corresponding increase of 1 dB in the output power of the fundamental. If the two RF signals, f_1 and f_2, are applied to a nonlinear amplifier, one finds that a number of additional signals, called distortion products, internally generated, are delivered to the load (Figure 3.15). The distortion products include sum and difference terms, $f_1 + f_2$ and $f_2 - f_1$, respectively, and the

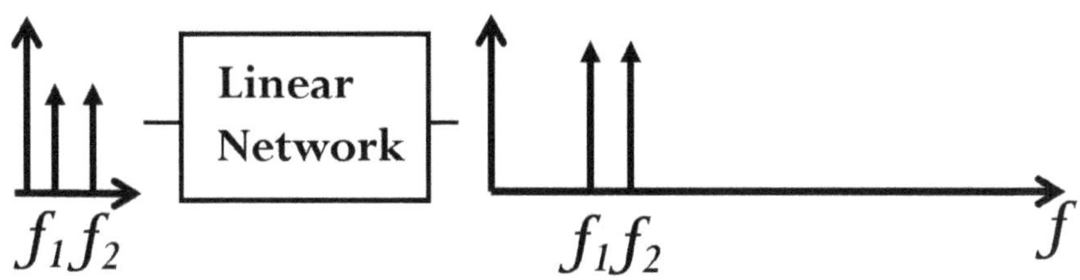

Figure 3.14 Processing of signals at frequencies f_1 and f_2 by an amplifying linear network. *Source:* Ref. [28].

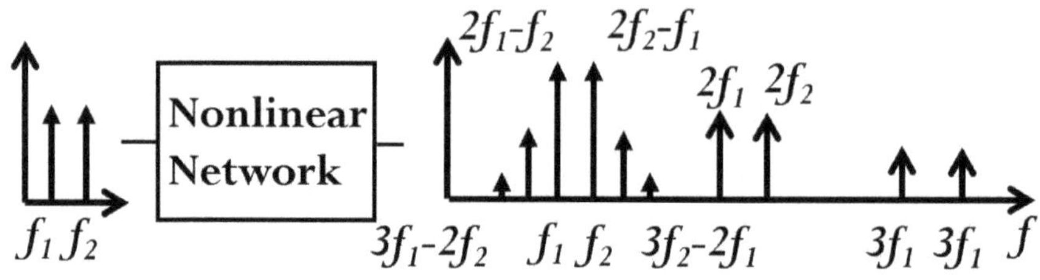

Figure 3.15 Processing of signals at frequencies f_1 and f_2 by a nonlinear network. *Source:* Ref. [28].

second harmonics $2f_1$ and $2f_2$, etc. In addition, specific mixing products will be generated as the frequencies

$$IMD = \pm m f_1 \mp n f_2. \tag{3.17}$$

It is apparent from Equation (3.17) that the *third-order intermodulation distortion (IMD)* products (i.e., those for which $m + n = 3$) occur at the frequency locations $2f_2$-f_1 and $2f_1$-f_2. Therefore, the second harmonic of f_2 interacts with fundamental f_1 and the second harmonic of f_1 interacts with fundamental f_2. As seen in Figure 3.15, the distortion products are spaced equally with respect to the two fundamental frequency terms and exhibit the behavior that, when the power levels of f_1 and f_2 are changed by *1 dB*, the amplitude of of both third-order IMD products changes, in turn, by *3 dB* (Figure 3.16). The amplitude of the third-order products is seen to be lower than that of the fundamental terms. This difference, which is dependent on the input power, is called the *rejection ratio* (RR) [28]. On the other hand, as the input power is increased, for every 1 dB of increase, the third-order intermodulation RR decreases by 2 dB. Thus, the continued increase in input power level leads to a distortion product power that equals that of the fundamental frequency. This power is referred to as the *third-order intercept point* (IP3).

Figure 3.16 shows a plot of the output power at the fundamental versus the input power, which reveals a linear characteristic with unity slope. It also shows, together with the fundamental, a plot of a third-order intermodulation distortion product, which reveals that the third-order intermodulation distortion product has a linear characteristic with a slope of *three*. Since the slope of the distortion product is steeper than that of the fundamental, the third-order distortion characteristic intersects the fundamental curve. This intersection is known as the network's *third-order*

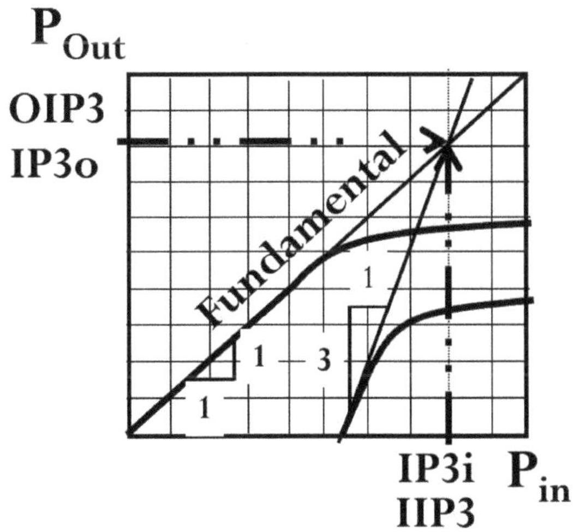

Figure 3.16 Output power P_{out} for one fundamental frequency signal versus input power P_{in}. *Source:* Ref. [28].

intercept point. When referenced to the network's input power, the intercept point is denoted as IP3i or IIP3; when referenced to the output power, it is denoted as IP3o or OIP3 [28].

The gain, IIP3, and OIP3 are related. In particular, the output intercept point is equal to the input intercept point plus the small-signal gain in dB, that is,

$$OIP\,3 = \; IIP\,3 + \; Gain. \tag{3.18}$$

In reality, the intercept concept is an ideal construct since it is not possible for the input power of a real system to be large enough to reach the theoretical intercept point due to the fact that compression sets in before that occurs. Once compression begins, further increases in the fundamental frequency power and in the distortion products are limited. It is found, in practice, that the 1-dB compression point usually occurs at close to 10 dB below the third-order intercept point. Thus, intercept point measurements must be performed at power levels at which the device is not compressed [29].

The third-order rejection ration (RR_{3o}), another distortion parameter, is measured using a spectrum analyzer at a fixed input power level (Figure 3.17) [28].

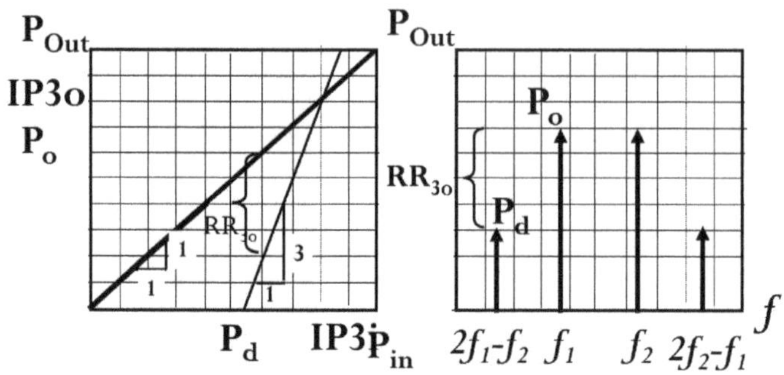

Figure 3.17 (a) Fundamental P_o and third-order distortion P_d output powers versus input power P_{in}. (b) Output frequency spectrum at a fixed input power level. *Source:* Ref. [28].

RR_{3o}, as shown in Figure 3.17, is the difference, in dB, between the output powers at the fundamental P_o and at the distortion P_d, taken at one point on the third-order distortion characteristic. Referenced to the output, the RR_{3o} is defined on the output power versus input power plot. To determine whether undesired signals in a receiver environment will interfere with it, the rejection ratio RR_{3o} for a given power level is needed.

The rejection ratio RR_{3o} is related to the output third-order intercept point as follows [28]:

$$IP_{3o} = P_o + \frac{RR_{3o}}{2} \tag{3.19}$$

or

$$OIP\,3 = P_o + \frac{RR_{3o}}{2}. \tag{3.20}$$

The third-order rejection ratio RR_{3o} is just the dB difference between the measured output power of one of the fundamental tones and one of the distortion products

$$RR_{3o} = P_o - P_d. \tag{3.21}$$

Substituting Equation (3.21) into Equation (3.20) results in the third-order output intercept point of the two-port

$$OIP\,3 = P_o + \frac{RR_{3o}}{2} = P_o + \frac{P_o - P_d}{2} = 1.5P_o - 0.5P_d. \tag{3.22}$$

Measuring the fundamental output power P_o and the distortion power P_d determines the third-order intercept point referenced to the output.

Furthermore, rearranging the equation for OIP3, the third-order input intercept point IIP3 is determined as

$$IIP3 = OIP3 - Gain. \tag{3.23}$$

Knowing the third-order output intercept point, the power level of the distortion product is determined for other input power levels as follows:

$$OIP3 = 1.5P_o - 0.5P_d \rightarrow P_d = 3P_o - 2OIP3. \tag{3.24}$$

The reader might wonder: Why spending all this effort discussing third-order products? Why is the third-order intercept point an important performance parameter? Well, there are a number of reasons for this, which are related to the environment in which the receiver operates. In the first place, all RF signals whose frequencies happen to coincide with those of the selected receiver channels will, in principle, enter and be processed by the receiver. In the second place, there will exist, in general, in addition to these, adjacent and close-in signals that will be processed through the nonlinear circuits in the receiver's front end, such as the amplifier and mixer circuits. In the third place, since signals in nearby channels enter and are also processed, they create third-order distortion products within the passband that are coincident with the RF front-end passband. In the fourth place, these distortion products may occur at the same frequency as the desired *signal*, and, therefore, in the fifth place, if these distortion products possess a power level slightly less than, or at a power anywhere above the desired signal, they will interfere with its reception! Figure 3.18, representing the spectrum of a perfectly linear receiver, gives an indication of this. In particular, the spectrum shows the desired signal $\mathrm{f_d}$ and two undesired signals $\mathrm{f_{U1}}$ and $\mathrm{f_{U2}}$, with the closest undesired signal $\mathrm{f_{U1}}$ assumed to be located one channel spacing away from the desired signal, say, 35 kHz, and the second undesired signal $\mathrm{f_{U2}}$ located two channel spacings away from the desired signal, say, 70 kHz. Assumption of a perfectly linear receiver implies that interference from third-order distortion products is impossible.

When the receiver in question is nonlinear, its output spectrum would look like that in Figure 3.18(b). In this case, the two undesired signals f_{U1} and f_{U2} cause a third-order intermodulation distortion product that *coincides* with the desired signal. As a result, if the power of the distortion product is greater than that of the desired signal, the reception of the information carried by the latter at f_{d} would be impossible. This is due to the fact that the coincidence of desired and third-order product frequencies would prevent, even filtering,

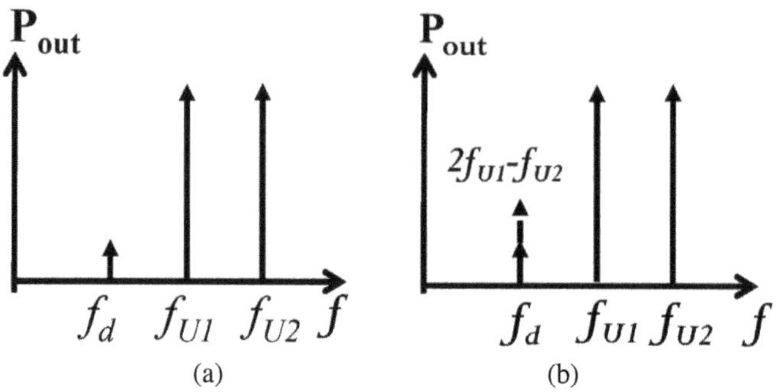

Figure 3.18 Typical output spectrum of (a) linear receiver and (b) nonlinear receiver. *Source:* Ref. [28].

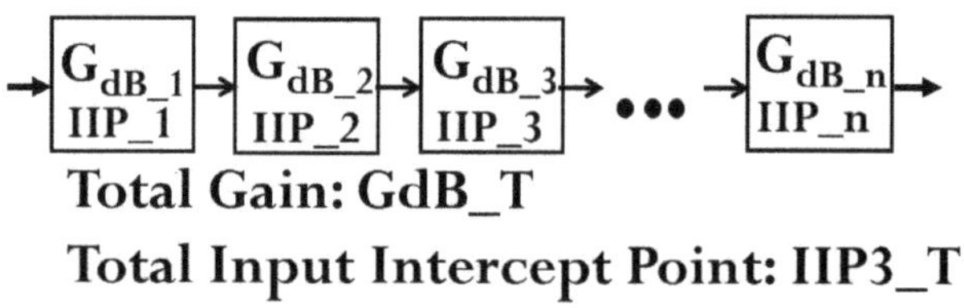

Figure 3.19 Receiver consisting of cascaded networks. *Source:* Ref. [28].

from enabling a clear reception. The intercept point, therefore, must be high enough to provide *a third-order RR* capable of protecting or guaranteeing that the largest interfering signals incident at the receiver input terminals would not give rise to products exceeding the power of the desired signal [28].

3.4.6 IP3 for Cascade of Networks

We now address the calculation of the overall third-order intercept point of a receiver system consisting of a cascade of building blocks, where each one is characterized by a gain and third-order input intercept point (Figure 3.19). The results of this calculation may be employed in analyses where the gains and intercept points are known, or for design, when trials are undertaken to optimize the parameters that will produce an overall desired target performance.

It will be apparent that each building block in the system has an effect on the total system distortion and that this may occur even when it adds no distortion of its own. In particular, it may be shown that IIP_T, the total *numeric* intercept point of the cascaded system, is given by the reciprocal of equation

$$\frac{1}{IIP3\_T} = \left[\left(\frac{1}{IIP3\_1} \right)^q + \left(\frac{g_1}{IIP\_2} \right)^q + \left(\frac{g_1 g_2}{IIP\_3} \right)^q + \cdots \left(\frac{g_1 g_2 \cdots g_{n-1}}{IIP3\_n} \right)^q \right] \tag{3.25}$$

where $q = \frac{m-1}{2}$ and m is the slope of the response ($m = 3$ for third order). It must be taken into account, when performing the calculations with this equation, that the gains and input intercept points (g_i's and *IIP3_i*'s) refer to numerical entities, rather than dB of dBm [28].

3.5 Definition of Dynamic Range

To calculate the DR, the MDS is subtracted from the 1-dB compression point referenced to the input, Figure 3.20, which is given by

$$DR = CP_{1dBin} - MDS_{dBm} = CP_{1dBin} + 174 \quad dBm - 10\log(B) - NF_{RX}. \tag{3.26}$$

In Equation (3.26), the units of the 1-dB compression point CP1dBin are in dBm. Often times, however, the receiver DR is modified from the above definition, which employs the MDS at the low end, to one that utilizes sensitivity, on the low end. In that case, an alternative definition gives a value less than the definition of the previous equation by an amount equal to the

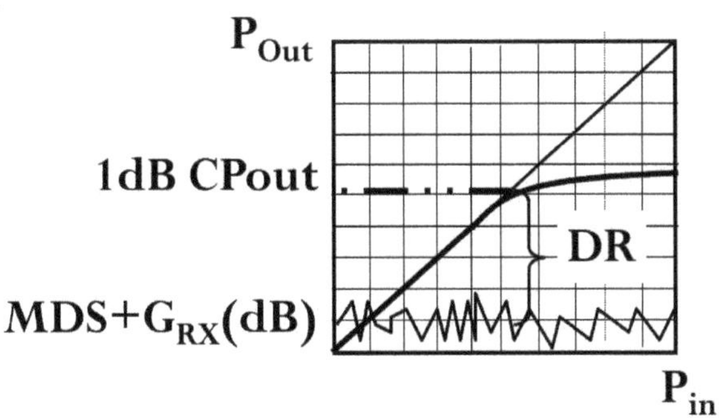

Figure 3.20 Dynamic range definition. *Source:* Ref. [28].

required carrier-to-noise ratio [27]

$$
\begin{aligned}
DR &= CP_{1dBin} - \mathit{Sensitivity} \\
&= CP_{1dB\,\mathrm{in}} + 174 \quad dBm - 10\log(B) - NF_{RX} - \frac{C\cdot}{N}
\end{aligned}
\tag{3.27}
$$

The power defining the top end of the DR of a receiver system may be modified from the 1-dB compression point. In particular, one may take for this purpose the highest power that the receiver can handle and still meet a particular level of performance. When this approach is used, the DR is defined as the difference between this power and the sensitivity power level [28].

Another way of defining the DR is by taking it as the difference between the largest signal that the receiver can reject while simultaneously receiving a smaller signal at the sensitivity level. In this case, this largest signal causes the loss of the receiver's ability to detect a signal at the sensitivity level, an effect called *receiver blocking* [28] which is discussed next.

3.5.1 Noise Figure of Blocks in Cascade

The presence of large undesired signals within the receiver's RF passband drives certain stages into gain compression. The impact of this gain compression is to reduce the signal power delivered to the load for all signals, including the desired signal. Depending on which stage in the cascade making up the receiver experiences the drop in gain, the drop in overall system gain, in turn, would cause the overall system NF to increase. In this case, Equation (3.28) may be utilized to predict the system NF as a function of the distributed gains (Figure 3.21).

$$
F_T = F_1 + \frac{F_2 - 1}{G_1} + \frac{F_3 - 1}{G_1 G_2} + \ldots + \frac{F_n - 1}{G_1 G_2 \cdots G_{n-1}}. \tag{3.28}
$$

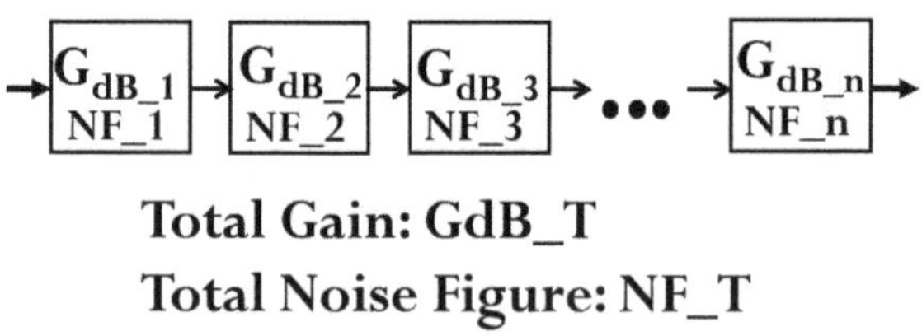

Figure 3.21 Cascaded noise figure. *Source:* Ref. [28].

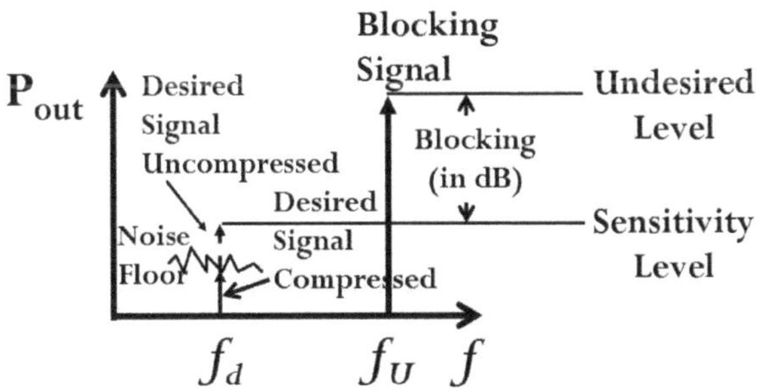

Figure 3.22 Receiver blocking. *Source:* Ref. [28].

If the NF degrades, so do the MDS and sensitivity levels. It may be the case, however, that depending on the stage experiencing the NF degradation (i.e., increase), the overall NF of the system may not significantly degrade. In general, if gain compression is experienced, the amount of desired signal delivered to the load is reduced. Therefore, both effects degrade the SNR.

Figure 3.22 shows the spectrum at the input of a receiver containing a desired signal at its sensitivity level (Figure 3.22).

When a large undesired (spurious) signal is received within the RF passband, the receiver gain drops due to compression, which, in turn, causes a reduction in the amplification of the desired signal, thereby lowering the SNR. Furthermore, due to this reduction, the desired signal drops below the sensitivity level. Under this circumstance, because the sensitivity level requires a particular SNR for the information in the received signal to be recoverable at the desired performance level, the SNR reduction causes a loss in the amount of information extracted. The amount of power required to cause the degradation of receiver sensitivity due to compression effects is referred to as the blocking power, and it is normally characterized with an undesired signal located 1 MHz away from the desired signal. The difference between the power of the undesired signal and the sensitivity level defines the parameter denoted as *blocking* and is expressed in dB (Figure 3.22) [28].

3.5.2 Spur-Free Dynamic Range

The SFDR parameter is analogous to the third-order intermodulation distortion rejection ratio RR_{3o}, which is defined as the power difference between the power at the fundamental and that at the distortion product, and is given by

$$RR_{3o} = P_o - P_d \tag{3.29}$$

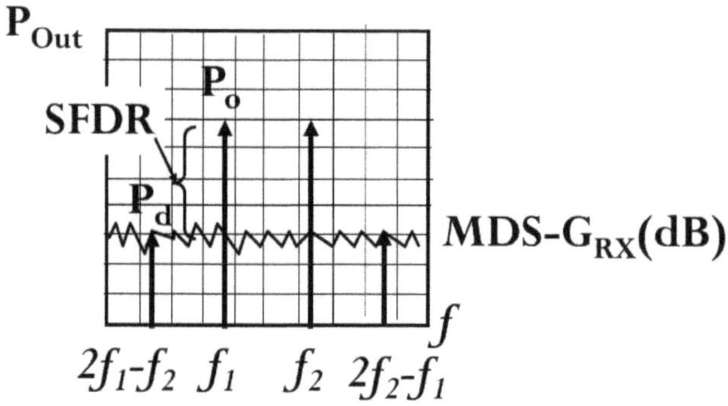

Figure 3.23 Sketch of spur-free dynamic range. *Source:* Ref. [28].

where the RR_{3o} is a function of a specific output power, P_o. From Equation (3.29), one sees that a change in fundamental power produces a change in RR. Since, as shown earlier, a 1-dB change in fundamental power is accompanied by a 3-dB change in the power of the third-order distortion product and because the fundamental has a slope of unity and the distortion product has a slope of 3, the RR exhibits a slope of 2 versus output power of the fundamental. Thus, a 1-dB change in the power of the fundamental changes the RR by 2 dB. The spur-free DR is defined as the difference between the power of the fundamental and the noise power when the distortion products are equal to the noise power (Figure 3.23).

The SFDR is, then, written as [28]

$$\begin{aligned} SFDR &= \frac{2}{3}\left[OIP3 - MDS - G_{RX}\right] \\ &= \frac{2}{3}\left[OIP3 + 174dB - 10\log(B) - NF_{RX} - G_{RX}\right] \end{aligned} \quad . \tag{3.30}$$

Next, an example exhibiting the tradeoffs met in a receiver design exercise as a result of architectural decisions is presented (Figure 3.24). This example compares the sensitivity and the spur-free DR when an RF filter follows the RF amplifier versus when the RF filter precedes the RF amplifier.

In the first place, we calculate the total gain for the receiver G_{RX}; this is done by adding together the gains of all the stages,

$$G_{RX} = -2 + 15 - 5 + 10 - 3 + 11 = 26 \quad dB \tag{3.31}$$

In the second place, we combine the losses in front of the gain stages with those of the gain stages, as shown in Figure 3.24; doing this to the six-stage cascade (Figure 3.24(a)) combines into a three-stage network (Figure 3.24(b)). Carrying out this combination, the loss in front of a gain stage decreases the gain dB for dB while it increases the NF and the input intercept point dB for dB. For NF, this is always the case whenever the additional noise is not induced by external sources into the lossy stage; for the intercept point, it is always true if the loss does not generate distortion products of its own. We continue by converting all dB values of gain, NF, and input intercept point to numerical values. This results in the following numerical gains, g_1 = 31.62; g_2 = 10; g_3 = 12.6, and the noise factor of each stage, f_1 = 1.78, f_2 = 3.2, and f_3 = 3.2. Computing the numerical input intercept points of each stage, we get: IIP3_1=0.126, IIP3_2=0.1, and IIP3_3=12.6. Then, the noise factor of the receiver is given as

$$F_T = 1.78 + \frac{3.2 - 1}{31.62} + \frac{3.2 - 1}{31.2(10)} = 1.86. \tag{3.32}$$

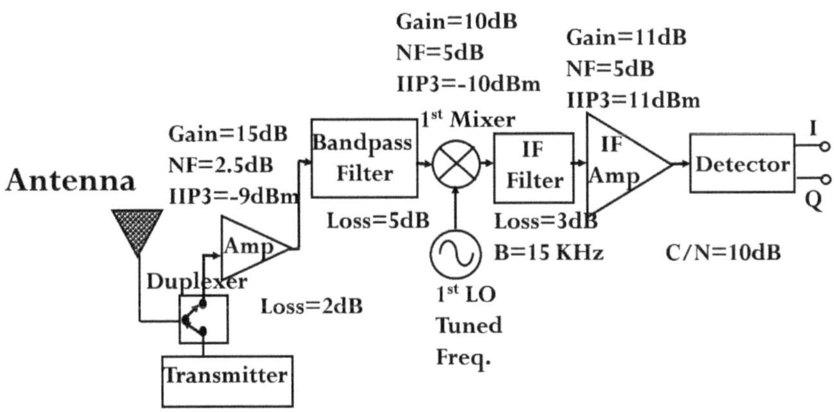

(a)

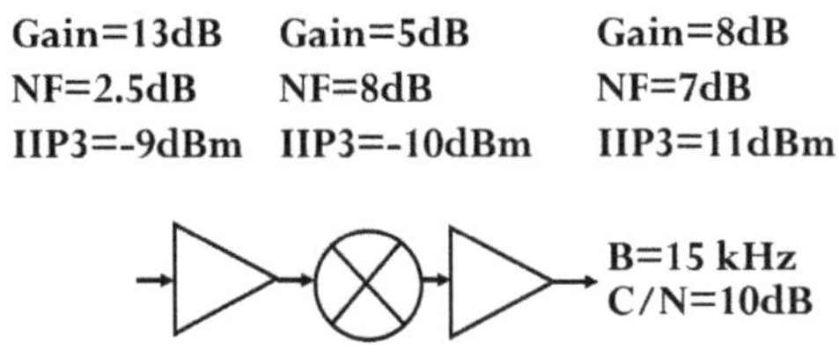

(b)

Figure 3.24 (a). Receiver block diagram. (b) Six-stage network combined into three-stage network. *Source:* Ref. [28].

Upon converting this to NF to dB, we get NF = 2.7dB. Similarly, calculating the third-order input intercept point, we obtain

$$\frac{1}{IIP\_T} = \left[\left(\frac{1}{0.126}\right)^{1} + \left(\frac{31.62}{0.2}\right)^{1} + \left(\frac{31.62 \cdot 10}{12.6}\right)^{1}\right] = 191 \quad (3.33)$$

which in dBm is

$$IIP\_T = 10\log(0.005) = -22.8\ dBm. \quad (3.34)$$

The sensitivity accompanying these is

$$\begin{aligned} MDS &= -174dBm + 10\log B + N_{RX} \\ &= -174dBm + 10\log(15000)dB + 2.7dB = -129.5dBm \end{aligned} \quad (3.35)$$

$$\begin{aligned} Sensitivity\ (dBm) &= \mathrm{MDS}(dBm) + \frac{C}{N} = -129.5\ dBm + 10 \\ &= -119.5\ dBm \end{aligned} \quad (3.36)$$

To conclude, given $IIP3 = OIP3 - G_{RX} \rightarrow OIP3 = IIP3 + G_{RX}$, the spur-free DR is calculated from

$$\begin{aligned} SFDR &= \frac{2}{3}\left[OIP3 - MDS - G_{RX}\right] = \frac{2}{3}(-22.8 + 129.5) \\ &= 71dB \end{aligned} \quad (3.37)$$

3.6 Circuit Signal-to-Noise Ratio

In terms of circuit quantities, the SNR is defined as the ratio at a given port of the signal power to noise power

$$SNR = \frac{P_S}{P_N} = \frac{V_s^2}{V_n^2} \quad (3.38)$$

where V_s and V_n are the rms signal and noise voltages, respectively. The SNR is normally expressed in decibels, i.e.,

$$SNR(dB) = 10\log_{10}\frac{P_S}{P_N} \quad (3.39)$$

and it may be interpreted as follows: A large SNR signifies that the signal overpowers the noise; so the impact of the noise on performance is less. What

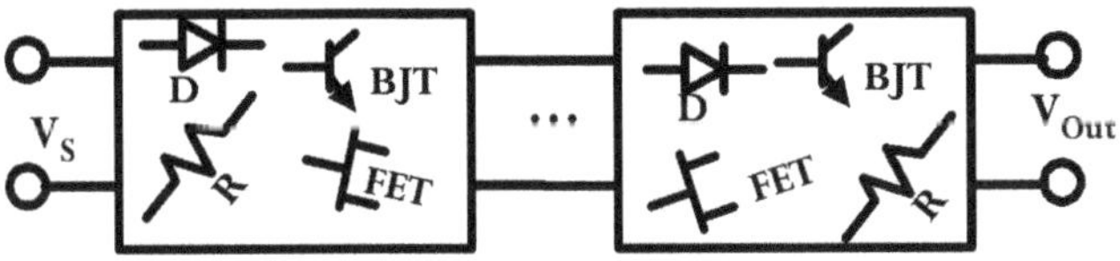

Figure 3.25 Each stage of a system adds noise to the signal passing through.

SNR value may be tolerated depends on the application at hand. For instance, at the input of an amplitude modulation (AM) detector, the minimum is SNR = 10dB; at the input of a frequency modulation (FM) detector, the minimum is SNR = 12dB; and at the input of a TV receiver, the minimum is SNR = 40dB [19].

It is instructive to "follow" the SNR as a signal passes through a cascade of amplifier stages; it is seen that the SNR continually decreases because each stage adds noise (Figure 3.25). In most systems, however, the amplified output noise is due primarily to: 1) noise present along with the input signal; 2) noise contributed by the first two stages (such as the RF amplifier and mixer stages in a receiver) [19].

3.6.1 Definition of Available Noise Power

The maximum power that can be extracted from a source is denoted as *available power* P_a [19]. For a source with an internal impedance Z_s=R+jX, the maximum power that will be delivered to a load will occur when this load is conjugate-matched to the source, i.e., Z_L=R-jX. When the open-circuit voltage of a source is V, then the maximum power transfer theorem results in

$$P_a = \frac{V^2}{4R}. \tag{3.40}$$

The noise introduced at the input to a system is, in general, the most important in determining the overall system SNR [19].

3.6.2 Network Noise Figure

The signal applied at the receiver input is processed through it simultaneously with the input noise. As a result, they are also both simultaneously amplified or attenuated by the same amount as they pass through the various successive stages of the system. It would be ideal if the systems we deal with would introduce no noise beyond that present at the input as, in such a case, a constant SNR would be present throughout the system. Unfortunately, it is observed in practice that it is impossible to maintain a constant SNR. This is so because noise is introduced by any additional resistors, transistors, and other devices; the signal comes across in its journey from input toward the output. This results in an SNR that continuously decreases throughout the system [28]. If amplification is available at the front end of the system and this is sufficiently high so as to make the extra noise introduced in the following stages, then the SNR decrease may be negligible. Studying the SNR at successive nodes in a system enables one to determine precisely which are the significant contributors to the noise and, thus, helps in discerning a strategy for designing circuits and systems that minimize overall noise. The NF, F, is a measure of how noisy a system (e.g., a stage) is [19].

The SNR is related to the NF F as follows:

$$\begin{aligned} NF &= \frac{\textit{input } SNR}{\textit{output } SNR} = \frac{P_{si}/P_{ni}}{P_{so}/P_{no}} \\ &= \frac{P_{si}}{P_{so}} \cdot \frac{P_{no}}{P_{ni}} = \frac{1}{G_a} \cdot \frac{P_{no}}{P_{ni}} = \frac{G_a P_{ni} + P_{ne}}{G_a P_{ni}} = 1 + \frac{P_{ne}}{G_a P_{ni}} \end{aligned} \tag{3.41}$$

In Equation (3.41), the available power gain of the system is G_a, the input signal and output noise powers are, respectively, P_{si} and P_{ni}, and the output signal and noise powers are, respectively, P_{so} and P_{no}. In addition, the noise power introduced by the system is P_{ne}. The NF, expressed in decibels, is given by

$$NF_{dB} = 10 \log_{10} NF. \tag{3.42}$$

To calibrate our intuition, we note that in a noise-free network, the input and output SNRs are equal and, therefore, NF = 1 or $NF_{dB} = 0$.

Another equivalent of NF, namely, the ratio of the actual noise power available from a noisy network to that which would be available if the network were noiseless may be defined [19].

$$NF = \frac{P_{no}}{G_a P_{ni}}. \tag{3.43}$$

3.6.3 Single-Frequency (Spot) Noise Figure

We saw earlier that the system bandwidth enters into the NF calculation. When interest is on the NF at a single frequency, it is denoted as the *spot NF*. This parameter is defined as the ratio of the total available noise power per unit bandwidth at the output port to that produced at the input frequency by the input termination, assuming a noise temperature of 290 K. For a conjugately matched network terminating the input port, the available power from the standard-temperature source in a 1-Hz bandwidth is equal to kT_0. Thus, the spot NF is expressed as [19]

$$NF = \frac{P_{no}}{G_a(f)kT_0}. \tag{3.44}$$

In Equation (3.44), the value of P_{no} is measured over a BW Δf that is more than 1 Hz, in which case the *NF* equation is

$$NF = \frac{P_{no}}{G_a(f)kT_0\Delta f}. \tag{3.45}$$

3.6.4 Equivalent Noise Temperature

The development of low noise receivers was elicited by the advent of radio astronomy, which resulted in the development of masers and parametric amplifiers with NF as low as 1.1 [16] and these devices required extremely low temperatures and special cooling systems for operating effectively. In these types of receivers, it was the external noise, captured by the antenna, as opposed to the internal receiver noise, which was the principal noise source, and, consequently, the receiver NF no longer played any significant role. The *noise temperature* parameter was thus introduced.

To any port in a network, one can associate a noise temperature by defining it as follows: Take the ratio of the available power P_a in a small frequency interval Δf to the product $k_B\Delta f$, giving the equivalent temperature:

$$\boldsymbol{T_e = \frac{P_o}{k_B\Delta f}} \tag{3.46}$$

3.6.5 Effective Noise Temperature of a Network

A thermal source with temperature T and connected to a noiseless network with bandwidth Δf and available gain $G_a(f)$ produces an available noise power [19]

$$\boldsymbol{P_{ns} = k_BT\Delta f} \text{ (Watts)} \tag{3.47}$$

which, upon amplification, manifests itself at the output as

$$\boldsymbol{P_{no} = G_a(f)k_BT\Delta f} \tag{3.48}$$

This situation is modeled by the schematic in Figure 3.26, where the network's own added noise power is P_{ne}.

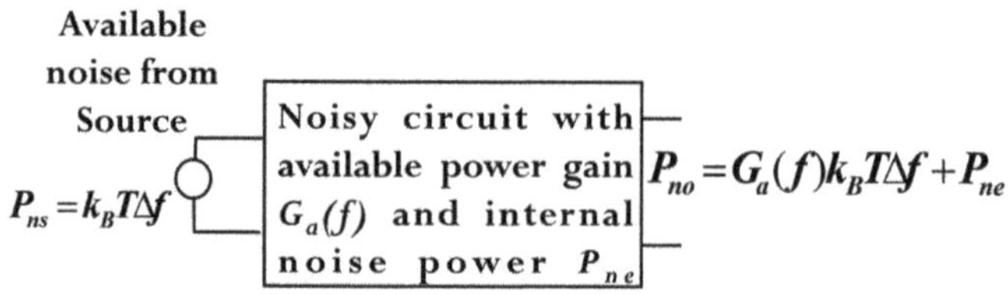

Figure 3.26 Model for addition of network's own power to input noise P_{ns}. *Source:* Ref. [19].

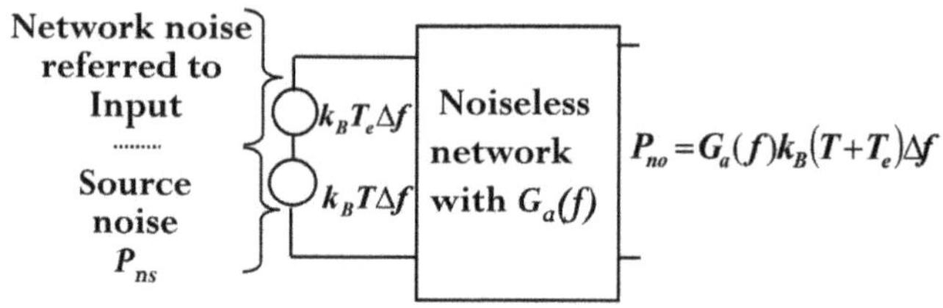

Figure 3.27 Noisy circuit modeled by its noise referred to the input plus noiseless circuit. *Source:* Ref. [19].

With the same input as before, the output noise power will be

$$P_{no} = G_a(f)k_BT\Delta f + P_{ne}. \tag{3.49}$$

If the noise circuit is replaced by a noiseless one with the same available power gain $G_a(f)$, the output noise P_{ne} may be modeled as an extra noise source in the input (Figure 3.27), where the value of the power P_{ne} is controlled by the temperature T_e,

$$T_e = \frac{P_{ne}}{G_a(f)k_B\Delta f} \tag{3.50}$$

and

$$P_{no} = G_a(f)k_B\left(T + T_e\right)\Delta f. \tag{3.51}$$

The temperature T_e sets the effective input noise of the circuit, which is denoted by the *effective input noise temperature* [19]. When analyzing a cascade of building blocks, the concept of effective temperature becomes useful since each individual building block is characterized by its own T_e. Next, the NF of cascaded building blocks is addressed.

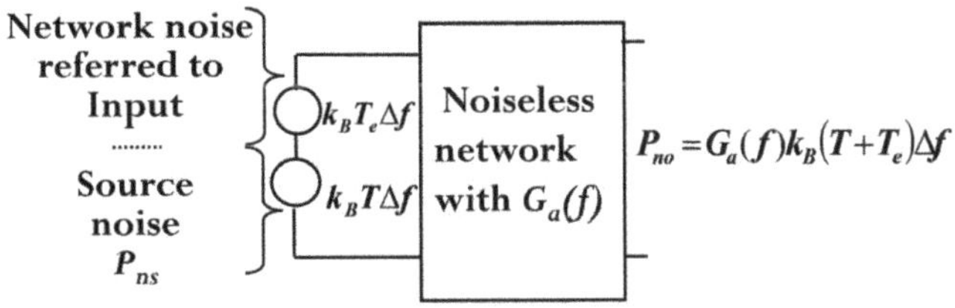

Figure 3.28 Cascaded building blocks. *Source:* Ref. [19].

3.6.6 Computing the Overall NF of Cascaded Circuits

The overall NF of a cascade of circuits is predicated on the overall bandwidth B of the overall system (Figure 3.29). Given a source temperature T_0, required for the standard definition of NF, in a small frequency band Δf, the output noise power produced by a single stage is [19]

$$P_{no} = G_a(f) k_B (T_0 + T_e) \Delta f. \tag{3.52}$$

Expressing Equation (3.56) in terms of $NF = P_{no}/G_a(f) k_B T_0 \Delta f$ yields

$$P_{no} = NF \cdot G_a(f) k_B T_0 \Delta f. \tag{3.53}$$

Solving for NF results in

$$NF = \frac{T_0 + T_e}{T_0} \tag{3.54}$$

or

$$T_e = T_0 (NF - 1). \tag{3.55}$$

The NF, referred to the standard temperature, T_0, may, thus, be expressed in terms of the *equivalent noise temperature* of the circuit and vice versa. The available power of two cascaded circuits (Figure 3.29) is given by using

$$P_{no} = G_a(f) k_B (T + T_e) \Delta f \tag{3.56}$$

to obtain

$$P_{no} = G_{a1} G_{a2} k_B T_0 \Delta f + G_{a1} G_{a2} k_B T_{e1} \Delta f + G_{a2} k_B T_{e2} \Delta f. \tag{3.57}$$

Here, the first term to the right of the equal sign is the noise $k_B T_0 \Delta f$ produced by the source, T_0, that reaches the output after being amplified by the gains G_{a1} and G_{a2}. The second term, to the right of the equal sign, is the noise

$k_B T_{e1} \Delta f$ reaching the output after being amplified by the gains G_{a1} and G_{a2}, where T_{e1} is the effective temperature of circuit 1 referred to the input. The third term, to the right of the equal sign, is the noise $k_B T_{e2} \Delta f$ reaching the output after being amplified by the gain G_{a2}, where T_{e2} is the effective temperature of circuit 2 referred to the input. If one compares (3.57) to (3.56) and (3.55), repeated here for convenience,

$$T_e = \frac{P_{ne}}{G_a(f) k_B \Delta f} \tag{3.58}$$

and

$$P_{no} = G_a(f) k_B \left(T + T_e\right) \Delta f \tag{3.59}$$

results in an expression for the effective input temperature of the two circuits blocks in cascade, namely,

$$T_{e1,2} = \frac{k_B G_{a2}\left(G_{a1} T_{e1} + T_{e2}\right) \Delta f}{G_{a1} G_{a2} k_B \Delta f} = T_{e1} + \frac{T_{e2}}{G_{a1}}. \tag{3.60}$$

In Equation (3.60), $T_{e1,2}$ is the effective input temperature that captures all the output noise introduced by the two noisy circuits in cascade. For *k* noisy circuit blocks in cascade, the effective temperature is [19]

$$T_{e1,N} = T_{e1} + \frac{T_{e2}}{G_{a1}} + \ldots + \frac{T_{eN}}{G_{a1} G_{a2} \ldots G_{a(k-1)}}. \tag{3.61}$$

By using $T_e = T_0(NF - 1)$, the following expression for the overall NF is obtained [28]:

$$NF_{1,k} = NF_1 + \frac{NF_2 - 1}{G_{a1}} + \ldots + \frac{NF_k - 1}{G_{a1} G_{a2} \ldots G_{a(k-1)}}. \tag{3.62}$$

Examination of Equation (3.62) reveals that the first stage in the cascade dominates the overall *NF*, unless its gain G_{a1} is small or NF_2 is large.

It may turn out that in the problem in question, the input noise temperature is *not* 290 K. In that case, the true measure of the SNR reduction is the actual NF:

$$NF_{act} = \frac{T_S + T_e}{T_S} \tag{3.63}$$

where $T_S \neq T_0$. The actual NF is related to the standard NF by setting $T_S = T_0$ to give

$$NF_{act} = 1 + (NF - 1)\left(\frac{T_0}{T_S}\right). \tag{3.64}$$

3.6.7 Noise Figure of a Mixer

As usual, the NF of a mixer is defined as the ratio of the input SNR to the output SNR. In this case, the input is taken as the RF port and the output is taken as the IF port [29]. As we have mentioned previously, the mixing process results in the IF containing not only the converted RF signal but also the image of the RF signal. This is depicted in Figure 3.29.

The signal to be converted by a mixer may be single-sideband (SSB) or double-sideband (DSB) [101]. Therefore, its NF is ambiguous, unless the scenario being dealt with is specified (Figure 3.30). In the case of a DSB signal v_{DSB}, we have

$$v_{DSB}(t) = A_{DSB}\left[\cos\left(\omega_{LO} - \omega_{IF}\right)t + \cos\left(\omega_{LO} + \omega_{IF}\right)t\right] \quad (3.65)$$

which, upon being down-converted and filtered, gives the IF signal

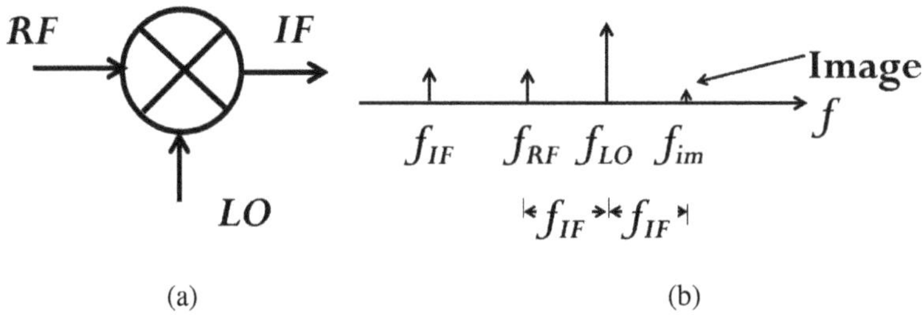

Figure 3.29 (a) Mixer symbol. (b) Mixing spectrum including image. *Source:* Ref. [19].

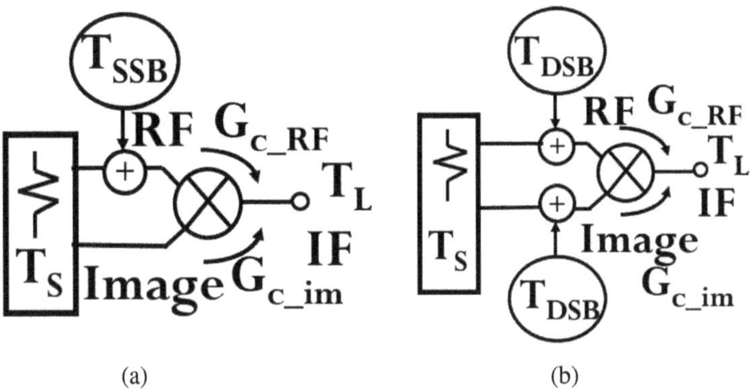

Figure 3.30 Noise scenarios in mixers. (a) Single-sideband. (b) Double-sideband. *Source:* Ref. [29].

$$\begin{aligned} v_{IF}(t) &= \frac{A_{DSB}G}{2}\cos(-\omega_{IF}t) + \frac{A_{DSB}G}{2}\cos(\omega_{IF}t) \\ &= A_{DBS}G\cos(\omega_{IF}t). \end{aligned} \tag{3.66}$$

The corresponding input and output powers of the DSB signal are

$$P_{i\_DSB} = \frac{A_{DSB}^2}{2} + \frac{A_{DSB}^2}{2} = A_{DSB}^2 \tag{3.67}$$

and

$$P_{o\_DSB} = \frac{A_{DSB}^2 G^2}{2}. \tag{3.68}$$

Now, the output noise is given by

$$N_o = \frac{(k_B T_0 B + N_{Mixer})}{G_{c\_RF}} \tag{3.69}$$

where G_{c_RF} is the mixer conversion loss, and the NF is

$$F_{DSB} = \frac{P_i/N_i}{P_o/N_o} = \frac{2}{G^2 G_{c\_RF}}\left(1 + \frac{N_{Mixer}}{k_B T_0 B}\right). \tag{3.70}$$

In the case of SSB, we have

$$v_{SSB}(t) = A_{SSB}\cos(\omega_{LO} - \omega_{IF})t \tag{3.71}$$

which, upon being down-converted and filtered, gives the IF signal

$$v_{IF}(t) = \frac{A_{SSB}G}{2}\cos(\omega_{IF}t). \tag{3.72}$$

The corresponding input and output powers of the SSB signal are

$$P_{i\_SSB} = \frac{A_{SSB}^2}{2} \tag{3.73}$$

and

$$\begin{aligned} P_{o\_SSB} &= \frac{A_{SSB}^2 G^2}{4}\cos^2(\omega_{IF}t) = \frac{A_{SSB}^2 G^2}{4}\left[\frac{1}{2}(1 + \cos(2\omega_{IF}t))\right] \\ &= \frac{A_{SSB}^2 G^2}{8} \end{aligned} \tag{3.74}$$

after neglecting the out-of-band second harmonic. Now, the output noise is given by

$$F_{SSB} = \frac{P_{i\_SSB}/N_{i\_SSB}}{P_{o\_SSB}/N_{o\_SSB}} = \frac{4}{G^2 G_{c\_RF}} \left(1 + \frac{N_{Mixer}}{k_B T_0 B}\right). \tag{3.75}$$

So, the single-sideband (SSB) NF is

$$F_{SSB} = 2F_{DSB}. \tag{3.76}$$

The SSB NF is twice the DSB NF.

Another way of arriving at the relationship between NF_{SSB} and NF_{DSB} for a mixer is as follows.

Obviously, there is an ambiguity in specifying the NF of a mixer; therefore, it is necessary to define which scenario one is referring to [29]. In the SSB temperature definition, it follows from observing Figure 3.30(a) that the output noise temperature T_L is [29]

$$T_L = (T_S + T_{SSB})\, G_{c\_RF} + T_S G_{c\_im}. \tag{3.77}$$

T_S and T_{SSB} are the noise temperatures of the source and SSB terminations, and G_{c_RF} and G_{c_im} are the conversion gains, respectively, from the RF and image ports to the IF port

$$T_{SSB} = \frac{T_L - T_S\left(G_{c\_RF} + G_{c\_im}\right)}{G_{c\_RF}}. \tag{3.78}$$

In the DSB noise temperature definition, it follows similarly, from Figure 3.30(b), that the output noise temperature T_L is [29]

$$T_L = (T_S + T_{DSB})\, G_{c-RF} + (T_S + T_{DSB})\, G_{c\_im} \tag{3.79}$$

which, solving for T_{DSB}, yields

$$T_{DSB} = \frac{T_L - T_S\left(G_{c\_RF} + G_{c\_im}\right)}{G_{c\_RF} + G_{c\_im}}. \tag{3.80}$$

For the case in which $G_{c\_RF} = G_{c\_im}$, one finds that $T_{SSB} = 2T_{DSB}$ exactly. The NF, while being applied to a three-port device, namely, the mixer, was originally introduced to characterize two-ports. This is why applying it to a three-port component, e.g., a mixer, gives rise to the ambiguity elicited by presence of the image-frequency termination [29]. This issue has been clarified by the IEEE, which provided the definition: "For heterodyne systems, [the output noise engendered by the input termination] includes only that noise from the input termination which appears in the output via the principal-frequency transformation of the system, and does not include spurious contributions such as those from an image frequency transformation." [29]

This statement has been interpreted in two ways, namely the following.

1) The source-termination noise at frequencies other than RF should not be included when calculating the noise at the output caused by the source. This means that the termination noise at the image should be ignored and, thus, gives rise to the output noise temperature $G_{c\_RF}(T_{SSB}+T_0)$ and the noise from the termination alone as $G_{c\_RF}\ T_0$. The NF is then the ratio of these quantities

$$NF_{SSB1} = 1 + \frac{T_{SSB}}{T_0}. \tag{3.81}$$

2) The image noise is treated as a noise source within the mixer so that the output temperature becomes $G_{c\_RF}(T_{SSB}+T_0) + G_{c\_im}\ T_0$. For identical conversion gains, this assumption yields

$$NF_{SSB2} = 2 + \frac{T_{SSB}}{T_0}. \tag{3.82}$$

In this scenario, a noiseless mixer (T_{SSB} = 0) has a NF of 3 dB. In practice, NF_{DSB} is employed in direct-conversion receivers where the image is the signal itself.

3.7 Summary

This chapter has addressed the parameters describing a communications system performance. We first discussed the transmitter's maximum output power, modulation accuracy, and adjacent and alternate channel power. Then, we discussed the parameters pertaining to the receiver, in particular, its sensitivity, NF, selectivity, IR, DR, and SFDR. Next, we dealt with the topics of noise and nonlinearity. These included an introduction of a variety of noise sources, in particular, thermal noise and shot noise, followed by the topic of SNR, including available noise power, NF, noise temperature, NF of cascaded linear networks, and mixer NF. We then continued with a discussion of the calculation of the noise for a cascade of building blocks. In particular, we reviewed the effects of nonlinearity, namely, gain compression and intermodulation, and IP3 of cascaded nonlinear stages.

3.8 Problems

1. Could the SNR as a signal travels from input to output in a circuit, increase? Hint: Do a patent search.
2. A temperature T_0 = 290 K sets the so-called standard *NF*. For T = T_0 = 290 K, what is the value of $k_B T_0 B$ in dBm/Hz?
3. Consider the receiver in Figure P3. What is its sensitivity? To what voltage is the sensitivity equivalent if the antenna impedance is 50 Ω?

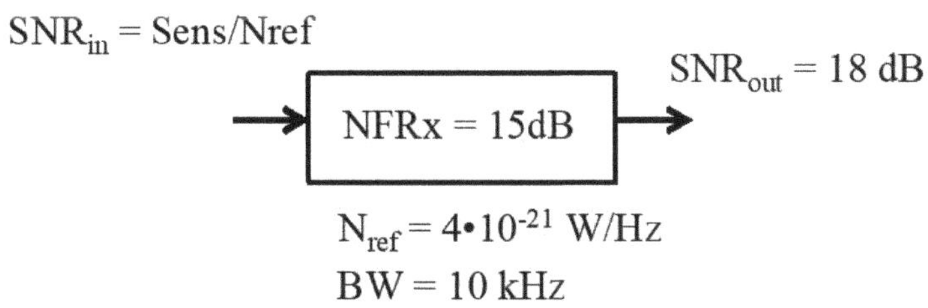

Figure P3.1 Receiver sensitivity calculation.

4. Consider a receiver system with the following parameters. Determine its total gain, NF, and IP3.

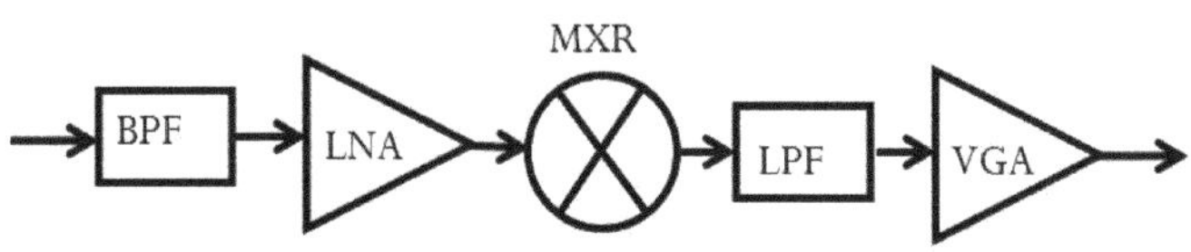

Gain (dB)	-1	15	10	-5	65
NF (dB)	1	1.5	8	5	15
IP3 (dBm)	100	0	16	100	20

Figure P3.2 Overall receiver performance calculation.

5. Implement the receiver of Figure 3.2 in SystemVue and compare your results with those obtained in Problem 4.
6. Given the receiver front end in Figure P3.3, calculate its T_e and NF.

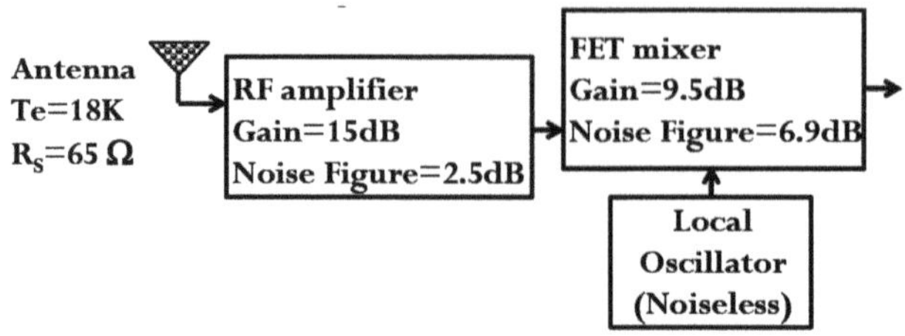

Figure P3.3 Simplified receiver front end.

Assumptions: (1) The radiation resistance and effective temperature due to external radiation of the antenna are, respectively, R_S = 65 Ω and T_e = 18 K.

4

Circuit Topologies for Signal Modulation and Detection

4.1 Introduction

This chapter deals with the principles of circuits and techniques utilized in effecting modulation and detection in wireless systems. The first topic deals with a system-level study of amplitude modulation (AM) and frequency modulation (FM) modulators and demodulators, including their principles of operation. The AM-related wireless system constituents discussed include the full carrier modulator, the single-sideband (SSB) suppressed carrier modulator, the double-sideband (DSB) suppressed carrier modulator, and the envelope detector. In what pertains to FM modulator/demodulator, we discuss the voltage-controlled oscillator (VCO) as an FM modulator, and the phase-locked loop (PLL) based FM detector. This is followed by a discussion on digital modulation, including binary modulation, binary amplitude-shift keying (BASK), binary frequency-shift keying (BFSK), binary phase-shift keying (BPSK), quadrature phase-shift keying (QPSK), M-ary quadrature amplitude modulation (QAM), orthogonal frequency division multiplexing (OFDM), direct sequence spread spectrum (DS/SS), and frequency hopping spread spectrum (FH/SS). The geometric representation of digital modulation schemes and the complex envelope form of a modulation signal are also introduced.

4.2 AM Modulation Approaches

The purpose of an AM modulator is to realize the waveform

$$v_{AM}(t) = a_0[1 + mg(t)]\cos\omega_0 t \qquad (4.1)$$

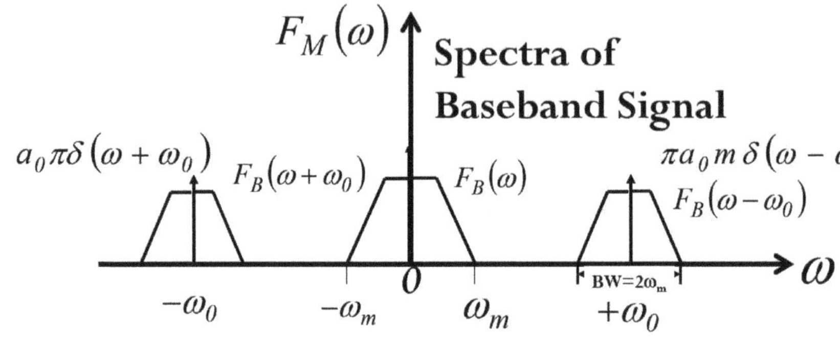

Figure 4.1 AM signal spectra. The baseband signal is centered at zero frequency and contains the maximum frequency ω_m. *Source:* Ref. [33].

where m represents the modulation index. The spectrum of the AM signal is given by [30] (4.2), which reveals that it is symmetric with respect to the

$$F_M(\omega) = \frac{a_0}{2}\left[2\pi\delta\left(\omega-\omega_0\right)+2\pi\delta\left(\omega+\omega_0\right)+F_B\left(\omega-\omega_0\right)+F_B\left(\omega+\omega_0\right)\right] \tag{4.2}$$

origin and that both sidebands, namely, one centered at $+\omega_0$ and one centered at $-\omega_0$, carry the same information (Figure 4.1).

This means that one of the sidebands is redundant and that, as a result, there is an inherent transmission inefficiency since the power is divided between the two sidebands. The presence of the two sidebands, in turn, dictates that the modulated signal, centered on ω_0, occupies a bandwidth given by

$$BW_{AM} = 2\omega_m. \tag{4.3}$$

However, since the maximum baseband signal frequency is ω_m, this means that there is also an excess bandwidth being used. In order to reduce the power and bandwidth inefficiencies of the AM process, another AM modulation scheme, which eliminates one of the sidebands, namely, *SSB*, AM, or SSB modulation, was invented. This is addressed next.

4.2.1 Generation of Single-Sideband AM Signals

The generation of SSB signals is predicated upon two approaches, namely, one that uses filtering to eliminate the extra sideband and one that uses a topological arrangement, the *balanced modulator*, to effect cancelation of the extra sideband.

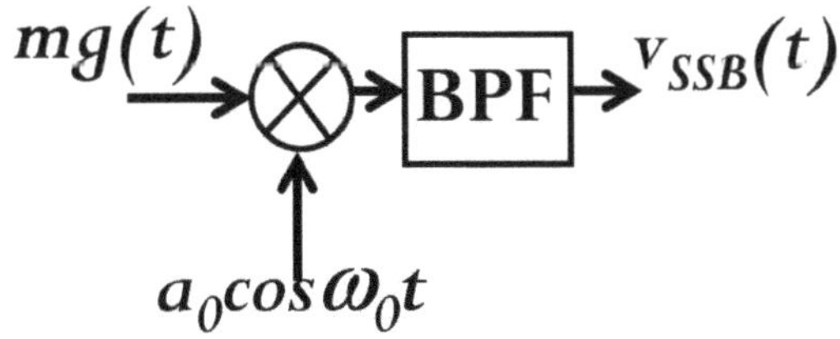

Figure 4.2 Illustration of SSB AM by the filtering approach. *Source:* Ref. [30].

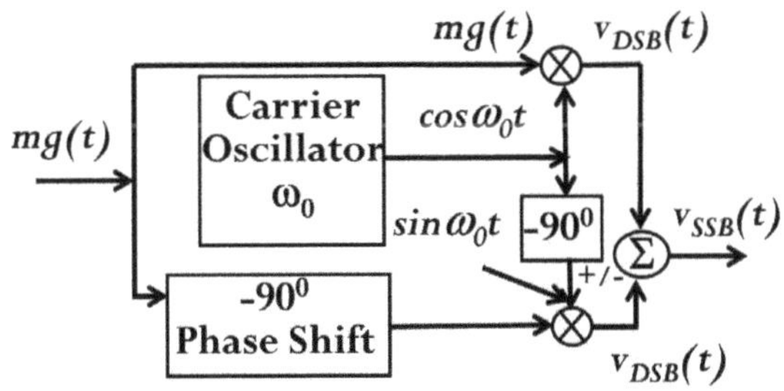

Figure 4.3 Illustration of the balanced modulator approach to SSB AM signal. *g(t)* is the modulation signal. *Source:* Ref. [30].

The approach that relies on filtering is conceptually illustrated in Figure 4.2; one of the sidebands is eliminated by direct filtering.

The approach to SSB AM that relies on the balanced modulator is conceptually illustrated in Figure 4.3.

In this approach, the baseband signal *g(t)* and its negative multiply the carrier and the result subtracted (Figure 4.3) [30].

4.3 AM Demodulation Approaches

Demodulation is the process of extracting the baseband signal, *g(t)*, from the modulated signal. One of the approaches to do so, namely, the *envelope detector* is addressed next.

4.3.1 Envelope Detector

This approach relies on filtering the modulated signal with a low-pass filter (LPF) to extract the baseband signal; this is illustrated in Figure 4.4.

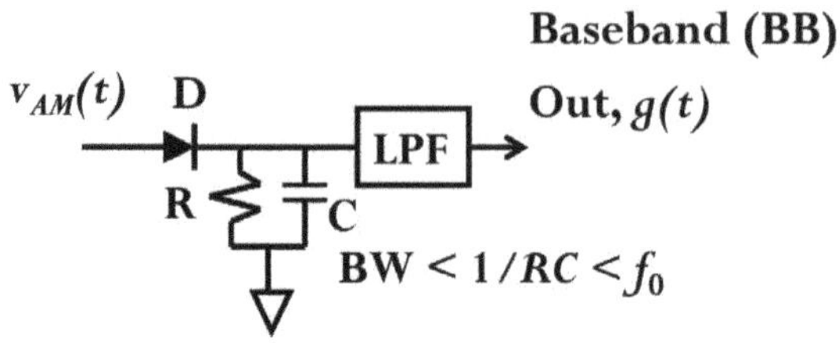

Figure 4.4 Envelope detector circuit plus low-pass filter. *Source:* Ref. [32].

In the circuit of Figure 4.4, the AM signal is rectified by the diode D, which, due to its nonlinearity, produces a spectrum containing the modulating signal and its harmonics. Now, by designing the RC filter such that its inverse time constant is greater than the modulating signal bandwidth but smaller than the carrier frequency, it is not responsive to the high-frequency carrier but only to the modulating signal and its harmonics. The harmonics, falling beyond the filter cutoff frequency, are then attenuated and only the modulating signal substantially remains [32].

4.4 FM Approaches

We will discuss here one FM modulation approach, namely, the *direct* approach.

4.4.1 Direct FM Modulator

In the direct FM modulation approach, the frequency of a VCO is modulated by applying the baseband signal to its control voltage. In particular, if the VCO has a free-running frequency, f_0, given by

$$f_0 = \frac{1}{2\pi\sqrt{L_0C_0}} \tag{4.4}$$

then by changing the value of the frequency-determining capacitors (varactors) through the application of the baseband signal, the VCO output frequency becomes

$$f(t) = f_0 + kg(t). \tag{4.5}$$

As the amplitude of the baseband signal increases and decreases, the carrier frequency follows due to variation of the tuned circuit capacitors (Figure 4.5) [30].

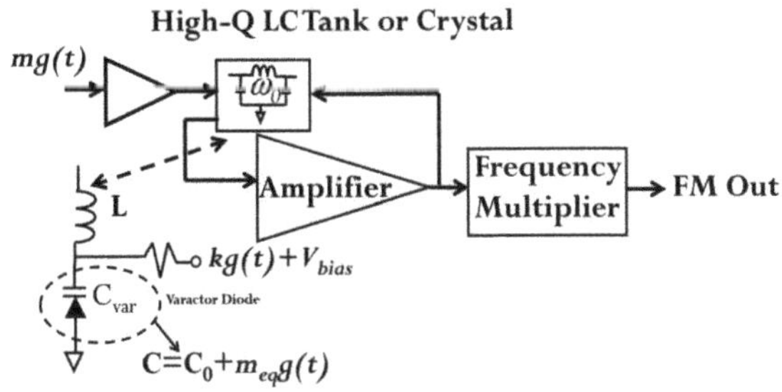

Figure 4.5 Voltage-controlled oscillator as FM modulator. *Source:* Ref. [30].

4.5 FM Demodulation Approaches

The approaches to FM demodulation derive from an examination of the equation of an FM waveform, namely [30],

$$v_{FM}(t) = A\cos\varphi(t) = A\cos\left[\omega_0 t + K\int g(t)dt\right]. \tag{4.6}$$

In particular, it is seen that, by differentiating the instantaneous frequency,

$$\omega_i(t) = \frac{d\varphi(t)}{dt} = \omega_0 + Kg(t) \tag{4.7}$$

the desired baseband signal *g(t)* may be obtained by canceling ω_0. Furthermore, if the frequency-modulated voltage waveform (Equation (4.7)) is differentiated, namely,

$$\begin{aligned}\frac{dv_{FM}(t)}{dt} &= -A\sin\varphi(t)\frac{d\varphi}{dt} \\ &= -A\left[\omega_0 + Kg(t)\right]\sin\varphi(t)\end{aligned} \tag{4.8}$$

then an AM-like signal is obtained, whose envelope is given by

$$A\left[\omega_0 + Kg(t)\right] = A\omega_0\left[1 + \frac{Kg(t)}{\omega_0}\right]. \tag{4.9}$$

Equation (4.9) reveals that we may also obtain *g(t)* by applying an envelope detection scheme to $dv_{FM}(t)/dt$. An alternate, intuitive, approach to FM demodulation relies on a PLL; this will be presented next.

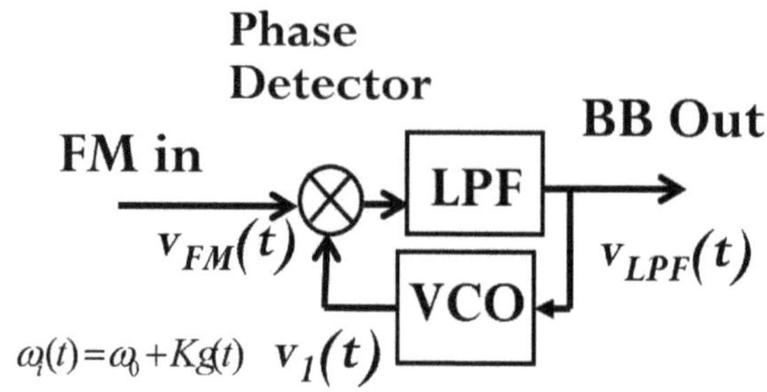

Figure 4.6 FM demodulation by phase-locked loop. *Source:* Ref. [30].

4.5.1 FM Demodulation by Phase-Locked Loop

A PLL is a closed-loop control system that follows the variations in the phase and frequency of an incoming signal (4.10).

$$v_{FM}(t) = A\cos(\omega_0 + \Delta\omega)\,t. \tag{4.10}$$

The mechanism of FM demodulation is as follows. The output of the VCO (Figure 4.6) may be represented by

$$v_1(t) = A_l\cos(\omega_0 t + \varphi_1(t))\,t \tag{4.11}$$

with its instantaneous frequency being given by

$$\omega_1 = \omega_0 + \frac{d\varphi_1(t)}{dt}. \tag{4.12}$$

This VCO output signal is applied to the phase comparator (detector), where it is compared to the input signal. The output of the phase comparator yields an output voltage proportional to phase difference. Subsequently, low-pass filtering the phase difference produces the voltage $V_{LPF}(t)$, which, when fed back to the VCO, controls its output frequency. The control of the VCO, $V_{LPF}(t)$ however, turns itself to be the demodulated FM signal, *g(t)* [30].

4.6 The Digital Modulation Technique

In the digital modulation technique, as its name implies, the baseband information is *digitized* and then utilized to vary (modulate) the carrier amplitude, frequency, or phase. Thus, in digital modulation, one begins by converting the baseband signal into a stream of digital bits.

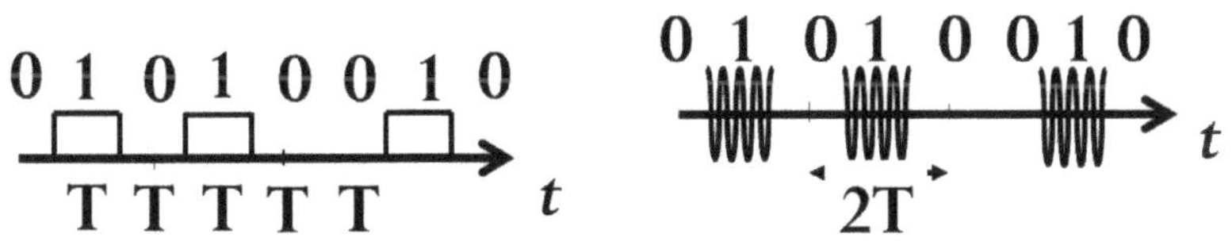

(a) (b)

Figure 4.7 Amplitude-shift keying. (a) Baseband signal represented by stream of bits. (b) Corresponding modulation of carrier amplitude with (a). *Source:* Ref. [31].

4.6.1 Amplitude-Shift Keying Modulation

In the binary modulation approach, the baseband signal is quantized using two levels, $N = 2$, so that it is represented by a binary stream of 1's and 0's. For example, if the digital representation of the baseband signal is the train of 1's and 0's in Figure 4.7(a), then the modulated carrier amplitude would look like that in Figure 4.7(b). This modulation is referred to as amplitude-shift keying (ASK).

4.6.2 Frequency-Shift Keying Modulation

In the frequency-shift keying (FSK) binary modulation approach, the transmitted frequency is dictated by the binary sequence of 1's and 0's representing the baseband signal. In particular, the carrier frequency is alternated between two frequencies f_1 and f_0 which are transmitted over intervals of time T (Figure 4.8).

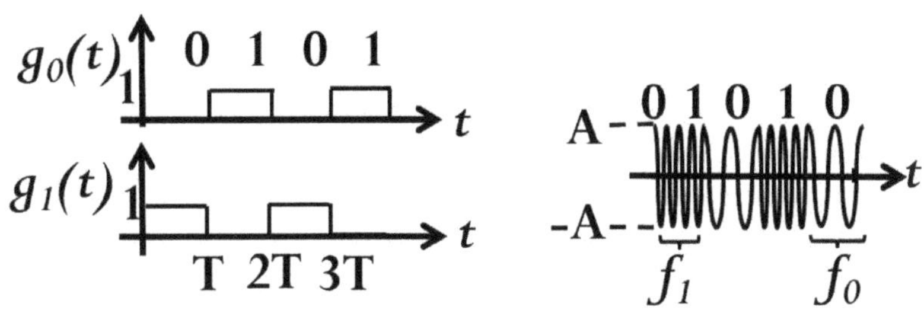

(a) (b)

Figure 4.8 Frequency-shift-keying. (a) Baseband signal represented by stream of bits. (b) Corresponding modulation of carrier frequency. *Source:* Ref. [31].

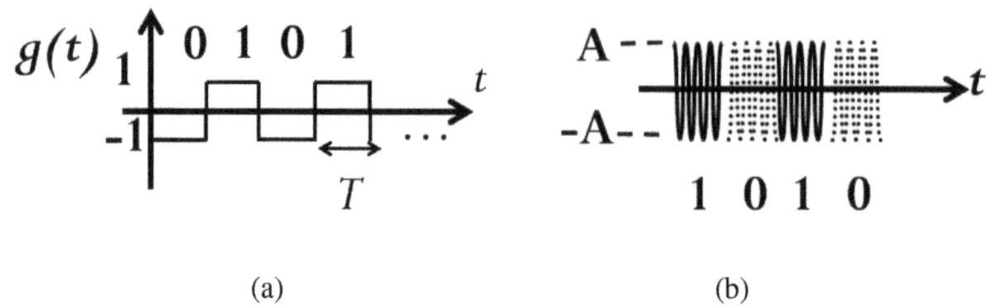

Figure 4.9 Phase-shift keying. (a) Binary representation of modulating baseband signal. (b) Corresponding modulation of carrier frequency. *Source:* Ref. [31].

4.6.3 Phase-Shift Keying Modulation

In the phase-shift keying (PSK) modulation approach, the phase of the carrier is shifted 180° in accordance with whether the modulating signal is a 1 or a 0; the 1's and 0's being the stream of bits representing the baseband signal (Figure 4.9).

4.7 Modulation Signal Representation by Complex Envelope Form

It is customary in the field to represent the waveforms utilized in digital modulation in complex envelope form [25, 27], that is,

$$M(t) = I(t) + jQ(t) = A(t)e^{j\varphi(t)}. \tag{4.13}$$

This equation represents the waveform in terms of two Cartesian components, namely, an in-phase one, *I(t)*, and a quadrature *Q(t)*, which in turn, are represented as

$$I(t) = \sum_k I_k p_I (t - kT_s - \tau) \tag{4.14}$$

and

$$Q(t) = \sum_k Q_k p_Q (t - kT_s - \tau). \tag{4.15}$$

These are pulses located at points in time $t{=}kT_s + \tau$ along the time axis, and I_k and Q_k are sequences of discrete variables corresponding to the baseband digital data. $1/T_s$ is the *symbol rate* and $p_I(t)$ and $p_Q(t)$ represent finite energy pulses, such as rectangular or Gaussian. τ is a possible delay, and *A(t)* and $\varphi(t)$ are the envelope amplitude and phase, respectively [25, 27].

The amplitude and phase in Equation (4.13) are given by

$$A(t) = \sqrt{I^2(t) + Q^2(t)} \tag{4.16}$$

and

$$\varphi(t) = \tan^{-1}\frac{Q(t)}{I(t)}. \tag{4.17}$$

The symbol duration, T_s, of an *M-ary* keying modulation is related to the bit duration T_b of the original binary data stream as

$$T_s = \log_2 M \cdot T_b. \tag{4.18}$$

This approach to representing digital modulation in terms of *I* and *Q* components is usually employed whenever more than two levels, i.e., "*m*-ary" as opposed to "bin-ary," are utilized to quantize the baseband signal. Below, we discuss *M*-ary modulation.

4.7.1 M-ary Modulation—MPSK

It is usual to represent digital *M*-ary modulation schemes via *constellations*, namely, *geometric* representations of digital modulation signals in the Cartesian plane.

Suppose there are *M* possible signals in terms of which the modulation signal set *S* may be represented [25]

$$S = \{s_1(t), s_2(t), \ldots, s_M(t)\}. \tag{4.19}$$

Then, a signal set of size *M* makes it possible to transmit a maximum of $log_2 M$ bits of information per symbol. Now, because of their quadrature representation, the elements of *S* may be visualized as constellations, i.e., points in a vector space. The baseband signal may be quantized into *M* levels, each of which is represented by an *M-ary* stream of 1's and 0's. Now, given a basis of *N* orthogonal (perpendicular) waveforms, we can expand a function in their vector space as a linear combination of them. In particular, once a basis is determined, any point in that vector space can be represented as a linear combination of the basis signals [25, 27],

$$\{\phi_j(t) \mid j = 1, 2, \ldots, N\} \tag{4.20}$$

such that

$$s_i(t) = \sum_{j=1}^{N} s_{ij}\phi_j(t). \tag{4.21}$$

Since they are orthogonal to one another, the basis signals must satisfy

$$\int_{-\infty}^{\infty} \phi_i(t)\phi_j(t)dt = 0 \quad i \neq j \tag{4.22}$$

and are defined such that they have unit energy,

$$E = \int_{-\infty}^{\infty} \phi_i^2(t)dt = 1. \tag{4.23}$$

4.7.2 Binary Phase Shift Keying Modulation—BPSK

Using a carrier of constant envelope, the BPSK modulation consists in switching its phase between two signals m_1 (bit 1) and m_2 (bit 0), where the two phases are 180° apart. Typical BPSK signals are expressed as [57]

$$s_{BPSK}(t) = \sqrt{\frac{2E_b}{T_b}} \cos\left(2\pi f_c t + \theta_c\right) \quad 0 \le t \le T_b \quad \text{(binary 1)} \tag{4.24}$$

or

$$s_{BPSK}(t) = -\sqrt{\frac{2E_b}{T_b}} \cos\left(2\pi f_c t + \theta_c\right) \quad 0 \le t \le T_b \quad \text{(binary 0)} \tag{4.25}$$

which, in turn, may be written in compact form as

$$s_{BPSK}(t) = m(t)\sqrt{\frac{2E_b}{T_b}} \cos\left(2\pi f_c t + \theta_c\right). \tag{4.26}$$

In this expression, it is seen that $m(t)$ modulates the carrier $cos(2\pi f_c t)$, and it can be produced with a balance modulator. If the rectangular pulse is expressed as $p(t)=rect((t-T_b/2)T_b$, for BPSK signals $s_1(t)$ and $s_2(t)$ given by

$$s_1(t) = \sqrt{\frac{2E_b}{T_b}} \cos\left(2\pi f_c t\right) \quad 0 \le t \le T_b \tag{4.27}$$

and

$$s_2(t) = -\sqrt{\frac{2E_b}{T_b}} \cos\left(2\pi f_c t\right) \quad 0 \le t \le T_b \tag{4.28}$$

where E_b and T_b are, respectively, the energy and period per bit, and $\phi_i(t)$ consists of the single waveform,

$$\phi_1(t) = \sqrt{\frac{2}{T_b}} \cos\left(2\pi f_c t\right) \quad 0 \le t \le T_b \tag{4.29}$$

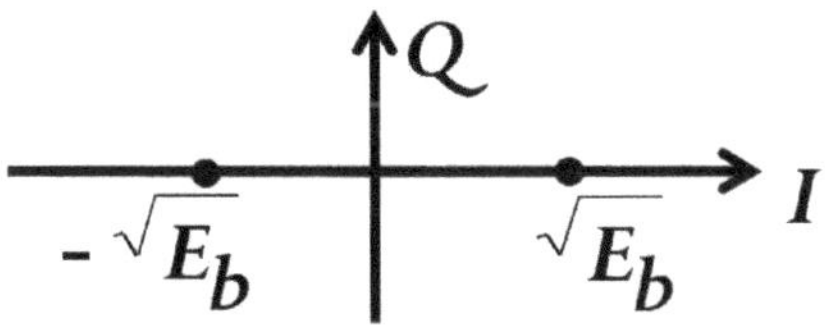

Figure 4.10 Constellation diagram for BPSK. *I: in-phase*; *Q: quadrature*.

Based on this basis signal, the BPSK signal set can be represented as

$$S_{BPSK} = \left\{ \sqrt{E_b}\phi_1(t), -\sqrt{E_b}\phi_1(t) \right\}. \tag{4.30}$$

Geometrically, Equation (4.30) may be represented as in Figure 4.10, the *constellation diagram*. This diagram represents the complex envelope of each possible *symbol state*.

The number of basis signals needed to represent a complex modulation signal is denoted as *dimension* of the vector space. A question that naturally arises in this context of waveform expansion in terms of basis functions is this: How many basis signals are required? The answer is: There should be as many basis signals as there are signals in the modulation signal set [25, 27]. In particular, the number of basis signals will always be less than or equal to the number of signals in the set. In the case of BPSK, there is one basis signal but two signals in the set. Where the number of basis signals equals that of signals in the modulation signal set, then all the signals in the set are orthogonal to one another.

In general, the properties of a modulation scheme may be inferred from its constellation. Such properties include: (1) the bandwidth occupied by the modulation signals, which decreases as the number of signals/dimension increases (in particular, the more densely packed a constellation, the more bandwidth-efficient it is); (2) the bandwidth occupied by a modulated signal, which *increases* with the dimension N of the constellation; (3) the probability of bit error, which is proportional to the distance between the closest points in the constellation [25, 27].

For a BPSK-modulated signal, it can be shown that its power spectral density (Figure 4.11) is [25]

$$P_{BPSK}(f) = \frac{E_b}{2}\left[\left(\frac{\sin \pi (f - f_c) T_b}{\pi (f - f_c) T_b} \right)^2 + \left(\frac{\sin \pi (-f - f_c) T_b}{\pi (-f - f_c) T_b} \right)^2 \right]. \tag{4.31}$$

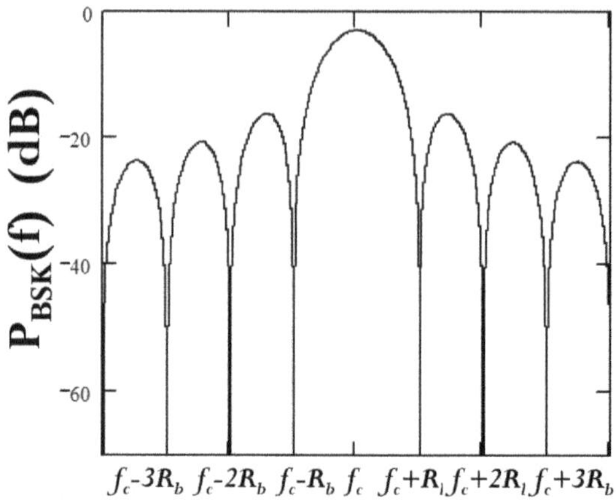

Figure 4.11 Power spectral density for BPSK. The bandwidth, $BW=2R_b=2/T_b$ *Source:* Ref. [25].

As may be surmised from Figure 4.11, the bandwidth containing the energy in the signal is a function of the shape of the pulse [57]. In the case of a rectangular pulse, 90% of signal energy is found within a bandwidth $BW \sim 1.6R_b$, whereas for a raised cosine filtering pulse with $\alpha = 0.5$, 100% of signal is found within a $BW=1.5R_b$ [25].

4.7.3 Quadrature Phase Shift Keying Modulation—QPSK

As its name suggests, *quadrature* PSK varies the carrier phase according to four equally spaced values, namely, 0, $\pi/2$, π, and $3\pi/2$, where each value of phase corresponds to a unique pair of baseband (information) bits given by [25, 27]

$$s_{QPSK}(t) = \sqrt{\frac{2E_b}{T_b}} \cos\left(2\pi f_c t + (i-1)\frac{\pi}{2}\right) \quad 0 \le t \le T_s \quad i = 1,2,3,4 \tag{4.32}$$

or

$$\begin{aligned} s_{QPSK}(t) = &\sqrt{\frac{2E_b}{T_b}} \cos\left[(i-1)\frac{\pi}{2}\right] \cos(2\pi f_c t) \\ &- \sqrt{\frac{2E_b}{T_b}} \sin\left[(i-1)\frac{\pi}{2}\right] \sin(2\pi f_c t \end{aligned} . \tag{4.33}$$

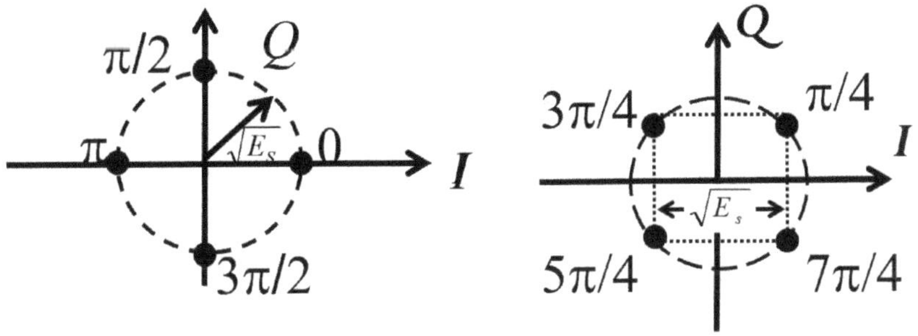

Figure 4.12 Constellations for QPSK.

The pertinent basis functions are

$$\phi_1(t) = \sqrt{\frac{2}{T_s}} \cos(2\pi f_c t) \tag{4.34}$$

and

$$\phi_2(t) = \sqrt{\frac{2}{T_s}} \sin(2\pi f_c t) \tag{4.35}$$

so that the QPSK set is

$$S_{QPSK} = \left\{ \sqrt{E_s} \cos\left[(i-1)\frac{\pi}{2}\right] \phi_1(t), -\sqrt{E_s} \sin\left[(i-1)\frac{\pi}{2}\right] \phi_2(t) \right\} \quad i = 1, 2, 3, 4 \tag{4.36}$$

whose corresponding constellation is shown in Figure 4.12. The constellations are equivalent; they differ by a 45° rotation.

The power spectral density for QPSK is as illustrated in Figure 4.13, which is given by [25]

$$P_{QPSK}(f) = E_b \left[\left(\frac{\sin 2\pi (f - f_c) T_b}{\pi (f - f_c) T_b} \right)^2 + \left(\frac{\sin 2\pi (-f - f_c) T_b}{\pi (-f - f_c) T_b} \right)^2 \right]. \tag{4.37}$$

In this spectrum, the bandwidth BW=R_b is given by the null-to-null spacing, which turns out to be half of that of a BPSK signal [25, 27].

4.7.3.1 Modulator Circuit for QPSK

To effect QPSK modulation, the baseband signal is converted into a unipolar bit stream, *m(t)*, with rate R_b. Then, the unipolar bit stream is converted into a

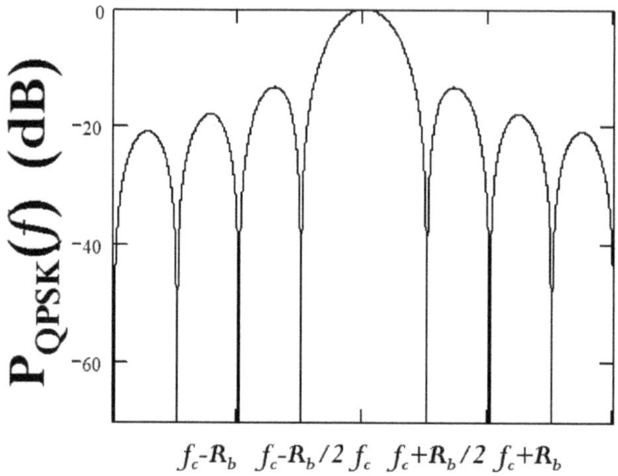

Figure 4.13 Power spectral density for QPSK. The bandwidth, BW=R_b. *Source:* Ref. [25].

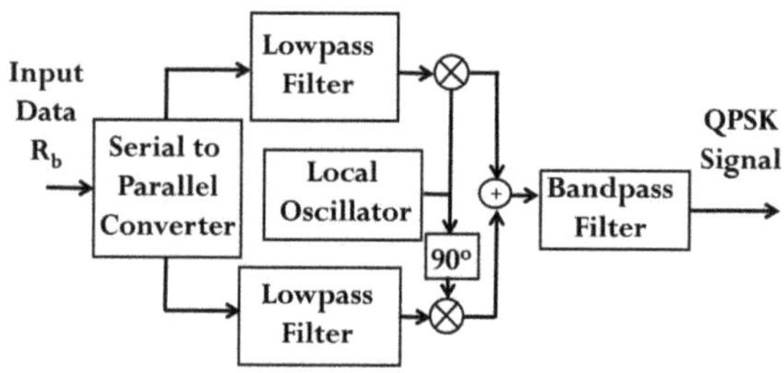

Figure 4.14 Modulator schematic for QPSK. *Source:* Ref. [25].

bipolar non-return-to-zero (NRZ) sequence. This stream is then divided into in-phase *mI(t)* and quadrature *mQ(t)* components, with each one having half the initial bit rate, i.e., $R_b/2$. Once the two separate binary streams are created, they are used to modulate two carriers so as to create two BPSK streams. Upon adding these, the QPSK signal results. Since the extent of the allowed spectrum is limited to a designated band, the QPSK power spectrum must be confined, and this is achieved by subsequently passing the signal through a bandpass filter (BPF). These operations are embodied in the schematic shown in Figure 4.14 [25, 27].

4.7.3.2 Circuit for QPSK Demodulation

To effect QPSK demodulation (Figure 4.15), the incoming QPSK signal is first filtered to remove out-of-band noise and adjacent channel interference

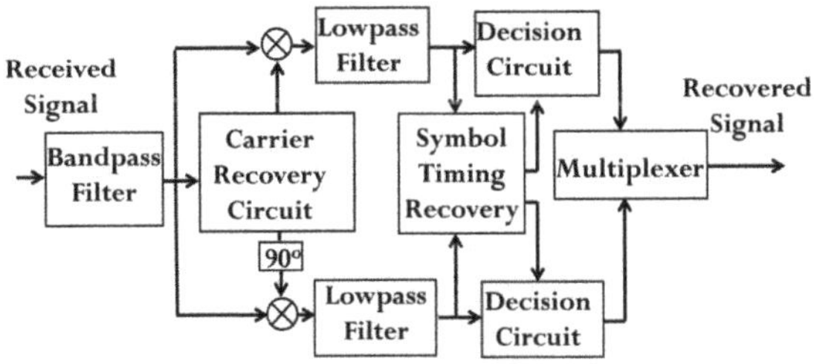

Figure 4.15 Circuit for QPSK demodulation. *Source:* Ref. [25].

noise accompanying it. Then, the filtered output is divided into two parts, which are coherently demodulated using I and Q carriers. The source of the coherent carriers is the received signal itself; they are extracted using carrier recovery circuits as in the BPSK demodulation. Following this, the outputs out of the demodulators are passed through decision circuits which generate I and Q streams. The final operation is to multiplex the two stream components to reproduce the original binary stream [25, 27].

4.7.4 Binary Frequency-Shift Keying Modulation Circuit

In BFSK approach, the frequency of a carrier of constant amplitude is switched between two values (bits 1 and 0) according to the digital representation of the baseband signal. Then, depending on the changes in frequency, there may be a continuous or a discontinuous phase between bits. Mathematically, the BFSK signal is given by [25]

$$s_{BFSK}(t) = \sqrt{\frac{2E_b}{T_b}} \cos\left(2\pi f_c + 2\pi \Delta f_c\right) t \quad 0 \leq t \leq T_b \quad (\mathit{binary} \quad 1) \tag{4.38}$$

or

$$s_{BFSK}(t) = \sqrt{\frac{2E_b}{T_b}} \cos\left(2\pi f_c - 2\pi \Delta f\right) t \quad 0 \leq t \leq T_b \quad (\mathit{binary}\, 0) \tag{4.39}$$

which may be synthesized according to

$$S_{BFSK}(t) = \sqrt{\frac{2E_b}{T_b}} \cos\left(2\pi f_c t + 2\pi k_f \int_{-\infty}^{\infty} m(\xi) d\xi\right). \tag{4.40}$$

Examination of Equation (4.40) reveals that even though *m(t)* is discontinuous, as a result of the integration, the phase is continuous [25].

4.7.4.1 Circuit for BFSK Modulation

Direct FM is usually utilized to implement BPSK modulation. In this scheme, the baseband signal is applied to an oscillator in order to modulate its frequency [25].

4.7.4.2 BFSK Demodulation via a Coherent Detector

The BFSK demodulation is carried out via a coherent detector, as illustrated in Figure 4.16. Its operation, following the numbers at the pertinent points in the schematic, is as follows. At point **1**, two correlators are applied to coherent reference signals generated locally. At point **2**, the outputs of the correlators are subtracted and compared by a threshold comparator. At point **3**, the comparator output triggers a decision circuit; if the difference signal has a value greater than the threshold, it is classified as "1"; otherwise, it is classified as a "0." The probability of bit error is given by [57]

$$P_{e,BPSK} = Q\left(\sqrt{\frac{E_b}{N_0}}\right). \tag{4.41}$$

4.7.4.3 BFSK Demodulation via a Noncoherent Detector

In the noncoherent detector approach to BFSK demodulation, a system as that illustrated in Figure 4.17 is utilized. Its operation will now be explained with reference to the numbers at the various points in the figure. At point **1**, the incoming signal is divided into two paths, a lower path where it is passed through a filter that is matched to the frequency f_H, and an upper path where it is passed by a filter matched to the frequency f_L. At point **2**, the matched filters function as BPFs centered at f_H and f_L, and their outputs are passed through envelop detectors. At point **3**, the difference of the outputs of the

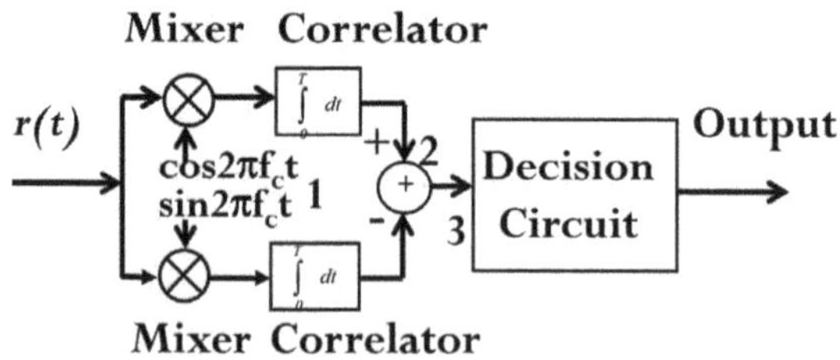

Figure 4.16 BFSK demodulator implemented by coherent detector. *Source:* Ref. [25].

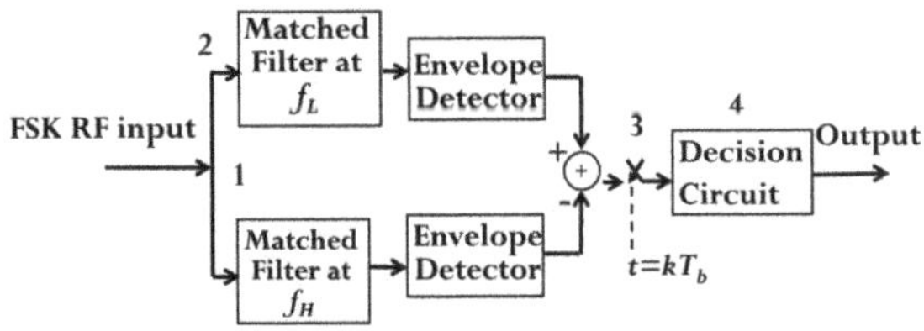

Figure 4.17 BFSK demodulation via a noncoherent detector. *Source:* Ref. [25].

envelope detectors is sampled at times $t=kT_b$. At point **4**, depending on the magnitude of the sampled value, a decision circuit decides whether to classify the received bit as a logic "1" or "0." The probability of bit error rate is [25]

$$P_{e,FSK,NC} = \frac{1}{2}\exp\left(\frac{-E_b}{2N_0}\right). \tag{4.42}$$

4.7.5 M-ary Quadrature Amplitude Modulation Approach

In this modulation approach, the amplitude of the carrier is allowed to vary together with the phase. The mathematical representation of the signal is

$$\begin{aligned} s_i(t) = &\sqrt{\frac{2E_{\min}}{T_s}}a_i\cos(2\pi f_c t) + \sqrt{\frac{2E_{\min}}{T_s}}b_i\sin(2\pi f_c t) \\ &0 \le t \le T \quad i = 1, 2, \ldots, M \end{aligned} \tag{4.43}$$

where E_{min} is the energy of the signal with the lowest amplitude, and a_i and b_i are a pair of independent integers according to the location of a signal point. The prototypical constellation for the case of, e.g., M = 16, is shown in Figure 4.18.

4.7.6 Orthogonal Frequency Division Multiplexing

In the OFDM approach, the goal is to enable the power-efficient signal transmission by a large number of users in the same channel (Figure 4.19). In particular, in this technique, multiple sub-carriers are utilized to transmit signal bits in parallel to achieve very high throughput. The resulting spectra from different sub-channels partly overlap and pulse shaping is not necessary. OFDM is usually combined with QAM or M-PSK [30].

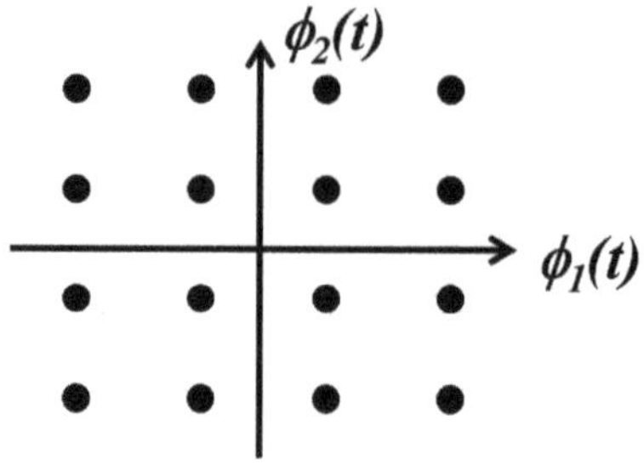

Figure 4.18 Constellation of M-ary QAM (M=16).

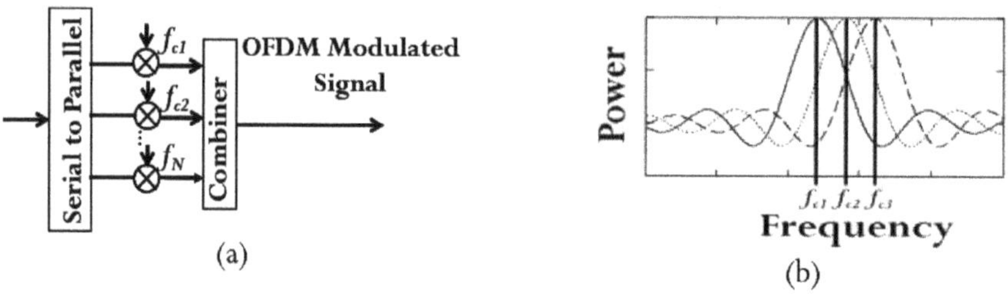

Figure 4.19 (a) Schematic of OFDM modulator. (b) Carriers modulated by rectangularly shaped data pulses are densely spaced and orthogonal in frequency. *Source:* Ref. [30].

4.7.7 Direct Sequence Spread Spectrum Modulation Approach

An illustration to convey the concept behind the DS/SS approach is given in Figure 4.20. Its purpose is to greatly enlarge or spread, (A) $\rightarrow$ (B), the carrier spectrum relative to the information rate [32]. This action endows the spread spectrum signal with an anti-jamming capability by forcing the jammer to deploy the transmitted power over a much wider bandwidth (C) than what would be necessary for a narrower band system [32]. In particular, for a given jammer power, the jamming power spectral density (D) is reduced in proportion to the ratio of the spread bandwidth Bs to the un-spread bandwidth B.

4.7.7.1 Modulation and Demodulation Circuits for Direct Sequence Spread Spectrum (DS/SS)

The modulator and demodulator circuits for DS/SS are illustrated in Figure 4.21. In order to produce an SS signal, the input data is fed into a data encoder and a data modulator, which results in the signal message $m(t)$. $m(t)$ is then fed into a mixer, where it is multiplied by the product of the

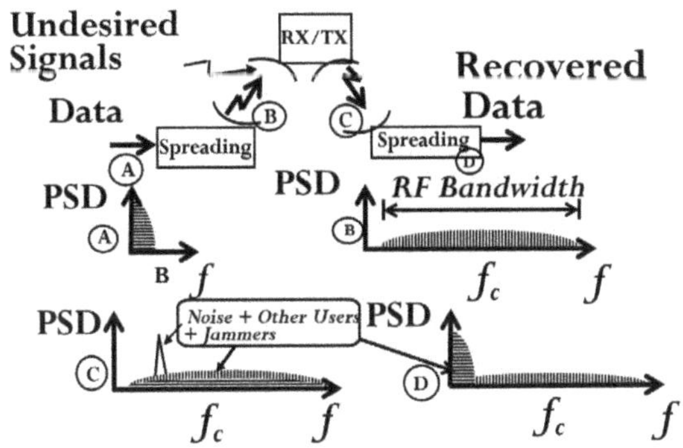

Figure 4.20 Concept of direct sequence spread spectrum modulation. *Source:* Ref. [33].

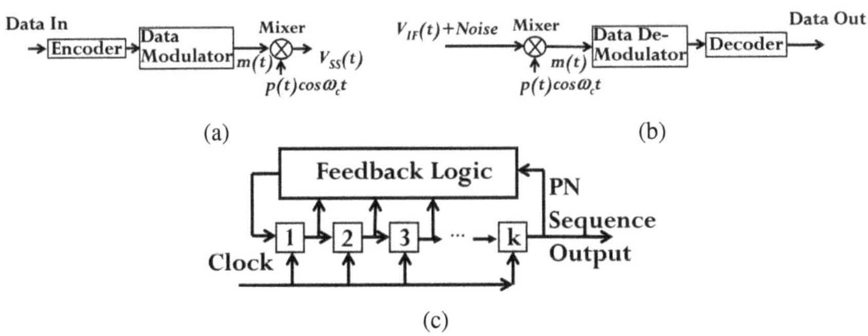

Figure 4.21 Direct sequence spread spectrum. (a) Modulator. (b) Demodulator. (c) PN sequence generator diagram. *Source:* Ref. [33].

carrier and a pseudo-random sequence (a random-like signal of $\pm 1p(t)$) which results in [32]

$$V_{SS}(t) = \sqrt{\frac{2E_s}{T_s}} m(t) p(t) \cos(2\pi f_c t + \theta) \tag{4.44}$$

with

$$m(t) = \sum_{m=-\infty}^{\infty} a_m p_m (t - mT_b) \tag{4.45}$$

and

$$p(t) = \sum_{n=-\infty}^{\infty} c_n p' (t - nT_c) . \tag{4.46}$$

The technique of code division multiple access (CDMA), in which each user possesses a pseudo-noise (PN) sequence and multiple users share the same bandwidth employs the DS/SS system. From each user's perspective, signals pertaining to all other users appear as noise. To produce the PN, a feedback shift register with k stages is employed (Equation (4.45)); see Figure 4.30(c). The sequence length generated by a k-stage shift register is $N = 2^k - 1$ [32]. The degree to which interference rejection of a DS/SS system is captured by the so-called *processing gain*, which is defined by

$$PG_{DS} = \frac{T_b}{T_c} = \frac{R_c}{R_b}. \tag{4.47}$$

4.7.8 Frequency Hopping Spread Spectrum Modulation/Demodulation

In the FH/SS modulation approach (Figure 4.22), the carrier frequency is periodically changed [32]. In particular, the baseband signal is multiplied by a carrier whose frequency is varied according to a pseudo-random sequence.

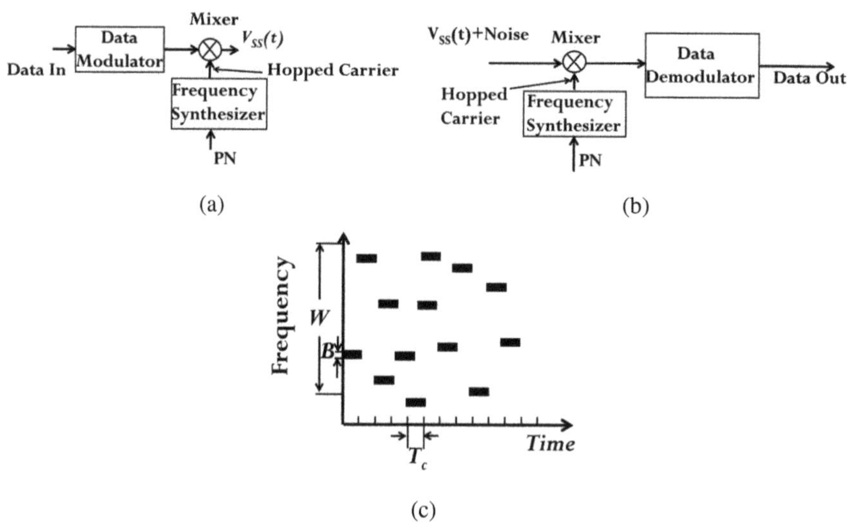

Figure 4.22 Frequency hopping spread spectrum approach. (a) Modulator schematic. (b) Demodulator schematic. (c) Frequency hopping versus time. T_c is the duration of time in a given frequency band. *Source:* Ref. [33].

The processing gain of FH is defined as

$$PG_{FHS} = \frac{W}{B}. \tag{4.48}$$

Here, *W* is the range of frequencies produced by hopping and *B* is the bandwidth of the band being hopped (the baseband bandwidth) [32].

4.8 Summary

In this chapter, a number of topics surrounding modulation and demodulation, with emphasis in unveiling to the reader pertinent *system-level* schematics/block diagrams, have been dealt with. First, we introduced system-level block diagrams of AM and FM modulators and demodulators and explained their respective principles of operation. In particular, under the topic of AM modulator/demodulator, the full carrier modulator, the SSB suppressed carrier modulator, and the envelope detector were introduced. Under the topic of FM modulator/demodulator, we introduced the VCO as FM modulator, the indirect FM modulator, and the PLL-based FM detector. Under the topic of digital modulation, the principles of a number of modulation schemes, namely, binary modulation, ASK, BFSK, BPSK, QPSK, minimum shift keying (MSK), M-ary QAM, OFDM, DS/SS, and FH/SS were discussed. The geometric representation of digital modulation schemes and the complex envelope form of a modulation signal were also introduced.

4.9 Problems

1. What is the bandwidth required for transmitting 100 AM channels?
2. How does the PLL FM demodulator work? Explain.
3. In DS/SS modulation, what tradeoff is made? Are there any cell phone products based on this technique? List them.
4. Why are DS/SS and FH employed in secure communications?
5. Explain how OFDM works.
6. Perform a literature search and describe the carrier aggregation (CA) technique.

5

Transmitter and Receiver Architectures

5.1 Introduction

Whether we are dealing with a communications or a RADAR system, both contain at least one transmitter and one receiver. This may be surmised from the block diagrams in Figures 5.1 and 5.2, where we show how both systems are *conceptually* similar, except that the RADAR is tailored to receive the return/echo of the signal it transmits, but the communication system is tailored to receive a signal produced by an independent information source.

An in-depth examination of traditional transmitters and receivers has already been presented in [33]. In this chapter, we will take on the subjects of transmitter and receiver independently. Receivers, in particular, will be addressed in the context of the new opportunities enabled by 5G, with the goal of understanding the design choices that accompany them.

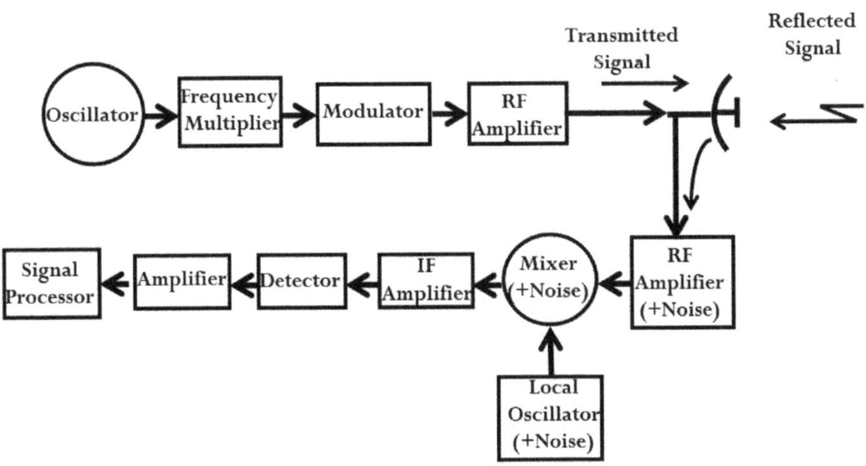

Figure 5.1 Schematic of RADAR system.

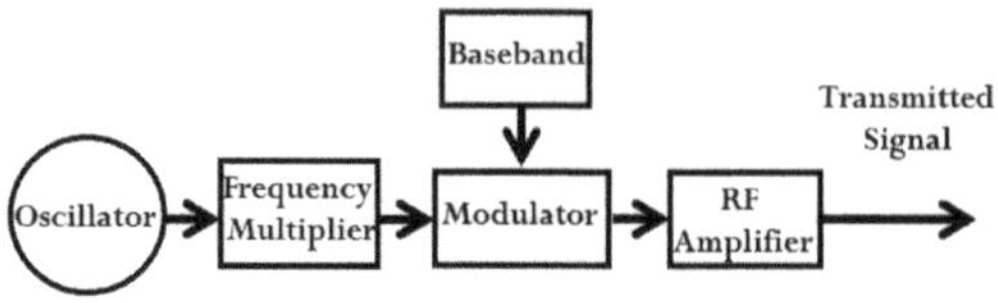

(a)

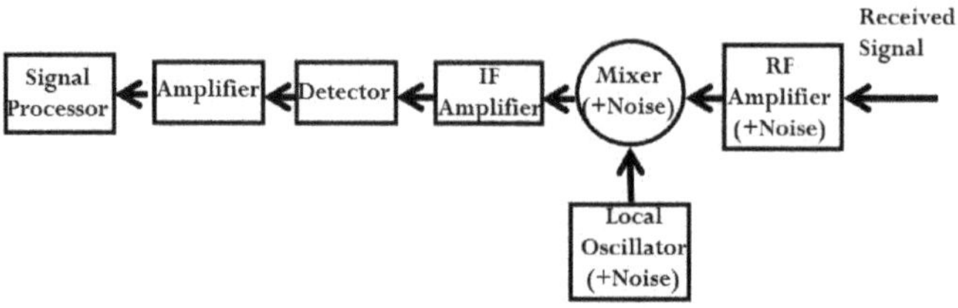

(b)

Figure 5.2 Communications system. (a) Transmitter. (b) Receiver.

5.2 The Transmitter

In general, the nature of the transmitter architecture employed is a function of two important factors, namely, the wanted and unwanted emission requirements, and the number of oscillators and external filters needed. When the transmitter and the receiver share the same platform, it is advisable to choose the architecture and frequency planning of the transmitter together with those of the receiver so as to allow the sharing of hardware and possibly power.

Whether dealing with a transmitter or a receiver, up until recently, one encountered two principal architectures, namely, the *heterodyne* and the *homodyne*.[1] The heterodyne architecture is predicated upon mixing the baseband signal with an offset-frequency local oscillator (LO) in a mixer circuit to generate the radio frequency (RF) signal to be transmitted [15, 19, 27, 28, 34, 35, 36]. It is found that, in general, the frequency translation processes may be performed more than once.

5.2.1 Heterodyne Transmitter Architecture

A prototypical schematic of a heterodyne transmitter is illustrated in Figure 5.3 [27]. The schematic may be conceptually divided into three sections, namely, the baseband (BB), the intermediate frequency (IF), and the RF.

---

[1]Hetero: different; Dyne: to mix.

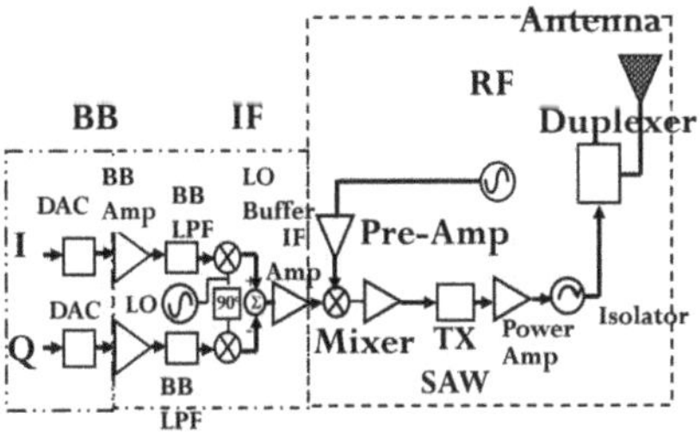

Figure 5.3 Block diagram of super-heterodyne transmitter.*Source:* Ref. [33].

In the baseband section, in-phase (I) and quadrature (Q) digital representations of the baseband signal to be transmitted are converted to analog signals by digital-to-analog converters (DACs). Then, following BB filtering, the I and Q BB signals are up-converted into IF signals by applying them to mixers in an I/Q modulator. In this circuit, which effects frequency up-conversion, the Q channel IF signal is phase-shifted by an additional 90° with respect to the I channel. The output of the I/Q modulator is a composite IF signal, which is then amplified by amplifier IF Amp (normally, a variable gain amplifier (VGA)). The IF VGA is followed by an up-converter mixer whose output is the RF signal. The RF signal is further amplified by an RF VGA, and then by a driver amplifier to a power level high enough to drive the power amplifier (PA). An RF bandpass filter (BPF; SAW filter) is inserted between the driver and the PA to select the desired RF signal and suppress other mixing products generated by the RF up-converter. While it would appear that placing the RF BPF immediately after the up-converter would be a better choice, this configuration would be inconvenient since the whole block, from the analog-to-digital converter (ADC) to the IF Amp driving the mixer, may be integrated on a single semiconductor chip, which would not be the case if the filter were moved before the pre-Amp.

In the RF section, depending on the application, a Class AB or Class C PA is employed [24]. While the Class C PA is capable of higher power efficiency than the Class AB PA, it can only be used for constant envelope modulation schemes such as frequency modulation (FM) and Gaussian minimum-shift keying (GMSK) [24].

The gain and nonlinearity performance of the PA are very sensitive to its load. Because of this, an isolator is usually employed between the PA and the antenna to reduce the influence of variations in the antenna environment [27, 34], which alters its input impedance and, therefore, the PA load

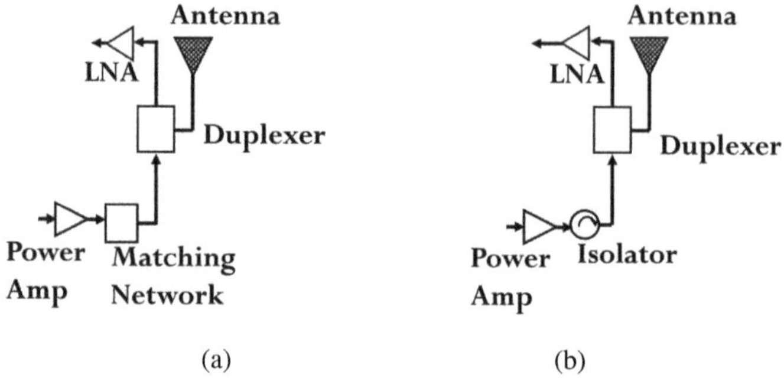

Figure 5.4 PA/antenna interface. (a) Employing duplexer-PA isolator. (b) Employing duplexer-PA matching network. *Source:* Ref. [33].

(Figure 5.4(a)). The isolator may, under certain circumstances, be replaced with a tunable matching network (Figure 5.4(b)).

The duplexer (Figures 5.3 and 5.4) is a filter that separates the transmit (TX) and receive (RX) bands, in the case of frequency-domain duplexing (FDD) or when the transmitter and receiver share the same hardware and power, e.g., a *transceiver*. When the TX and RX bands coincide, the duplexer is replaced by an RF switch to perform time-division duplexing (TDD). The duplexer filter suffers from a loss of several dB, whereas the switch introduces a typical loss of the order of 1 dB [34]. The loss in these components is responsible for a lower effective PA efficiency since additional bias current would be necessary to compensate for this loss [66]. The LO synthesizer of the RF section provides the LO power to the RF up-converter and effects channel tuning.

5.2.2 The Homodyne Transmitter Architecture

A prototypical schematic of a homodyne transmitter architecture is illustrated Figure 5.5.

As in the heterodyne transmitter architecture, the homodyne transmitter has a baseband section and an RF section but no IF section. In the baseband section, the digital information to be transmitted, after being rendered into I and Q signals, is fed into DACs and then passed through low-pass filters (LPFs); the filters suppress adjacent channel and alternate-channel emission levels and eliminate aliasing products [66]. This is followed by the I and Q BB signals being fed to an I/Q modulator where they are both up-converted to RF and then added [34].

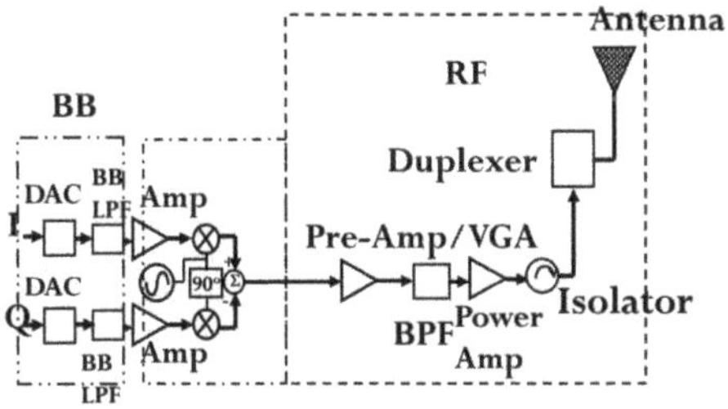

Figure 5.5 Schematic of a homodyne (zero IF/direct-conversion) transmitter.

In the RF section of the homodyne transmitter, the direct conversion, or zero-IF (no IF) up-conversion to the RF for transmission frequency, is *directly* effected. In particular, the composite RF signal is amplified all the way up to the RF PA. After this, the signal is fed to a BPF to suppress out-of-band signals, especially those in the receiver band, noise, and spurs emissions [27, 34]. When compared to the super-heterodyne transmitter, the main difference is that the homodyne transmitter eliminates the need for the LO synthesizer. However, the outstanding feature of the homodyne transmitter is that its transmission contains much less spurious products than the super-heterodyne transmitter [34]. The homodyne transmitter is not perfect; its advantages are accompanied by a number of drawbacks which must be dealt with for a successful implementation; these are reviewed below.

5.2.2.1 Drawbacks of Homodyne transmitter architecture

5.2.2.1.1 LO disturbance and its corrections

LO disturbance is caused by coupling from the output of the PA to the LO, which, thus, disturbs the purity of its output frequency (Figure 5.6(b)). The reason for the disturbance is that since the waveform at the PA output is modulated and its spectrum centered about the LO frequency, when the frequency variation reaches the LO circuit, it manifests as noise added to the LO frequency [34]. This noise, adding to it, proceeds to corrupt the purity of the LO signal. The mechanism of corruption is "injection pulling" or "injection locking," which describes the shift in the frequency of an oscillator toward that of an external stimulus (Figure 5.6(c)).

When the frequency of the injected noise is close to the oscillator center frequency, increases in the amplitude of the injected noise cause the LO frequency to broaden with the peak magnitude of the LO spectrum shifting toward the injected frequency until eventually the strongest signal produced

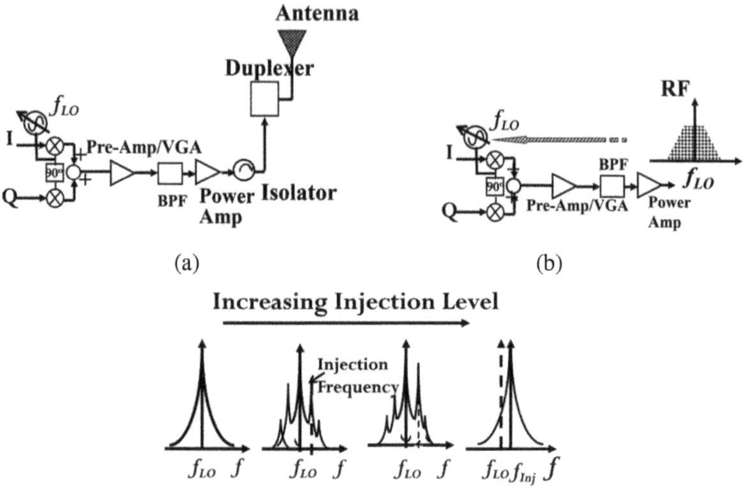

Figure 5.6 (a) Direct-conversion transmitter. (b) The coupling from the power amplifier output to the local oscillator causes LO frequency pulling. (c) Phenomena of frequency pulling by signal coupled from PA back to the LO. *Source:* Ref. [34].

by the LO circuit is that of the injected signal, thus reaching the "locking" condition [34, 35]. In practice, it is found that to avoid injection locking, it is necessary to maintain noise levels well under 40 dB [34].

To reduce LO pulling, several techniques have been advanced. In one of them (frequency planning), the output spectrum of the PA is set far away from the LO frequency. This can be achieved when using quadrature up-conversion, by producing the LO frequency at an offset, i.e., at a frequency away from the intended LO frequency, and then synthesizing the actual LO frequency by adding or subtracting the offset frequency [66]. An example of this is illustrated in Figure 5.7, which shows oscillators generating frequencies f_{LO_1} and f_{LO_2}. These frequencies are then mixed and filtered to produce $f_{LO\_1} + f_{LO\_2} = f_{LO}$ such that $f_{LO} \gg f_{LO\_1}$ and $f_{LO} \gg f_{LO\_2}$.

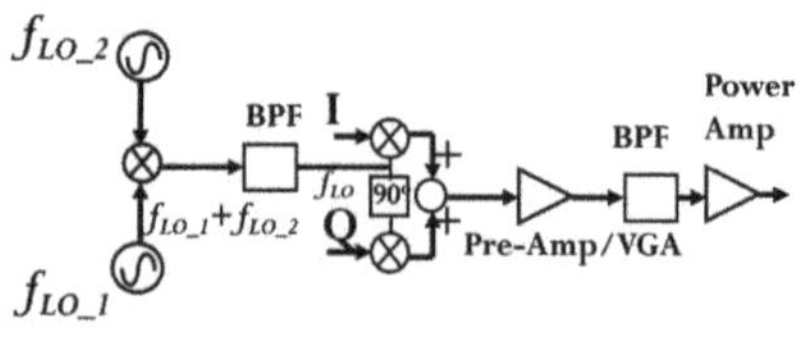

Figure 5.7 Direct-conversion transmitter with *offset* LO for diminishing frequency pulling.

If the filter does not suppress the mixing terms $nf_{LO\_1} \pm mf_{LO\_2}$, there is a potential difficulty that can be met with this approach, namely, that these will manifest as spurious frequencies at the amplifier output spectrum. Therefore, the selectivity of the filter whose output is f_{LO} becomes decisive [34].

In a second approach to circumventing the problem of LO pulling in transmitters, the baseband signal is up-converted in at least two steps so that the PA output spectrum is far from the frequency of the LOs (Figure 5.8). What happens in this situation is that the baseband I and Q channels undergo quadrature modulation at a lower frequency, f_{LO_1}, and the result is up-converted to $\mathrm{f}_{LO\_1} + \mathrm{f}_{LO\_2}$ by mixing and bandpass filtering. As pointed out by Razavi [34], one advantage that accompanies the two-step up-conversion approach over the direct one is that, since quadrature modulation is performed at lower frequencies, a superior I and Q matching may be obtained, thus resulting in a lower level of cross-talk between the two bit streams. In addition, the utilization of channel filters at the first IF helps in limiting spurs and noise originating in neighboring channels [34]. A drawback of the two-step transmitters, however, is that because the up-conversion process produces both wanted and unwanted sidebands with the same magnitude, the filter that follows the second up-conversion must reject the unwanted sideband by at least 50 – 60 dB [66], thus requiring the filter to be implemented in an expensive off-chip technology.

In addition to the previously discussed issues, there are other problems that accompany the homodyne transmitter if the requirements that it needs to meet are stringent such as those pertaining the output power range, for instance, as would be necessary for wideband-CDMA systems [34]. These issues include the high linearity requirements for the baseband filters and the modulator due to most of the gain being realized in the baseband section. It also includes high requirements on LO–RF isolation due to the LO frequency

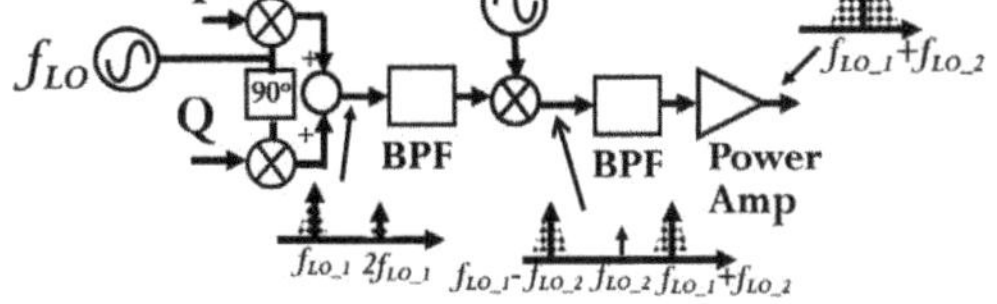

Figure 5.8 Heterodyne transmitter utilizing a two-step up-conversion. The first BPF suppresses the harmonics of the IF signal while the second removes the unwanted sideband centered around $f_{LO\_1} - \mathrm{f}_{LO\_2}$.

being set in the transmit band. Finally, it includes issues with I/Q phase mismatches because even a low error in the phase shifting network may lead to a severe degradation of the error vector magnitude (EVM) [34].

5.3 The Heterodyne Receiver Architecture

The prototypical heterodyne receiver architecture is illustrated in Figure 5.9. In this architecture, the received RF signal is translated to much lower frequencies, which makes it possible to reduce the required quality factor for the channel-select filter; so it is easier to implement [27].

Upon detection by the antenna, the RF front end subsequently processes the modulated high-frequency carrier signal. The architecture includes part of the duplexer as the pre-selector, a low noise amplifier (LNA), an RF BPF, an RF amplifier as the preamplifier of the mixer, and an RF-to-IF down-converter mixer.

The LNA provides amplification to improve the desired sensitivity level of the reception, and its gain can be adjusted to compensate for deficiencies in the dynamic range (DR) of the receiver.

The LNA is followed by an RF BPF, which is usually implemented in SAW filter technology. Its purpose is to dramatically reduce the power leakage that originates on the transmitter as well as the image and other interference. When the rejection to the transmitted power produced by the pre-selector is sufficiency large, or if the receiver is in a half-duplex system, then the SAW filter is not necessary [27].

The SAW filter is followed by an RF amplifier (Amp) or preamplifier preceding the mixer and is endowed with a large gain to reduce the influence of the noise figure (NF) of the mixer, and subsequent stages, on the overall receiver NF and sensitivity. This amplifier is normally necessary when the mixer is passive [27].

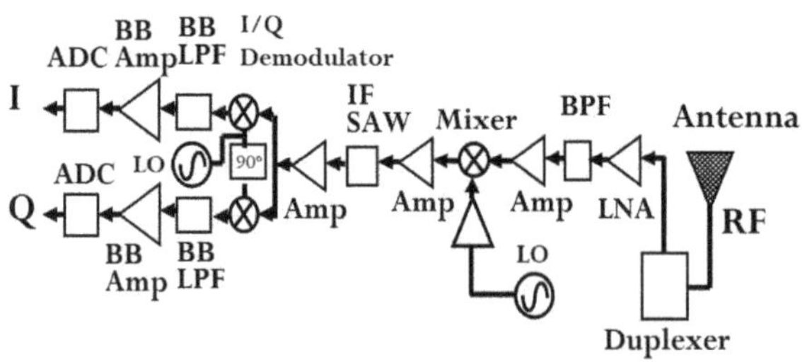

Figure 5.9 Block diagram of super-heterodyne receiver.

The mixer translates the RF signal into the IF. The signal thus translated is then fed to an IF amplifier whose output is fed into an IF BPF to effect channel selection and suppress unwanted mixing products. The implementation of this filter usually relies on high-selectivity surface acoustic wave (SAW) or crystal filter technologies. The overall receiver gain is set by the IF amplifier which, formed by multiple stages, may produce a variable/adjustable gain.

Following the IF block, we find the I/Q demodulator, which realizes the second frequency converter, whose function is to down-convert the signal frequency from IF to BB. The I/Q demodulator possesses two mixers and converts the IF signal into two baseband signals I and Q signals that are phase shifted 90° with respect to each other. The 90° phase shift is realized with a polyphase filter that shifts the phase between very-high frequency (VHF) LO signals going to the mixers in the I and Q channels.

The mixers in the I/Q modulator are followed by LPF in both the I and Q channels; its function is to filter out the unwanted mixing products and to further suppress interferers.

5.4 The Homodyne (Zero IF/Direct-Conversion) Receiver

The direct conversion receiver circumvents the image problem accompanying the heterodyne receiver front end. This is accomplished because it simply does not have an image band. In it, the received signal, after pre-selection by the duplexer, is amplified by an LNA and is further filtered by an RF filter (Figure 5.10).

In the context of a transceiver, due to the leakage from the transmitter and to control the LO-RF leakage-related self-mixing, the rejection of this filter needs to be higher than that required in the heterodyne receiver; this higher rejection also allows the relaxation of the second-order distortion requirement of the down-converter [34].

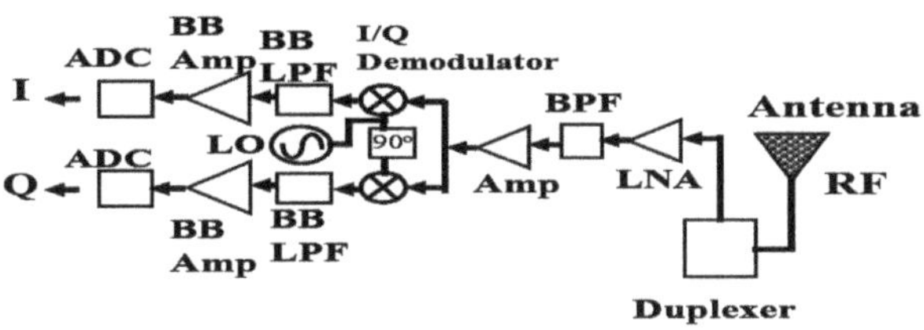

Figure 5.10 Homodyne receiver architecture.

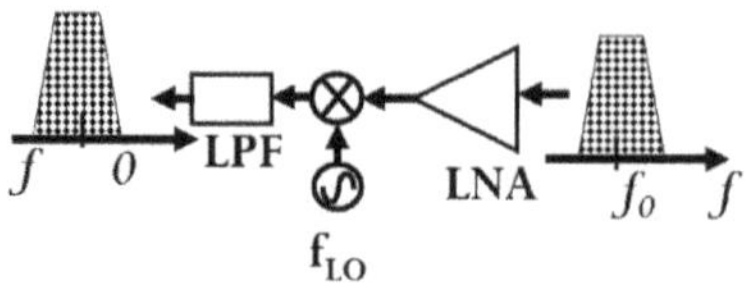

Figure 5.11 Simplified homodyne receiver.

The discussion on the homodyne receiver is continued with reference to Figure 5.11.

The key distinction between the homodyne and the heterodyne receiver is that the LO frequency in the homodyne is equal to the received input carrier frequency. As a result, channel selection requires only an LPF with relatively sharp cutoff characteristics [34]. Since homodyne down-conversion results in an overlapping of the positive and negative parts of the input spectrum, the circuit in Figure 5.11 operates properly only with double-sideband amplitude modulation (AM) signals. In the case of frequency and phase modulated signals, the down-conversion has to provide quadrature outputs in order to avoid loss of information, and this is effected by the circuit on Figure 5.12. This is because the two sides of FM or quadrature phase-shift keying (QPSK) spectra carry different information and must be separated into quadrature phases in translation to zero frequency [27, 34].

Several advantages may be attributed to the homodyne receiver over the heterodyne receiver. First, the image problem is circumvented because $f_{IF} = 0$; therefore, no image filter is required. Second, the LNA does not need to drive a $50 - \Omega$ load. And, third, the IF SAW filter and subsequent down-conversion stages are replaced with LPFs and baseband filters that are amenable to monolithic integration [34].

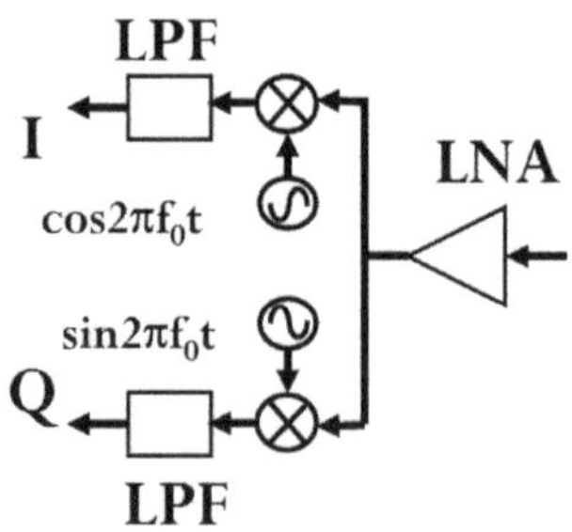

Figure 5.12 Homodyne receiver with quadrature down-converter.

5.5 Receiver Architectures in Light of 5G [37]

One of the fundamental factors, perhaps the most important, that must be considered in choosing a receiver architecture, whether for communications or RADAR applications, is cost. Cost, in turn, will be determined by the attainable levels of integration, of flexibility, and of power consumption [37].

For wireless communications, the fact that "evolving flexibility" usually entails compatibility with past generations of standards means that it is imperative to adopt architectures with the highest degree of flexibility. Toward this end, Bronckers, Roc'h, and Smolders have recently presented a rather timely examination of receiver architectures, which clarifies the field of choices available [37]. This examination addresses frequencies under 6 GHz, which includes more than 50 bands for LTE and WiFi, but which will also be common to 5G.

The receiver (RX) candidates considered were classified according to the framework illustrated in Figure 5.13. As could be surmised from our earlier discussions, what will differentiate one RX architecture from another is the hardware between the receiver antenna and the baseband. In that context, the framework for discussing the various candidate RX architectures was formulated in terms of the down-conversion and ADC functions. Thus, the RX functionality, in Figure 5.13, is visualized in three dimensions, with each dimension capturing one aspect of the overall RX functionality and design freedom.

The "dimension 1" deals with the IF. This captures the fact that an RX architecture may be chosen depending on its intended IF, which may range from IF = 0 to IF = RF. Traditionally, most RX architectures have adopted a solution in which the received signal is down-converted from its carrier frequency f_c to an IF $\neq$ 0, by one or several mixers. In this case, the

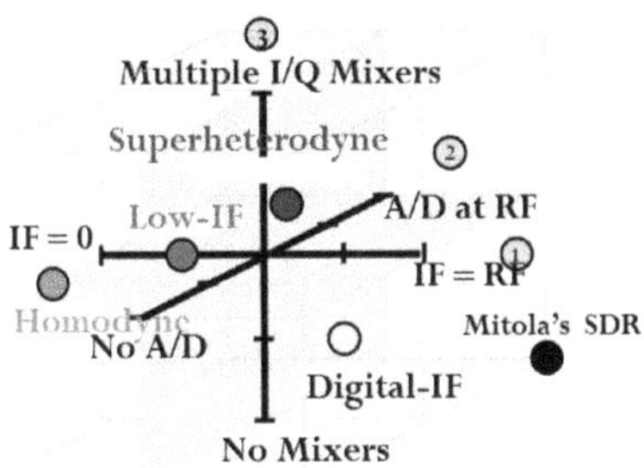

Figure 5.13 Proposed framework for discussing the most commonly used receiver architectures. *Source:* Ref. [37].

actual value of the IF, namely, [0, RF], is the key parameter embodied by dimension 1. A displacement from IF = 0 toward IF = RF embodies the circumstance in which a finite IF is considered. In this case, the increase in IF would afford the designer the possibility of including a second down-conversion which, by placing an ADC after the first IF, would, in turn, enable him to access greater bandwidth or perform additional digital signal processing (DSP). This may be understood as follows: The first down-conversion would result in a higher IF frequency and, thus, more bandwidth, than the second IF, which would have to be low enough (thus, smaller bandwidth). Unfortunately, the additional components accompanying multiple down-conversions, the faster ADC needed, and the additional design complexity involving the higher frequency would result in overall higher power consumption and cost.

The “dimension 2” deals with the position of an ADC within the “receive chain.” This could range from having no ADC, the traditional solution, to placing the ADC immediately after the antenna, sampling the RF frequency, or placing it at some intermediate point, sampling an IF. A displacement in the direction of no ADC toward ADC embodies the circumstance in which the ADC sampling frequency is increased. In this case, the designer is afforded the greater flexibility of performing more sophisticated DSP functions due to the greater digitized bandwidth available. While the move toward digital processing at higher frequencies would ease the issue of realizing tough filter specifications, this would also be accompanied by more demanding requirements for the analog component linearity. This is due to the impossibility of filtering out all the undesired signals present in a large bandwidth. Further, the greater sampling frequency would be accompanied by a greater ADC power consumption and cost.

The “dimension 3” deals with the complexity of the mixer employed to frequency-translate the RF frequency. This could range from having no mixers to employing multiple blocks of analog I/Q mixers. A displacement in the direction of no mixer toward multiple I/Q mixers would embody the implementation of analog mixers utilizing in-phase and quadrature branches or employing multiple IF stages. The two alternatives result in increased complexity due to the larger number of components utilized in the analog domain. On the other hand, this would also endow the designer with greater freedom during the frequency planning stage.

The “space” of RX architectures elicited by the framework depicted in Figure 5.15 are next reviewed.

5.5.1 Super-Heterodyne Receiver

A high-level, simplified, diagram of the super-heterodyne (also referred to as heterodyne or IF-receiver) RX is shown in Figure 5.14. This RX architecture, while offering high performance, also possesses limited flexibility, relatively high cost, and relatively high power consumption. It is instructive to examine the rationale behind each of its building blocks to develop an intuitive understanding for how it works.

As discussed previously, the dominant drawback of this RX is the inevitable down-conversion of both the desired signal and its image since both are located at the same distance, above and below, from the local oscillator LO1. As a result, filters must be included in the receive path to eliminate the mirror frequency prior to down-conversion. This is the function of filters BPF 1 and BPF 2 in Figure 5.14.

The performance specifications of BPF 1 and BPF 2 are a function of the IF chosen. In particular, when a high IF is chosen and, consequently, the distance between the desired signal and its image is large, the image frequency will be easier to reject. However, in this case, the filters may be of such specifications that they cannot be realized in integrated-circuit form.

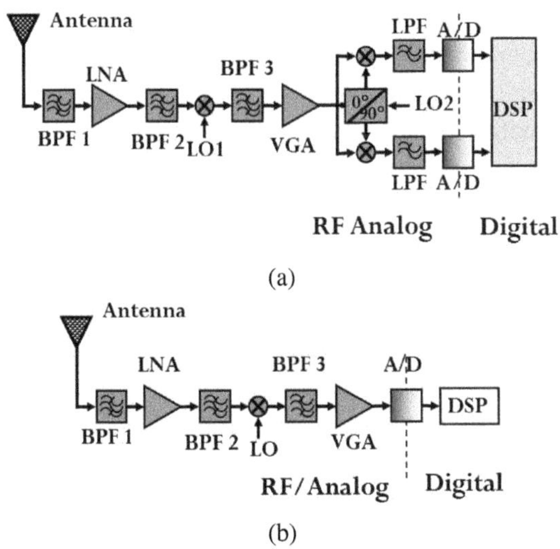

Figure 5.14 Heterodyne receiver block diagrams. (a) Double-conversion quadrature super-heterodyne receiver. (b) Single-conversion digital-IF receiver: Here, the down-converted received signal is converted into digital form at the IF stage. *Source:* Ref. [37].

The filters, then, may preclude the frequency flexibility needed to cover 5G applications.

Before the initial filtering, effected by BPF 1, it is first necessary to amplify the received signal with an LNA, whose purpose is to decrease the RX NF and, thus, increase its sensitivity. Thus, a second function of BPF 1 is to block the out-of-band signal energy from undesired signals that may result in RX desensitization.

After down-conversion to the IF, effected by the mixer, filter BPF 3 is employed to further separate the signal of interest from other nearby signals; this filter is, thus, called "channel select filter." The specifications of BPF 3, in turn, depend on its operating frequency, which coincides with the IF. Thus, this is another factor that the choice of IF influences. Considerations that must be taken into account, therefore, include Q-factor since, for the identical channel separation, a lower IF would require a lower-Q filter than higher IF. The high-Q requirement may, therefore, preclude IF filter integration. It becomes apparent, then, that selecting the IF of the RX entails a tradeoff between image rejection (IR) (before mixing) and channel selection (after mixing).

After effecting filtering by BPF 3, a VGA is then employed to condition the signal amplitude so that it may reach the optimal range for the subsequent ADC. If it is impossible to employ an affordable and low power-consuming ADC at the IF frequency following the first down-conversion, a second down-conversion may be effected to bring the IF closer to the baseband for information extraction (Figure 5.14(a)). In this case, it is necessary to perform the second down-conversion by employing analog quadrature mixers. These are necessary to separate the negative from the positive frequencies, as done in a homodyne receiver. This approach is the most often employed in modern heterodyne receivers. On the other hand, if the first down-conversion results in a low enough IF signal frequency, then it may be digitized directly (Figure 5.14(b)), thus realizing the "digital-IF receiver." In this case, while the issue of dealing with I/Q imbalance is eliminated, it is necessary to effect the ADC conversion at a frequency requiring an expensive and power-consuming ADC.

While, conceptually, a double-conversion receiver may be implemented by using multiple LOs, one for each frequency conversion step, in practice, it is found that it is impossible to prevent these from coupling in an IC environment. This is overcome by using a so-called "sliding-IF receiver" architecture, in which the desired LO frequencies are derived from a single oscillator by frequency division.

So far, it has been made clear that the heterodyne RX architecture involves overcoming serious tradeoffs that induce filtering requirements that are difficult to meet. Since these requirements will worsen when targeting 5G applications, architectures invoking IR techniques have been proposed. These include the Hartley and Weaver architectures. An in-depth treatment of these is given in [33]. Essentially, however, it may be said that these rely on applying quadrature (I/Q) down-conversion to the RF signal, followed by low-pass filtering. In particular, the Hartley architecture combines the down-converted signals after one of them is phase-shifted by 90°. Similarly, in the Weaver architecture, the phase-shifting is effected by a second pair of mixers. Due to the utilization of phase shifting and quadrature mixing, extra measures have to be taken to minimize sensitivities to amplitude imbalances and errors in the phase shift operation.

5.5.2 Homodyne Receiver

The homodyne receiver architecture, as pointed out in our earlier discussion, is also referred to as the direct-conversion or zero-IF receiver and is shown in Figure 5.15.

It is apparent that, upon comparison with the super-heterodyne architecture, it differs from it, in that it possesses no IF; that is, it embodies the IF = 0 case in the framework depicted in Figure 5.15. Thus, by avoiding the need for an "rf chain," this architecture achieves lower cost and power consumption at the expense of high performance. In particular, the homodyne RX's simplicity makes it germane to low-power and scalable applications, as it may be totally integrated on-chip while disposing of the need for high-performance BPFs. Thus, BPFs with performance that is not expensive to achieve, but sufficient, for 5G become feasible.

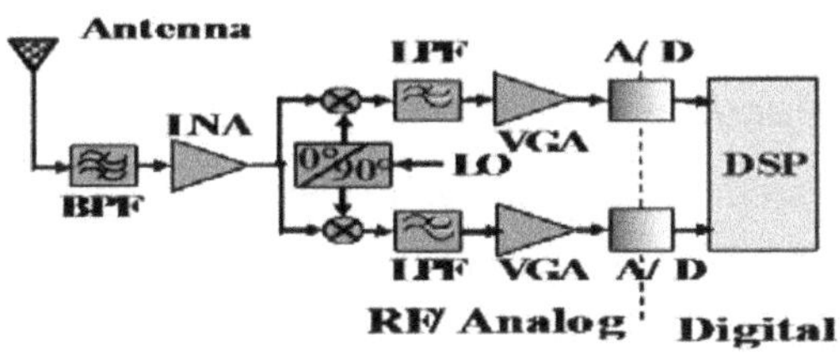

Figure 5.15 Block diagram of zero-IF/direct-down-conversion/homodyne receiver. *Source:* Ref. [37].

Following the received signal in Figure 5.15, we see that it is immediately down-converted to baseband, without first translating it to an IF. In this circumstance, then, the image and desired frequencies would overlap. To avoid this, a quadrature down-converter is employed. After this, the information is extracted at baseband frequencies via DSP. The function of the building blocks in the zero-IF RX architecture is now described. The output signal produced by the antenna is first passed through the filter BPF to preclude desensitization by blockers, after which it is amplified by an LNA in order to, again, set the RX sensitivity. The LNA is followed by the quadrature down-converter, after which the signal is applied to an LPF, which renders the desired signal at baseband, where it effects the selection of the channel. Again, the signal is applied to a VGA which brings its amplitude within the optimal range for the ADC. The ADC produces a digital representation of the signal, which is demodulated via DSP.

A comparison of the super-heterodyne and the zero-IF receivers reveals that the principal justification for using the latter over the former is the high level of integration it enables. In particular, by avoiding the need for high-Q bandpass filters (BPFs), channel selection is then effected by LPFs, which means that there is no need for an IR filter. On the other hand, zero-IF receivers are accompanied by the following drawbacks: (1) flicker (or $1/f$) noise because the down-converted signal contains DC; (2) difficulty suppressing even-order intermodulation; (3) DC offsets, which desensitize the RX due to mixer non-idealities such as finite port-to-port isolation, manifested as the leakage of the LO signal to the input of the mixer or to the LNA input; this then can mix with itself resulting in another contributor to the DC offset. A further manifestation of LO leakage is when it reaches the antenna, which may give rise to an in-band interferer. Approaches to overcome DC offsets include using a blocking capacitor or employing offset estimation and cancelation.

5.5.3 The Low-IF Receiver

The low-IF receiver architecture aims at realizing a compromise between the characteristics of the super-heterodyne and the zero-IF receivers (Figure 5.16); it is mostly employed for standards predicated upon narrow channels.

The compromise lies in that, like the homodyne, it has two down-conversion paths (one for positive and one for negative frequencies), but, in addition, it employs an IF, like the heterodyne. The bandpass filtering

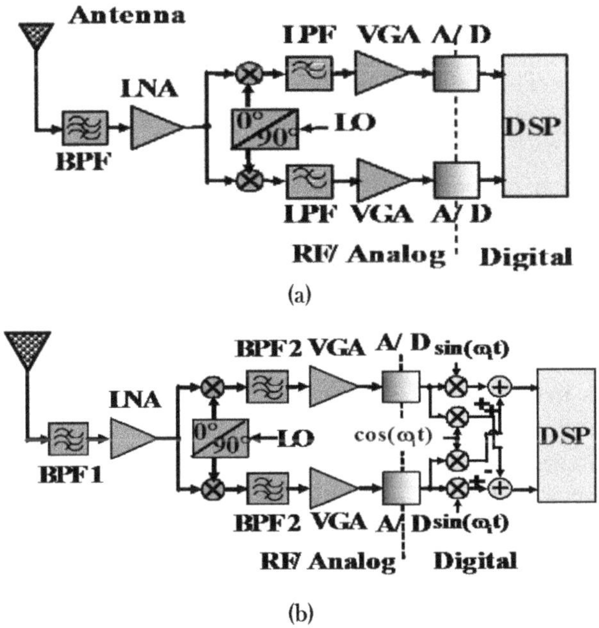

Figure 5.16 Block diagram of (a) low-IF receiver which utilizes a complex BPF and (b) low-IF receiver which employs real filters. *Source:* Ref. [37].

in the low-IF receivers may be either a complex BPF (a polyphase BPF) or a real BPF.

Examining Figure 5.16(a), we see that in the low-IF receiver, the desired signal and its image, while down-converted to the IF, are not superimposed due to the use of I/Q mixers. In particular, the desired signal is situated at negative frequencies, while its mirror image is located at positive frequencies. In practice, the IF ranges from half to several times the channel bandwidth.

The polyphase BPF approach works by suppressing the mirror signal, which is then conditioned by the VGAs to be within the optimal amplitude range of the ADCs. Once digitized by the ADCs, the signal is digitally processed with the result being down-converted to baseband by multiplication with the sine function. One virtue of employing the low IF is that it is easier to effect signal sampling after the first mixer stage than would be the case in the super-heterodyne receiver, where the frequency would still be too high. On the other hand, since both the signal and its image are digitized, the required bandwidth will be larger than that for the desired signal alone. When real BPFs are utilized, the filters are capable of separating both the desired and mirror signals from other adjacent signals. Following filtering, it is again the role of the VGAs to condition the signals to an optimal amplitude range for input to the ADCs. The digitized signal is then fed to the DSP where the last down-conversion is carried out, employing a positive frequency equal to the IF.

A positive aspect of the low-IF RX architecture is that it is free from the DC-offset issue. This is so because the desired signal is far from DC, as is the case in homodyne receivers. In addition, since the required filtering is close to DC, it has a similar level of integrability as the homodyne RX. Another advantage of this architecture is that, also due to the low frequency involved, there is enough flexibility to realize 5G requirements. The serious challenge that is met with low-IF receivers, however, lies in the difficult requirements pertaining to the suppression of the mirror image signal due to the fact that it can be stronger than the desired signal. In particular, since I/Q mixing is accompanied by the inevitable imbalances between the in-phase and quadrature paths, one must contend with IR levels that do not exceed 35 dB, unless extra measures such as tuning and calibration are exercised.

5.5.4 The Software-Defined Receiver

The software-defined receiver (SDR) architecture is an idealistic concept (Figure 5.17), in which a receiver is construed as consisting of only an antenna and an ADC; the implication is that all functions are realized digitally, i.e., by software.

Thus, any desired versatility, be it standard (including 5G), frequency band, bandwidth, DR, sensitivity, sampling rate, etc., will be accessible and realizable as prescribed by the software program "loaded" onto it. That the SDR has not been realized yet in practice is easy to see. In particular, its realization imposes extremely exacting requirements on the ADC which, despite being obtainable, would entail a prohibitive amount of power dissipation, not practical for mobile applications.

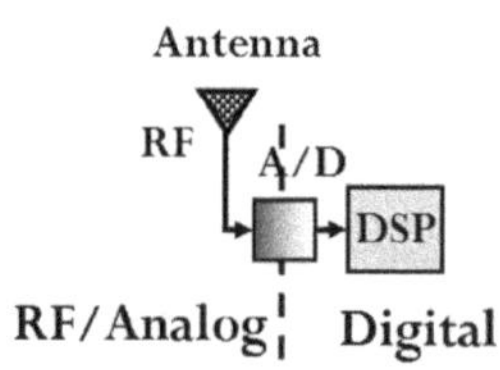

Figure 5.17 Block diagram of the "ideal" SDR receiver. *Source:* Ref. [38].

5.6 Summary

This chapter has dealt with the principles of transmitters and receivers, as they relate to wireless communications and RADARs. We discussed the advantages and disadvantages of the fundamental principles of heterodyne and homodyne architectures. We then, at the risk of repeating ourselves, reviewed the excellent paper by Bronckers, Roc'h, and Smolders which places into perspective wireless receiver architectures in the context of the needs of the upcoming/nascent 5G technology.

5.7 Problems

1. Compare and contrast the different receivers.

6

5G

6.1 Introduction

5G, the fifth generation of wireless communications standards, represents the latest in the evolution of mobile phone and network developments (Figure 6.1) [40].

The history of mobile communication may be said to have begun with the first generation (1G) implemented by analog cellular systems and launched in 1979 [45]. 1G has matured to today's 4G LTE/LTE-Advanced, which was launched around 2010 and is now deployed rapidly throughout the world. As may be surmised from Figure 6.1, a new mobile technology "generation" is usually introduced approximately every 10 years, but its life lasts about, and reaches its maximum number of users, in roughly 20 years after introduction.

Every generation is typically accompanied by the introduction of technological advances. For example, 2G introduced time division multiple access (TDMA), 3G introduced code division multiple access (CDMA), and 4G introduced orthogonal frequency division multiple access (OFDMA) (Figure 6.2).

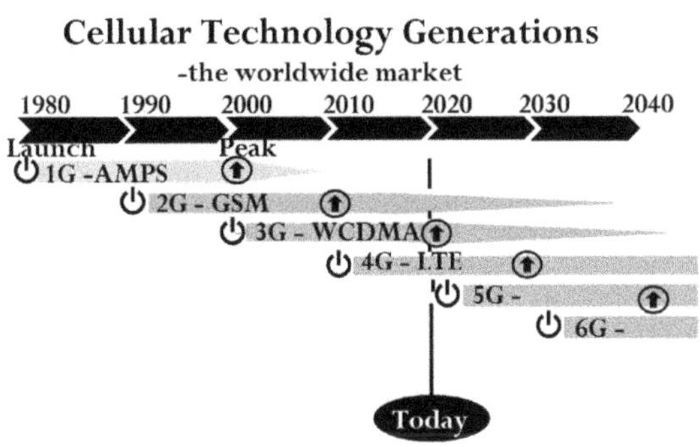

Figure 6.1 Evolution of networks and mobile technologies. *Source:* Ref. [40].

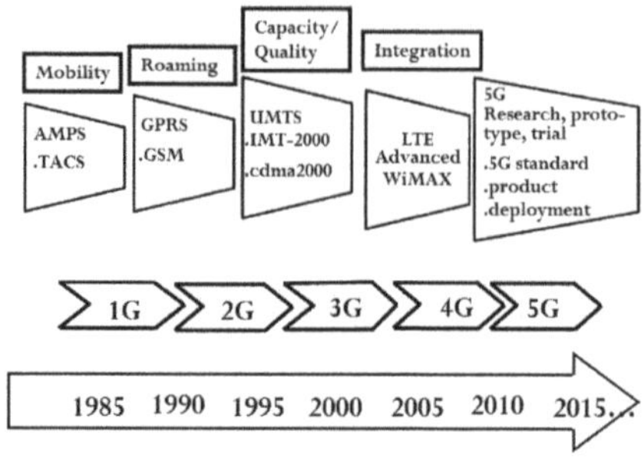

Figure 6.2 Historical evolution of mobile technology. *Source:* Ref. [45].

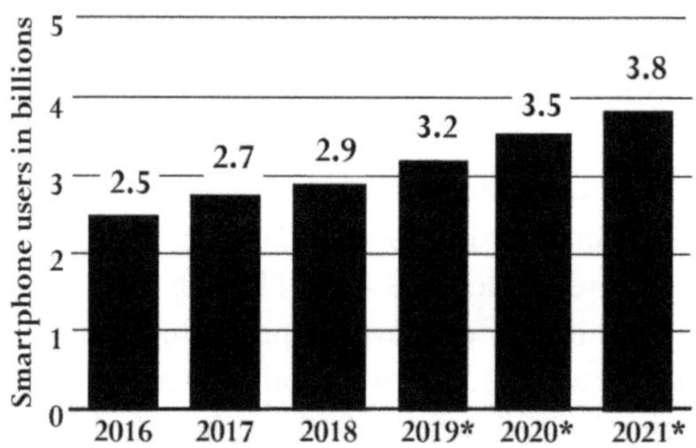

Figure 6.3 Evolution of networks and mobile technologies. *Source:* Ref. [41].

This evolution is motivated by, at least, two main reasons. In the first place, there is the rise in the use of smart phones, namely, from about 2.5 billion users in 2016 to close to 3.8 billion uses by 2021 [73] (Figure 6.3). And in the second place is the imminent saturation of 4G radio technology which, due to this explosive growth, has almost achieved Shannon capacity at the link-level. The expectation is, therefore, that to overcome the limitations of today's technology (4G LTE/LTE-Advanced), which would preclude growth, 5G will become a necessity starting from the year 2020 and onwards. 5G may enable higher capacity upon the introduction of a number of new technological advances and the use of higher frequency bands (above 6 GHz) [40, 45].

Table 6.1 provides a brief summary of features of the various mobile technology/standards generations.

Bringing 5G into reality requires meeting a variety of requirements to provide real performance superiority. These requirements are indicated in Figure 6.4 and Table 6.2 [44, 45].

Table 6.1 Features of wireless technologies. *Source:* Refs. [42–44].

First Generation	*Second Generction*	*Third Generation*	*Fourth Generation*	*Fifth Generation*
o Analog cellular (single band)	o Digital (dual-mode, dual band)	o Multi-mode, multi-band software defined radio	o Multi-mode, multi-band software defined radio	o All-IP-based model for wireless and movable network interoperability.
o Voice telecom only	o Voice + data telecom	o New services markets beyond traditional telecom: higher speed data, improved voice, multi-media mobility	o IP telephony, gaming services, high-definition mobile TV, video conferencing	o Enhanced mobile broadband (eMBB), ultra reliable low latency communications (URLLC), and massive machine type communications (mMTC)
o Macrocell only	o Macro/micro/ pico cell	o Data networks, Internet, VPN, WINternet		o Cloud-based networking
o Outdoor coverage	o Seamless indoor/outdoor coverage			
o Distinct from public switched telephone network (PSTN)	o Complementary to fixed PSTN			
o Business customer focus	o Business + customer focus	o Total communications subscriber, virtual personal networking		

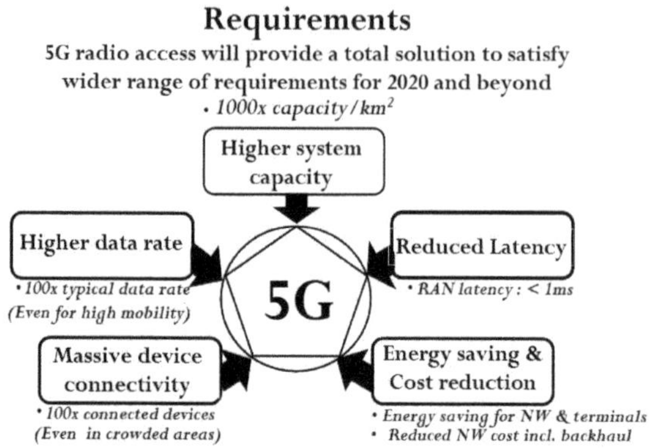

Figure 6.4 5G requirements. *Source:* Ref. [45].

Table 6.2 Target 5G features. *Source:* Refs. [40, 44, 45].

Data Rate	Latency	Energy and Cost	Device Types and Quantities
○ Aggregate data rate in bits/s per unit area: 1000× that of 4G ○ Worst data rate: 100 Mbps to 1 Gbps ○ Maximum data rate: tens of Gbps ○ Wireless download speeds of above 1 Gbps in local area network (LAN) and 500 Mbps in wide area network (WAN)	○ Air latency: 8–12 milliseconds ○ User-level latency of less than 1 ms over the radio access network (RAN)	○ Joules per bit and cost per bit will need to fall by at least 100× that of 4G	○ A single macrocell may need to support 10,000 or more low-rate devices along with its traditional high-rate mobile users

6.2 5G Systems Technologies

The technologies that are expected to enable the enhanced 5G network capacity include [44] 1) improving the spectral density by placing more active nodes per unit area and Hz, a technique referred to as *ultra-densification*, 2) exploiting the mmWave frequency spectrum to increase the communications bandwidth available, and 3) employing advances on massive multiple-input multiple-output (MIMO) receive/transmit antenna nodes to achieve greater spectral efficiency, i.e., more bits/s/Hz per node.

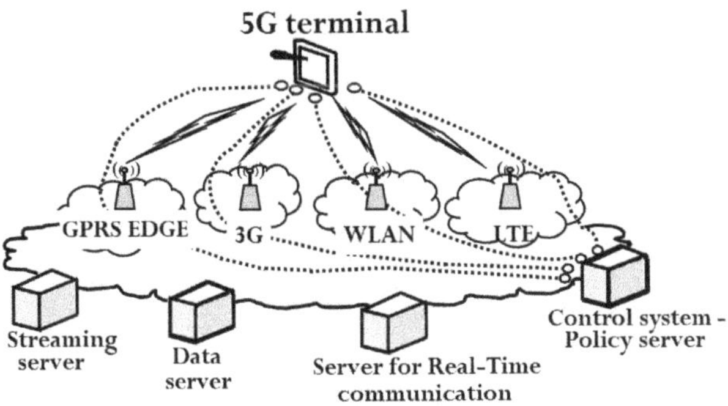

Figure 6.5 Functional architecture of 5G wireless technologies. *Source:* Ref. [40].

The range of frequencies that are occupied by mmWaves is demarcated by frequencies in the 30–300 GHz, which corresponds to wavelengths in the 1–10 mm range; this range is often extended to include the 20–30 GHz range.

A pictorial view of the structural design of 5G systems is shown in Figure 6.5, where it is seen that 5G embodies an all-internet protocol (IP)-based model for wireless and movable network interoperability [40].

6.2.1 5G Systems: mmWaves [44]

Several challenges have been identified pertaining to the utilization of mmWaves for 5G wireless communications. These are briefly addressed next.

6.2.1.1 Propagation issues

One of the consequences of using higher frequencies is that the antenna size decreases and its aperture scales as $\lambda^2/2\pi$, where $\lambda = c/f_c$ is the wavelength, c is the speed of light, and f_c is the carrier frequency. This results in free-space path loss (PL) increase and [46] between a transmit and a receive antenna, growing in fact as f^2. To diminish this loss, the adoption of antenna arrays has been employed.

6.2.1.2 Blocking

Blocking refers to the fact that since at mmWave signal propagation undergoes more specular propagation and lower diffraction than at microwave frequencies, sensitivity to blockages is more pronounced. This causes additional loss, of the order of 15–40 dB per decade, as compared to the free-space PL value of 20 dB/decade, whenever distance between the transmit/receive antenna increases. In turn, this blocking indices the more rapid link transition from usable to unusable, which is not possible to circumvent with countermeasures involving standard small-scale diversity.

6.2.1.3 Atmospheric and rain absorption

At mmWave frequencies, the absorption due to air and rain is not negligible, in particular, due to the 15 dB/km oxygen absorption within the 60 GHz band. This absorption, however, is negligible in the context of the urban cellular deployments envisioned, where it is anticipated that the distance between base stations (BS) would not exceed 200 m. The problem of propagation losses at mmWave frequencies, however, has been deemed to be surmountable by exploiting large antenna arrays to steer the energy and collect it in a coherent fashion, but it requires the production of narrow beams.

6.2.1.4 Large arrays, narrow beams

In contrast to traditional wireless systems at low frequencies, the design of wireless systems predicated upon mmWave the design and utilization of narrow beams. This need is elicited by the abruptness accompanying the interference behavior of highly directional pencil beams which, in turn, produce a higher sensitivity to beam misalignment. Beam interference, then, manifests as an on/off behavior in which most of the beams do not interfere, but, instead, events of strong interference are intermittently experienced.

6.2.1.5 Link acquisition

At mmWaves, the mandatory utilization of narrow beams is accompanied by the difficulty of establishing links between BSs and users, and this induces problems pertaining to the processes of initial access and handoff.

The approach to overcoming these problems involves the integration of both microwave and mmWave frequencies (Figure 6.6).

In this approach, mmWave frequencies are utilized for the transmission of payload data from small-cell BSs, while microwave frequencies are employed for beam control in macro BSs.

6.3 5G: Internet of Things [46, 47]

Due to the virtually unlimited network capacity potential afforded by mmWaves/5G, this technology becomes an enabler for the emerging Internet of Things (IoT) paradigm. In this context, a number of concepts have been advanced [47].

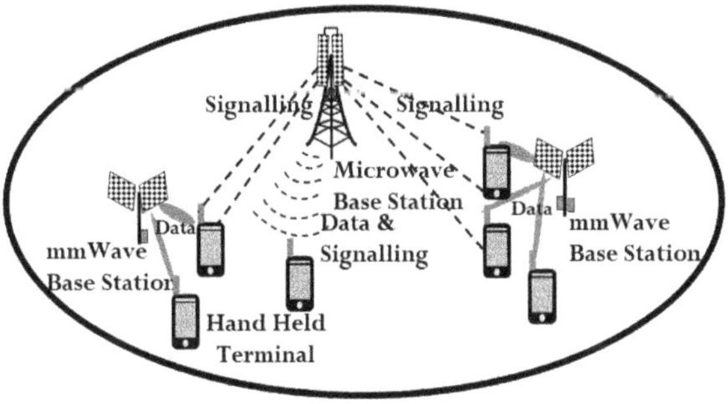

Figure 6.6 mmWave-enabled 5G network. *Source:* Ref. [40].

6.3.1 Device-to-Device Communications

mmWaves for mediating Device-to-Device (D2D) communications have been identified as a promising technology. The technique will facilitate avoiding the bottleneck between source and destination posed by the BS. Before finally accessing the 5G network, the communication traffic would go directly from one IoT device to nearby devices upon establishing *local* links. Communication via D2D links, in particular, is expected to reduce latency and power consumption, while increasing peak data rates. In addition, this would enhance spectrum reuse as a multitude of D2D links could be designed for sharing the same bandwidth, thus resulting in an increase of spectral reuse per node.

6.3.2 Simultaneous Transmission/Reception (STR)

The STR technique enables simultaneous transmission and reception. Since these would be carried out simultaneously and utilize the same frequencies, this would enable greater spectral efficiency. In particular, a doubling of the spectral efficiency would be immediately attained in D2D links. Furthermore, it would be easier to use STR to discover adjacent devices in D2D communication networks due to the possibility of monitoring uplink signals from adjacent nodes without having to suspend transmission momentarily.

6.4 Non-Orthogonal Multiple Access [45, 49]

Besides the direct approach of increasing system capacity by utilizing mmWave frequencies, novel signal processing techniques that further boost spectrum efficiency are available. Such is the case of non-orthogonal multiple

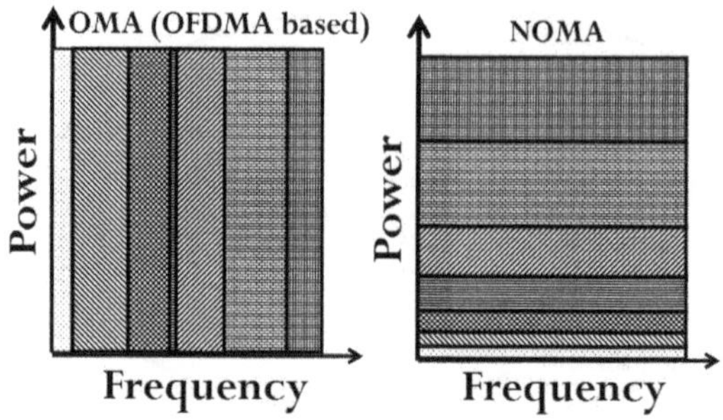

Figure 6.7 A pictorial comparison of OMA and NOMA. *Source:* Ref. [49].

access (NOMA), introduced as an intra-cell multi-user multiplexing scheme that exploits a new domain, namely, the power domain, which is not sufficiently utilized in previous generations.

In orthogonal multiple access (OMA) approaches, individual users are able to utilize orthogonal communication resources, whether in a selected time slot, frequency band, or code, in order to avoid multiple access interference. The traditional OMA approaches include frequency division multiple access (FDMA) utilized in 1G, TDMA utilized in 2G, CDMA utilized in 3G, and OFDMA utilized in 4G [46, 49]. Figure 6.7 contrasts the OMA and NOMA approaches.

In the NOMA scheme (Figure 6.8), on the other hand, multiple users are able to employ non-orthogonal resources concurrently so that a high spectral efficiency is attained at the expense of suffering some degree of multiple access interference at the receivers [49].

6.4.1 NOMA Approaches

In practice, there are two versions of NOMA, namely, power-domain multiplexing and code-domain multiplexing.

In the power-domain multiplexing approach to NOMA, high system performance is attained by allocating large power differences to different users, as per their pertinent channel conditions. This is accomplished by superimposing the information signals of multiple users at the transmitter. On the other hand, at the receiver, decoding of the signals is effected by employing the successive interference cancellation (SIC) or maximum-likelihood detection (MLD) techniques on received signals one by one until the desired user's signal is obtained. This provides an acceptable tradeoff between system throughput and user fairness for both downlinks and uplinks.

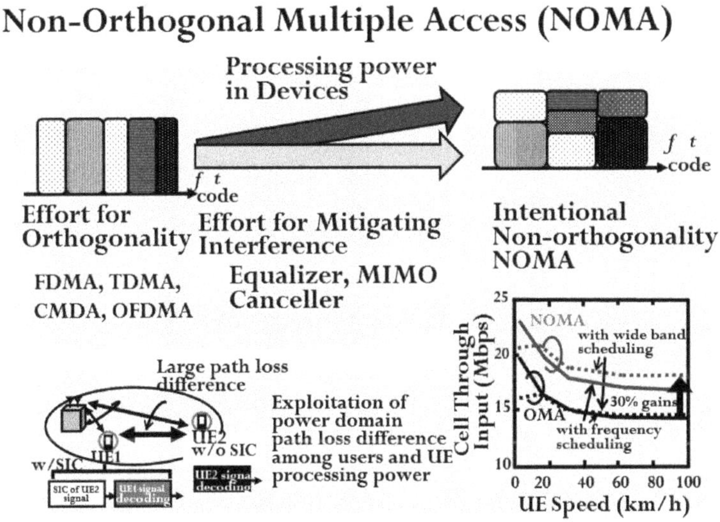

Figure 6.8 Non-orthogonal multiple access (NOMA). *Source:* Ref. [45].

In the code-domain multiplexing approach to NOMA, high system performance is achieved by allocating different codes to different users and multiplexing these over the same time-frequency resources, in particular, multi-user shared access (MUSA), sparse code multiple access (SCMA), and low-density spreading (LDS). NOMA is also advantageous for enabling the many more simultaneous connections required for either uplinks or downlinks in the context of massive device connectivity.

6.5 5G Evolution

The pervading consensus pertaining the evolution of 5G toward its maturity has as one of its main aspects the development of *massive MIMO* technology, which would be natural to exploit higher frequency bands [82]. Massive MIMO promises the availability of larger bandwidths and, consequently, higher data rates (Figure 6.9).

In particular, operation at higher frequencies will entail lower-size antenna elements which, in turn, will enable a larger number of such elements. Antenna arrays with a large number of elements could be located close to each other, yet avoid interference due to their ability to form very narrow beams. However, with the number of elements required to achieve a

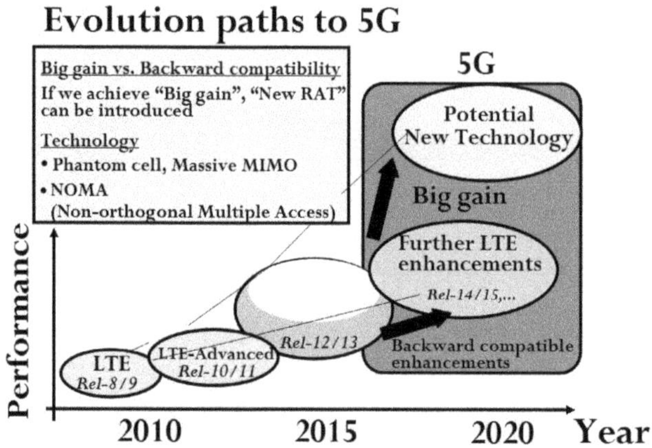

Figure 6.9 Evolution paths to 5G. *Source*: Ref. [45].

reach of kilometers in bands over 10 GHz being into the thousands, many technical challenges will have to be overcome to bring 5G to fruition [45].

6.6 Summary

In this chapter, we have dealt with various aspects of 5G systems technologies. In particular, we have discussed the role of mmWaves, propagation issues, atmospheric absorption, the need for large arrays and narrow beams, and the problem of link acquisition. Other aspects of 5G dealt with its relation to the IoT, and NOMA approaches to greater capacity. The chapter ends with a discussion of the projected evolution of 5G.

6.7 Problems

1. What technological hurdles must be overcome for ubiquitous adoption of 5G?
2. Projections are already being made for the capabilities of 6G. How is the performance of 6G expected to improve/extend over that of 5G? Please, investigate and provide a report.

7

MIMO

7.1 Introduction

It could be said that the concept of MIMO, multiple-input multiple-output systems was first explored by Foschini [51] and Foschini and Gans [84] when they studied the impact of multi-element arrays (MEAs) at both the transmitter and receiver, on the ability to approach the ultimate limits of the bandwidth-efficient delivery of high bit rate digital signals in wireless communication systems [52]. In particular, they showed that by constraining the channel bandwidth and the total transmitted power, and forming a channel using increased spatial dimensions, it was possible to get an extraordinarily large capacity. In this chapter, we begin our study of MIMO by first discussing the *channel capacity* of the simpler single-input single-output system (SISO) in terms of which MIMO will be interpreted. Then, we address the topics of MIMO channel models and propagation models. This is followed by a discussion of the singular value decomposition (SVD) approach and its application to the channel matrix, and the interpretation of MIMO in terms of SVD. Next, we present the water-filling algorithm to MIMO transmit antenna input power optimization and the topic of MIMO receive antenna signal processing. MIMO detection and transmission techniques, in particular, maximum ratio combining (MRC), zero-force beamforming (ZFBF), and minimum mean-square error (MMSE) are then discussed. The chapter ends with a discussion of precoding, massive MIMO architectures, and massive MIMO limitations.

7.2 The SISO Channel

7.2.1 The SISO Channel Model

We begin by presenting a sketch of the SISO channel model (Figure 7.1); this will require the following definitions.

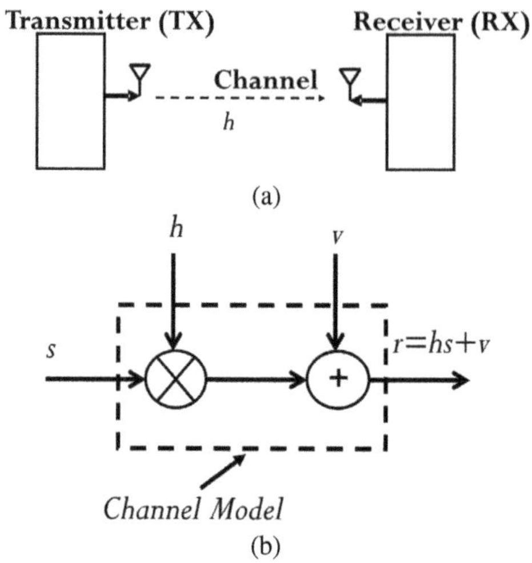

Figure 7.1 (a) Model of single-input single-output flat fading wireless channel; h is the channel gain. (b) System-level representation.

Table 7.1 SISO model parameters. *Source:* Ref. [51].

Parameter	*Description*
T	Time variable, assumed to evolve in discrete normalized steps, i.e., t_i, $i = 0, 1, 2, \ldots$
$s(t)$	Transmitted signal assumed to be narrowband such that the channel frequency response is perceived as flat throughout the band. This signal has a total power P_T.
$r(t)$	Received signal at each point in time. It exhibits an average output power P.
$v(t)$	Noise at the receiver. It consists of an AWGN complex signal with normal distribution, zero mean, and mean square deviation, $\sigma_v^2 = N$, where N is the average power.
$h(t)$	Channel gain, a complex scalar.

In the first place, it is assumed that the channel is a flat fading[1] wireless channel that introduces additive white Gaussian noise [85] (AWGN)[2] represented by the gain h. In the second place, the channel is considered to have constant parameters. The SISO model parameters are defined in Table 7.1 [51].

[1]Fading wireless channel: One whose gain fluctuates randomly.

[2]AWGN is a noise model utilized in the field of information theory to simulate random processes occurring in nature.

Accordingly, at each time point, occurring every T seconds, the SISO system is modeled by the following equation [51]:

$$r_i = hs_i + v_i \tag{7.1}$$

This equation represents a linear system with wide-sense stationary (WSS)[3] [54] random uniformly distributed input *s*, added random noise *v*, and random output *r*, having gain *h*, assumed to be constant. The response of such a linear system with random input is given in terms of the distribution properties of the random variables *S* and *V*, and the mean and variance of the output *R* [54, 55]. In particular, we are interested in the received noise and signal powers, which are given by their respective mean square values at the system output [55]

$$E\{V(t)\} = E\{V^2\} = \sigma_v^2 = N \tag{7.2}$$

and

$$\begin{aligned} E\{R(t)^2\} &= E\left\{\left(\int_0^{\infty} S(t-\lambda)\cdot h(t)\cdot d\lambda\right)\right\} = E\{R^2\} \\ &= E\{(hS+V)\cdot(hS+V^*)\} \\ &= E\{|hS|^2\} + 2hE\{S\}\cdot E\{V\} + E\{|V|^2\} \\ &= |h|^2 E\{|S|^2\} + 2hE\{S\}\cdot 0 + E\{|V|^2\} \\ &= |h|^2\sigma_s^2 + \sigma_v^2 = \sigma_r^2 \end{aligned} \tag{7.3}$$

7.2.2 The SISO Channel Capacity

The channel capacity is defined by Shannon [13]

$$C = \max_{P(S):E\{S^2 \le P_T\}} I(S,R) \tag{7.4}$$

where *I(S,R)* is the equivocation or mutual information between the input and output signals. The capacity, *C*, is thus defined as the maximum of the equivocation with respect to all possible signal distributions *p(s)*. In particular, the equivocation is expressed as

$$I(S,R) = H(R) - H(R|S) \tag{7.5}$$

[3]The mean of a random WSS process, $s(t)$, is $E\{s(t)\}$, a constant.

and subtracts that part of the received signal due to noise from the transmitted information. For a continuous density function, *H* is given by the conditional differential entropy, defined by [54]

$$H(X) = -\int_{-\infty}^{\infty} p(x)\text{In}\, p(x)dx. \tag{7.6}$$

Now, Shannon [13] has derived the form of the distribution function *p(x) of a real random variable x* that maximizes the entropy subject to the condition of having a constant standard deviation, σ_x, and it is given by

$$p(x) = \frac{1}{\sqrt{2\pi\sigma_x^2}} e^{\frac{-x^2}{2\sigma_x^2}} \tag{7.7}$$

To calculate *H*, with inserting Equation (7.7) into Equation (7.6), we get

$$H(x) = -\int_{-\infty}^{\infty} \frac{1}{\sqrt{2\pi\sigma_x^2}} e^{\frac{-x^2}{2\sigma_x^2}}\ \text{In}\ \frac{1}{\sqrt{2\pi\sigma_x^2}} e^{\frac{-x^2}{2\sigma_x^2}}\, dx \tag{7.8}$$

Evaluating the logarithm

$$-\ln \frac{1}{\sqrt{2\pi\sigma_x^2}} e^{\frac{-x^2}{2\sigma_x^2}} = \ln\sqrt{2\pi\sigma_x^2} + \frac{x^2}{2\sigma_x^2} \tag{7.9}$$

we then proceed to calculate *H* as follows:

$$\begin{aligned} H(x) &= \int_{-\infty}^{\infty} p(x)\left[\text{In}\sqrt{2\pi\sigma_x^2} + \frac{x^2}{2\sigma_x^2}\right] dx \\ &= \int_{-\infty}^{\infty} p(x)\left[\text{In}\sqrt{2\pi\sigma_x^2}\right] dx + \int_{-\infty}^{\infty} p(x)\left[\frac{x^2}{2\sigma_x^2}\right] dx \\ &= \text{In}\sqrt{2\pi\sigma_x^2}\int_{-\infty}^{\infty} p(x)dx + \frac{1}{2\sigma_x^2}\int_{-\infty}^{\infty} x^2 p(x)dx \\ &= \text{In}\sqrt{2\pi\sigma_x^2}\cdot(1) + \frac{1}{2\sigma_x^2}\int_{-\infty}^{\infty} x^2 \frac{1}{\sqrt{2\pi\sigma_x^2}} e^{-\frac{x^2}{2\sigma_x^2}}\, dx \\ &= \text{In}\sqrt{2\pi\sigma_x^2} + \frac{1}{2\sigma_x^2}\cdot\frac{1}{\sqrt{2\pi\sigma_x^2}}\cdot\int_{-\infty}^{\infty} x^2 e^{-\frac{x^2}{2\sigma_x^2}}\, dx \end{aligned} \tag{7.10}$$

Now, from a table of integrals, we find that [56]

$$\int_{-\infty}^{\infty} x^{2n} e^{-\alpha x^2} dx = \sqrt{\frac{\pi}{\alpha}}\cdot\frac{(2n-1)}{2\alpha} \tag{7.11}$$

which, comparing with $\int_{-\infty}^{\infty} x^{2n}e^{-\alpha x^2}dx$, we identify, $n = 1$ and $\alpha = \frac{1}{2\sigma_x^2}$, and substituting in Equation 7.11, we obtain

$$\int_{-\infty}^{\infty} x^{2n}e^{-\alpha x^2}dx = \sqrt{2\pi\sigma_x^2}\cdot\sigma_x^2 \tag{7.12}$$

so that Equation (7.10) becomes

$$\begin{aligned} H(x) &= \text{In}\sqrt{2\pi\sigma_x^2} + \frac{1}{2\sigma_x^2}\cdot\frac{1}{2\pi\sigma_x^2}\cdot\int_{-\infty}^{\infty} x^2 e^{-\frac{x^2}{2\sigma_x^2}}dx \\ &= \text{In}\sqrt{2\pi\sigma_x^2} + \frac{1}{2\sigma_x^2}\cdot\frac{1}{\sqrt{2\pi\sigma_x^2}}\cdot\sqrt{2\pi\sigma_x^2}\cdot\sigma_x^2 \\ &= \text{In}\sqrt{2\pi\sigma_x^2} + \frac{\sigma_x^2}{2\sigma_x^2} \\ &= \text{In}\sqrt{2\pi\sigma_x^2} + \frac{1}{2}\cdot 1 \\ &= \text{In}\sqrt{2\pi\sigma_x^2} + \frac{1}{2}\cdot\text{In}\, e \\ &= \text{In}\sqrt{2\pi\sigma_x^2}\cdot\text{In}\sqrt{e} \\ &= \text{In}\left[\left(\sqrt{2\pi\sigma_x^2}\right)\cdot(\sqrt{e})\right] \\ &= \text{In}\sqrt{2\pi\sigma_x^2} \end{aligned} \tag{7.13}$$

This result (7.13) is for a real random variable. To apply it to our case of a *complex* random variable with real and imaginary parts $X_c = Re(X) + i\text{Im(X)}$, that are independent identically distributed (i.i.d.) [97], $H(X_c) = H(\text{Re(X)}) + \text{H(Im(X))} = 2\text{In}\sqrt{2\pi \text{e}\sigma^2} = \text{In } 2\pi\text{e}\sigma^2$. This expression may now be applied to calculate the mutual information as follows:

$$\begin{aligned} I(S,R) &= H(R) - H(R|S) \\ &= H(R) - H(hS + V|S) \\ &= H(R) - H(hS|S) - H(V|S) \\ &= H(R) - 0 - H(V|S) \\ &= H(R) - H(V) \end{aligned} \tag{7.14}$$

And from Equation (7.12), we can write

$$H(R) = \text{In } 2\pi e\sigma_r^2 \tag{7.15}$$

and

$$H(V) = \text{In}\ 2\pi e\sigma_v^2 \tag{7.16}$$

and, thus,

$$\begin{aligned} C &= H(R) - H(V) \\ &= \text{In}\ 2\pi e\sigma_r^2 - \text{In}\ 2\pi \text{e}\sigma_\text{v}^2 \\ &= \text{In}\ 2\pi e \left(|h|^2\sigma_s^2 + \sigma_v^2\right) - \text{In}\ 2\pi \text{e}\sigma_\text{v}^2 \\ &= \text{In}\ \frac{|h|^2\sigma_s^2 + \sigma_v^2}{\sigma_v^2} \\ &= \text{In}\ \left(1 + \frac{\sigma_s^2}{\sigma_v^2} \cdot |h|^2\right) \\ C &= \text{In}\ \left(1 + \frac{P_r}{N} \cdot |h|^2\right) \textit{nats per transmission} \end{aligned} \tag{7.17}$$

Equation (7.17) gives the channel capacity in *nats* per transmission. If the base of the logarithm is changed to 2, namely,

$$C = \log_2 \left(1 + \frac{P_r}{N} \cdot |h|^2\right) \textit{bits per transmission} \tag{7.18}$$

the capacity is in bits per transmission [57]. Examination of Equation (7.18) indicates that the SISO channel capacity is a function of the signal-to-noise ratio (SNR) at the receiver, where the signal is $S_R = P_r \cdot |h|^2$ and the noise is *N*. The channel gain, *h*, may be determined experimentally [58]; so it may be assumed known.

The capacity in bits per second, rather than bits per transmission, is obtained by determining how many transmissions per second are made. Noting that a signal of bandwidth *B* repeats every 1/*B* seconds, the capacity in bits per second is

$$\begin{aligned} C &= \log_2 \left(1 + \frac{P_r}{N} \cdot |h|^2\right) / (1/B)\ \textit{bits per second} \\ &= B\log \left(1 + \frac{P_r}{N} \cdot |h|^2\right) \end{aligned} \tag{7.19}$$

If the noise *density* is given by N_0, in watts/Hz, then the noise power in the bandwidth is $N = N_0$B. This allows us to rewrite Equation (7.19) as [54, 57,]

$$C = B\log_2 \left(1 + \frac{S_r}{N}\right) = \log_2 \left(1 + \frac{S_r}{N_0 B}\right) \quad \textit{bits per second.} \tag{7.20}$$

Now, to compare channel capacities amongst various realizations, it is useful to compare their *bandwidth efficiency*, and this is given by [57]

$$\frac{C}{B} = \log_2\left(1 + \frac{S_R}{N_0 B}\right) \textit{bits second per Hz.} \tag{7.21}$$

7.3 The MIMO Channel Model

The MIMO channel model (Figure 7.2) consists of n_T transmit antennas and n_R receive antennas, where individual channels relating transmit antennas $i = 1, 2, \ldots, n_T$ to receive antennas $j = 1, 2, \ldots, n_R$ are described by complex gains h_{ij}. The MIMO system model parameters are defined in Table 7.2 [52]. Again, we assume the channel is a flat fading wireless channel introducing AWGN, and the gains, h_{ij}, while complex constants, individually, vary randomly among them.

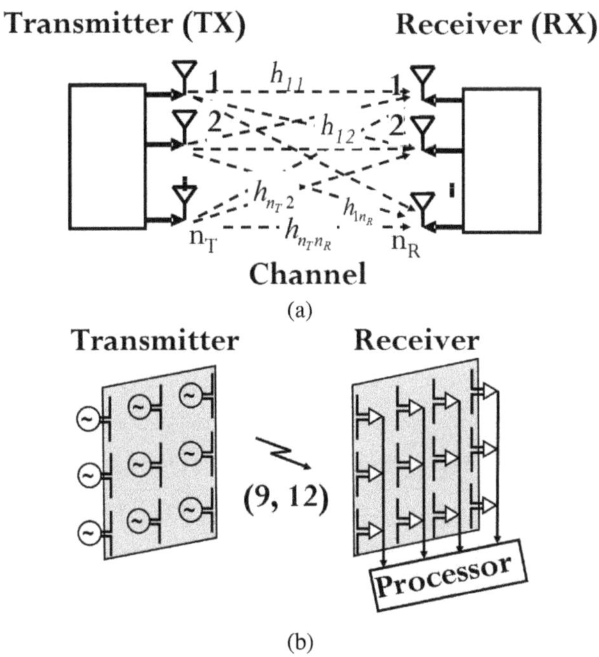

Figure 7.2 (a) Model of MIMO flat fading wireless channel; h_{ij} are the channel gains between transmitter *i* and receiver *j*. (b) Example of MIMO system with $n_T = 9$ and $n_R = 12$. *Source:* Ref. [51].

Table 7.2 MIMO model parameters. *Source*: Refs. [51, 52].

Parameter	*Description*
t	Time variable, assumed to evolve in discrete normalized steps, i.e., t_k, k=0, 1, 2, ...
n_T	Number transmit of antenna elements.
n_R	Number of receive antennas elements.
$s(t)$	Transmitted signal, assumed to be narrowband such that the channel frequency response is perceived as flat throughout the band. This signal has a total power P_r independent of the number of elements, n_T, with each of the antenna elements radiating a power P_r/n_T.
$r(t)$	Received n_R-dimensional signal at each point in time. Each entry of this vector corresponds to one of the n_R antenna elements, each of which exhibits a (spatial) average output power S_R.
$v(t)$	Noise at the receiver. It consists of an AWGN complex signal with normal distribution, zero mean, and mean square deviation, $\sigma_v^2 = N$, where N is the average power.
ρ	Average signal-to-noise-ratio (S_R/N) at output of each of the n_R antenna elements. It is independent of n_T.
$h_{ij}(t)$	Channel gain between transmitter i and receiver j, a complex scalar.

With reference to Figure 7.2(a), the equations governing the MIMO system may be written as

$$\begin{aligned} r_1 &= h_{11}S_1 + h_{12}S_2 + h_{13}S_3 + \cdots h_{1n_T}S_{n_T} + V_1 \\ r_2 &= h_{21}S_1 + h_{22}S_2 + h_{23}S_3 + \cdots h_{2n_T}S_{n_T} + V_2 \\ \cdots & \\ r_{n_g} &= h_{n_T1}S_1 + h_{n_T2}S_2 + h_{n_T}S_3 + \cdots h_{n_gn_T} + V_{n_g} \end{aligned} \tag{7.21}$$

which is interpreted as meaning that the signal received by antenna 1 is the sum of the signals s_i transmitted by antennas i=1, 2,$\cdots$, n_T, modified by the channel gains h_{ij}, plus the noise V_i introduced into each channel, and so on for each receive antenna. In matrix form, we can write Equation (7.21) as

$$\begin{bmatrix} r_1 \\ r_1 \\ \cdots \\ r_{n_g} \end{bmatrix} = \begin{bmatrix} h_{11} & h_{12} & \cdots h_{1n_r} \\ h_{21} & h_{22} & \cdots h_{2n_r} \\ & & \\ h_{n_T1} & h_{n_T2} & \cdots h_{n_gn_r} \end{bmatrix} \cdots \begin{bmatrix} S_1 \\ S_2 \\ \cdots \\ S_{n_T} \end{bmatrix} + \begin{bmatrix} V_1 \\ V_2 \\ \cdots \\ V_{n_g} \end{bmatrix} \tag{7.22}$$

or

$$\mathbf{r} = \mathbf{H} \cdot \mathbf{s} + \mathbf{v} \tag{7.23}$$

In general, the elements of the channel matrix, h_{ij}, representing the complex gain from transmitter *j* to receiver *i*, are complex and given by [93]

$$\begin{aligned} \mathbf{h}_{ij} &= \alpha + \mathbf{j}\beta \\ &= \sqrt{\alpha^2 + \beta^2} \cdot \mathbf{e}^{\mathbf{j}\tan^{-1}\left(\frac{\beta}{\alpha}\right)} \\ &= |\mathbf{h}_{ij}|\mathbf{e}^{j\phi_{ij}} \end{aligned} \tag{7.24}$$

The parameters adopted by the complex gain, in turn, model the *fading* phenomena [51, 52] of the propagation environment. These are captured by three main models, namely, the Rayleigh model, the Ricean model, and the Nakagami-m model. We present these models next.

7.3.1 MIMO Channel Propagation Models [13, 60, 61]

7.3.1.1 The rayleigh distribution model

The Rayleigh model utilizes a probability density distribution that captures the envelope of a Gaussian random noise. It models the amplitude of signals that reach a receiver, from their source of origin, upon scattering through multiple paths, thus capturing non-line-of-sight (NLOS) propagation. It is given by a Gaussian probability density function with zero mean and variance $\sigma^2 n$

$$\begin{aligned} p(r) &= \frac{r}{\sigma_n^2} e^{-\frac{r^2}{\sigma_n^2}} \quad r \geq 0 \\ E(r) &= \bar{r} = \sigma_n \sqrt{\pi/2} \\ E(r-\bar{r})^2 &= \sigma_r^2 = \left(2 - \frac{\pi}{2}\right) \sigma_n^2 \end{aligned} \tag{7.25}$$

7.3.1.2 The Ricean distribution model

The Ricean probability density function models the envelope of a sinusoid whose random amplitude includes the sum of a constant amplitude ***A*** and a Gaussian noise of variance σ. It models the situation when there is a line-of-sight (LOS) component between transmitter and receiver. It is given by

$$p(r) = \frac{r}{\sigma^2} e^{-\left[\frac{\left(r^2+A^2\right)}{2\sigma^2}\right]} I_0\left(\frac{Ar}{\sigma^2}\right) \quad r \geq 0 \tag{7.26}$$

where I_0 *()* is the modified Bessel function of 0th order and the first kind

$$\boldsymbol{I}_0(\boldsymbol{x}) = \frac{1}{2\pi}\int_0^{2\pi} e^{x\cos\theta}\boldsymbol{d\theta}. \quad (7.27)$$

For $A/\sigma \gg 1$, the Ricean distribution is approximated by

$$\boldsymbol{p}(\boldsymbol{r}) = \frac{1}{\sigma\sqrt{2\pi}} e^{-\left[\frac{\left(r^2+A^2\right)}{2\sigma^2}\right]} \quad (7.28)$$

7.3.1.3 The Nakagami-m distribution model

The Nakagami-m probability density function approximates the distribution that results from the sum of a large number of vectors exhibiting correlated components with different mean values and variances. It models a wide class of fading channel conditions while enabling better fits to empirical data that could be obtained with Rayleigh and Rice distributions. It is given by

$$p(\boldsymbol{r}) = \frac{2m^m}{\Gamma(m)\Omega^m} r^{2m-1} e^{-\frac{m}{\Omega}r^2} \quad m \geq 0.5, \Omega \geq 0 \quad (7.29)$$

where $\Gamma(m)$ is the Gamma function, *m* is the fading parameter that determines the shape of the distribution, and Ω controls the spread. The value of the parameter *m* captures the degree of fading and may adopt a minimum value of 0.5. *m*=0.5 represents the severest fading that can be modeled by the Nakagami-m distribution, being identical to a single-sided Gaussian distribution. For *m*=1, on the other hand, the Nakagami-m distribution matches the Rayleigh distribution. For *m*>1, the distribution captures the existence of a strong LOS component, representing least severe fading conditions.

7.3.2 The Singular Value Decomposition Approach [62, 63]

Examination of Equation (7.22) reveals that the average output power of each receive antenna simultaneously receives signal contributions from all the transmit antennas so that, in the context of Equation (7.21), it is not possible to relate the signal received by a given antenna to the signal sent by a given transmit antenna and, thus, calculate the improvement in signal-to-noise-ratio and, consequently, capacity, in a transparent way. The best way to understand the improvement in capacity brought about by the added multiplicity of transmit and receive antennas is to express Equation (7.22) in diagonal form so that the coupling terms in the channel matrix are eliminated and we have single equations similar to $r_i = hs_i + v_i$. This is accomplished by transforming Equation (7.23) as per the SVD approach, which we present below.

The SVD is a technique to, given a matrix **A** of dimension *m x n*, find square matrices **U**, of dimension *m* x *m*, and **V**, of dimension *n* x *n*, such that if $m > n$

$$\mathbf{U}^{\mathrm{H}}\mathbf{A}\mathbf{V} = \Sigma \tag{7.30}$$

where the matrix $\mathbf{U}^H$ is the conjugate transpose of the matrix **U**, and

$$\Sigma = \begin{bmatrix} \sigma_1 & 0 & \dots & 0 \\ 0 & \sigma_2 & \dots & 0 \\ & \dots & & \\ 0 & 0 & \dots & \sigma_n \\ 0 & 0 & \dots & 0 \\ & & \dots & \\ 0 & 0 & \dots & 0 \end{bmatrix} = \begin{bmatrix} D & 0 \\ 0 & 0 \end{bmatrix} \text{ is } m \times n. \tag{7.31}$$

In other words, $\sum$ is a matrix with elements $\sigma_{ij} = 0$ if $i \neq j$, and, in particular, **D** is a diagonal *k* x *k* matrix with elements $\sigma_1 \geq \sigma_2 \cdots \geq \sigma_k >$ for $1 \leq i \leq k$, where $k = \min\{m, n\}$ The elements σ_i are called *singular values* and are given by the strictly positive square roots of the eigenvalues of $\mathbf{A^H}$ **A**; the matrix $\sum \sum^H$ is diagonal, of dimension $n \times n$, with diagonal elements σ^2, which, it can be shown, are the eigenvalues of $\mathbf{A^H}$ **A**. It follows from Equation (7.30) that **A** may be expressed as

$$\mathbf{A} = \mathbf{U}\Sigma\mathbf{V^H} \tag{7.32}$$

Now, applying Equation (7.32) to Equation (7.23), with **A** renamed as **A** = **H** = **U** $\sum$ $\mathbf{V^H}$, we have

$$\boldsymbol{r} = \boldsymbol{U\Sigma V}^H \cdot \boldsymbol{s} + \boldsymbol{v} \tag{7.33}$$

then multiplying on the left by $\mathbf{U}^H$, we have

$$\begin{aligned} \boldsymbol{U}^H\boldsymbol{r} &= \boldsymbol{U}^H\boldsymbol{U}\Sigma\boldsymbol{V}^H \cdot \boldsymbol{s} + \boldsymbol{U}^H\boldsymbol{v} \\ &= \Sigma\boldsymbol{V}^H \cdot \boldsymbol{s} + \boldsymbol{U}^H\boldsymbol{v} \end{aligned} \tag{7.34}$$

so that we get

$$\boldsymbol{U}^H\boldsymbol{r} = \Sigma\boldsymbol{V}^H \cdot \boldsymbol{s} + \boldsymbol{U}^H\boldsymbol{v} \tag{7.35}$$

which, identifying $\mathbf{U}^H\mathbf{r} = \widetilde{r}$, $\mathbf{V}^H \cdot \mathbf{s} = \widetilde{\mathbf{s}}$ and $\mathbf{U}^H\mathbf{v} = \widetilde{v}$ yields

$$\tilde{\boldsymbol{r}} = \Sigma \cdot \widetilde{\boldsymbol{s}} + \widetilde{\boldsymbol{v}}. \tag{7.36}$$

This is an equation of the form,

$$\begin{bmatrix} \widetilde{\boldsymbol{r}}_1 \\ \widetilde{\boldsymbol{r}}_2 \\ \cdots \\ \widetilde{\boldsymbol{r}}_k \\ \widetilde{\boldsymbol{r}}_{k+1} \\ \cdots \\ \widetilde{\boldsymbol{r}}_m \end{bmatrix} = \begin{bmatrix} \sigma_1 & 0 & 0 & 0 & 0 & 0 & 0 \\ 0 & \sigma_2 & 0 & 0 & 0 & 0 & 0 \\ 0 & 0 & \cdots & 0 & 0 & 0 & 0 \\ 0 & 0 & 0 & \sigma_k & 0 & 0 & 0 \\ 0 & 0 & 0 & 0 & 0 & 0 & 0 \\ \cdots & \cdots & \cdots & \cdots & \cdots & \cdots & \cdots \\ 0 & 0 & 0 & 0 & 0 & 0 & 0 \end{bmatrix} \cdot \begin{bmatrix} \widetilde{\boldsymbol{s}}_1 \\ \widetilde{\boldsymbol{s}}_2 \\ \cdots \\ \widetilde{\boldsymbol{s}}_{\boldsymbol{k}} \\ \widetilde{\boldsymbol{s}}_{\boldsymbol{k}+1} \\ \cdots \\ \widetilde{\boldsymbol{s}}_n \end{bmatrix} + \begin{bmatrix} \widetilde{\boldsymbol{v}}_1 \\ \widetilde{\boldsymbol{v}}_2 \\ \cdots \\ \widetilde{\boldsymbol{v}}_{\boldsymbol{k}} \\ \widetilde{\boldsymbol{v}}_{k+1} \\ \cdots \\ \widetilde{\boldsymbol{v}}_n \end{bmatrix}. \tag{7.37}$$

Equation (7.37) is, obviously, a series of equations of the form

$$\begin{aligned} \widetilde{\boldsymbol{r}}_1 &= \sigma_1 \widetilde{\boldsymbol{s}}_1 + \widetilde{\boldsymbol{v}}_1 \\ \widetilde{\boldsymbol{r}}_2 &= \sigma_2 \widetilde{\boldsymbol{s}}_2 + \widetilde{\boldsymbol{v}}_2 \\ &\cdots \\ \widetilde{\boldsymbol{r}}_k &= \sigma_k \widetilde{\boldsymbol{s}}_k + \widetilde{\boldsymbol{v}}_k \end{aligned} \tag{7.38}$$

Now, from Equation (7.1), and Equation (7.21) for the SISO model, repeated here for convenience,

$$r_i = hs_i + v_i \tag{7.39}$$

and

$$\frac{C}{B} = \log_2 \left(1 + \frac{\boldsymbol{P_T}}{\boldsymbol{N}} \cdot |\boldsymbol{h}|^2\right) \text{ bits per second per Hz} \tag{7.40}$$

it becomes clear that to obtain the MIMO model, we can make the following associations: $\mathbf{r}_i \rightarrow \widetilde{\mathbf{r}}_i$, $\mathbf{h}_i \rightarrow \sigma_i$, and $\mathbf{v}_i \rightarrow \widetilde{\mathbf{v}}_i$ from where the bandwidth efficiency for MIMO *per channel* becomes

$$\left.\frac{C}{\boldsymbol{B}}\right|_i = \log_2 \left(1 + \frac{\boldsymbol{P_T}/\boldsymbol{n_T}}{\boldsymbol{N}} \cdot |\sigma_i|^2\right) \text{ bits per second perHz} \tag{7.41}$$

since the power per transmit antenna is the total power P_T divided by the number of antennas n_T. So, the capacity or bandwidth efficiency for the MIMO model is

$$\left.\frac{\boldsymbol{C}}{\boldsymbol{B}}\right|_{\text{MIMO}} = \sum_{i=1}^{k} \log_2 \left(1 + \frac{\boldsymbol{P_T}/\boldsymbol{n_T}}{\boldsymbol{N}} \cdot |\sigma_i|^2\right) \text{ bits per second per Hz} \tag{7.42}$$

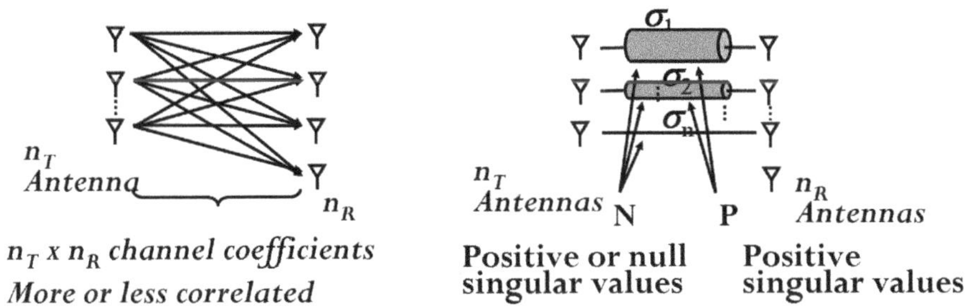

(a) (b)

Figure 7.3 Decomposition of a MIMO channel into independent sub-channels by decomposition into singular values. (a) Classical correlated channels. (b) Independent sub-channels representation. *Source:* Ref. [60].

For an easy comparison with SISO, we consider the case in which all the MIMO channels have identical gains σ. Then, the capacity becomes

$$\left.\frac{C}{B}\right|_{\text{MMO}} = k \cdot \log_2 \left(1 + \frac{P_T/n_T}{N} \cdot |\sigma|^2\right) \text{ bits per second per Hz} \quad (7.43)$$

which clearly shows that the MIMO capacity increases linearly with the number of antennas. The SVD transformation conceptually replaces the coupled transmit–receive antenna links into effectively *parallel* channels; see Figure 7.3. Figure 7.4 shows MIMO capacity calculations for various cases of probability density distributions, a number of transmit and receive antennas, and SNRs. Next, we go over the mechanics of effecting the SVD approach.

7.3.2.1 The mechanics of the SVD approach

The key to computing the SVD of an $m \times n$ matrix **A** is finding the matrices **U** and **V**. These matrices are defined as

$$\mathbf{U} = [\mathbf{u_1}, \mathbf{u_2}, \ldots, \mathbf{u_m}] \text{ and } \mathbf{V} = [\mathbf{v_1}, \mathbf{v_2}, \ldots, \mathbf{y_p}] \quad (7.44)$$

where the vectors $\mathbf{v}_i$ are the normalized eigenvectors of the matrix $\mathbf{A}^H\mathbf{A}$, and $\mathbf{u}_i$ are the normalized eigenvectors of the matrix $\mathbf{AA}^H$. This is best visualized by going over an example.

Example 7.1:

SVD decomposition.

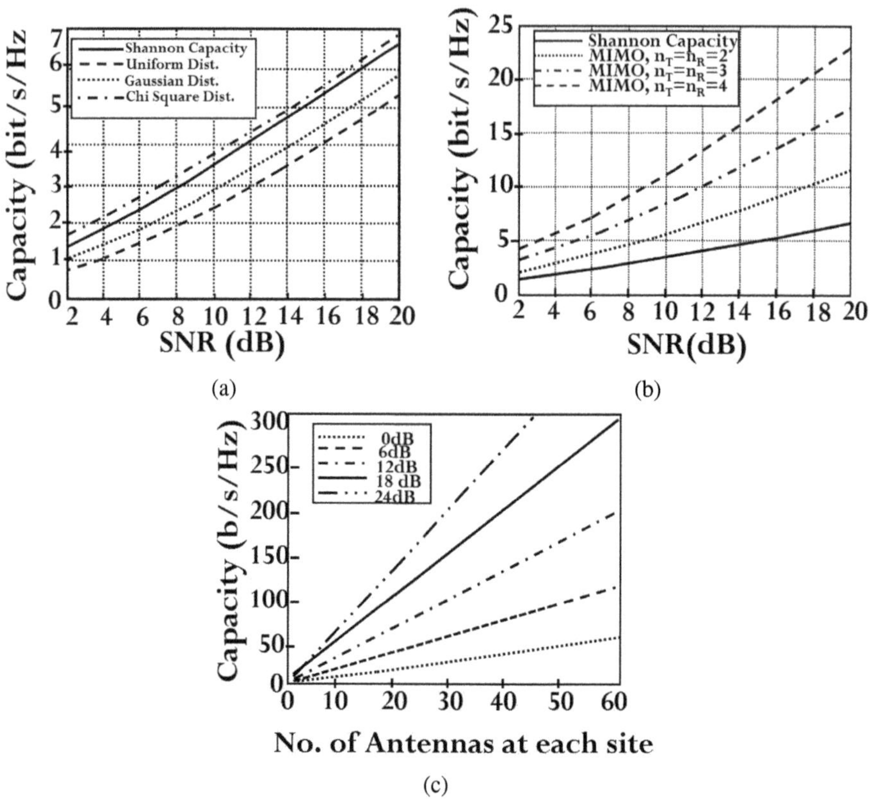

Figure 7.4 (a) SISO channel capacity with various distributions. (b) MIMO channel capacity.. (c) MIMO channel capacity. *Source:* Ref. [67].

Consider the matrix

$$\boldsymbol{A} = \begin{bmatrix} 1 & 2 & -1 \\ 2.3 & 4 & 4 \\ -2 & 5.1 & 1 \\ 0 & 0.8 & 6 \end{bmatrix}: \tag{7.45}$$

Find its SVD.

Step 1: Find **V**.

(a) To find **V**, we create $\mathbf{A}^H\mathbf{A}$.

$$A^H A = \begin{bmatrix} 1 & 2 & -1 \\ 2.3 & 4 & 4 \\ -2 & 5.1 & 1 \\ 0 & 0.8 & 6 \end{bmatrix}^H \cdot \begin{bmatrix} 1 & 2 & -1 \\ 2.3 & 4 & 4 \\ -2 & 5.1 & 1 \\ 0 & 0.8 & 6 \end{bmatrix} \tag{7.46}$$

$$A^H A = \begin{bmatrix} 1 & 2.3 & -2 & 0 \\ 2 & 4 & 5.1 & 0.8 \\ -1 & 4 & 1 & 6 \end{bmatrix} \cdot \begin{bmatrix} 1 & 2 & -1 \\ 2.3 & 4 & 4 \\ -2 & 5.1 & 1 \\ 0 & 0.8 & 6 \end{bmatrix} \tag{7.47}$$

$$A^H A = \begin{bmatrix} 10.29 & 1 & 6.2 \\ 1 & 46.65 & 23.9 \\ 6.2 & 23.9 & 54 \end{bmatrix} \tag{7.48}$$

(b) Find the eigenvalues of $\mathbf{A}^H\mathbf{A}$.

This can be done by hand calculation or using a numerical software tool. Using a numerical software tool, we find that the eigenvalues of $\mathbf{A}^H\mathbf{A}$ are

$$\text{eigenvectors } A^H A = \begin{bmatrix} 74.951 \\ 26.773 \\ 9.216 \end{bmatrix}. \tag{7.49}$$

(c) Find the normalized eigenvectors of $\mathbf{A}^H\mathbf{A}$.

This can be done by hand calculation or using a numerical software tool. Using a numerical software tool, we find that the eigenvectors of $\mathbf{A}^H\mathbf{A}$ are

$$\text{eigenvectors } A^H A = \begin{bmatrix} -0.083 & -0.188 & -0.979 \\ -0.645 & 0.759 & -0.092 \\ -0.76 & -0.623 & 0.184 \end{bmatrix} \tag{7.50}$$

Each column in Equation (7.50) represents one normalized (unit magnitude) eigenvector, namely,

$$\boldsymbol{v}_1 = \begin{bmatrix} -0.083 \\ -0.645 \\ -0.76 \end{bmatrix} \boldsymbol{v}_2 = \begin{bmatrix} -0.188 \\ 0.759 \\ -0.623 \end{bmatrix} \boldsymbol{v}_3 = \begin{bmatrix} -0.979 \\ -0.092 \\ 0.184 \end{bmatrix}. \tag{7.51}$$

d) Write **V**.

$$\boldsymbol{V} = \begin{bmatrix} -0.083 & -0.188 & -0.979 \\ -0.645 & 0.759 & -0.092 \\ -0.76 & -0.623 & 0.184 \end{bmatrix}. \tag{7.52}$$

Step 2: Find **U**.

a) To find **U**, we create $\mathbf{AA}^H$.

$$\boldsymbol{AA}^H = \begin{bmatrix} 1 & 2 & -1 \\ 2.3 & 4 & 4 \\ -2 & 5.1 & 1 \\ 0 & 0.8 & 6 \end{bmatrix} \cdot \begin{bmatrix} 1 & 2 & -1 \\ 2.3 & 4 & 4 \\ -2 & 5.1 & 1 \\ 0 & 0.8 & 6 \end{bmatrix}^H \tag{7.53}$$

$$\boldsymbol{AA}^H = \begin{bmatrix} 1 & 2 & -1 \\ 2.3 & 4 & 4 \\ -2 & 5.1 & 1 \\ 0 & 0.8 & 6 \end{bmatrix} \cdot \begin{bmatrix} 1 & 2.3 & -2 & 0 \\ 2 & 4 & 5.1 & 0.8 \\ -1 & 4 & 1 & 6 \end{bmatrix} \tag{7.54}$$

$$\boldsymbol{AA}^H = \begin{bmatrix} 6 & 6.3 & 7.2 & -4.4 \\ 6.3 & 37.29 & 19.8 & 27.2 \\ 7.2 & 19.8 & 31.01 & 10.08 \\ -4.4 & 27.2 & 10.08 & 36.64 \end{bmatrix} \tag{7.55}$$

b) Find the eigenvalues of $\mathbf{AA}^H$.

This can be done by hand calculation or using a numerical software tool. Using a numerical software tool, we find that the eigenvalues of $\mathbf{AA}^H$ are

$$\text{eigenvalues } \boldsymbol{AA}^H = \begin{bmatrix} 74.951 \\ 26.773 \\ 9.216 \\ 0 \end{bmatrix} \tag{7.56}$$

c) Find the normalized eigenvectors of $\mathbf{AA}^H$.

This can be done by hand calculation or using a numerical software tool. Using a numerical software tool, we find that the eigenvectors of $\mathbf{AA}^H$ are

$$\text{eigenvectors } \boldsymbol{AA}^H = \begin{bmatrix} -0.071 & 0.377 & -0.443 \\ -0.671 & 0.021 & -0.619 \\ -0.448 & 0.7 & 0.552 \\ -0.586 & -0.605 & 0.34 \end{bmatrix}. \tag{7.57}$$

Each column in Equation (7.56) represents one normalized (unit magnitude) eigenvector, namely,

$$\boldsymbol{u}_1 = \begin{bmatrix} -0.071 \\ -0.671 \\ -0.448 \\ -0.586 \end{bmatrix} \boldsymbol{u}_2 = \begin{bmatrix} 0.377 \\ 0.021 \\ 0.7 \\ -0.605 \end{bmatrix} \boldsymbol{u}_3 = \begin{bmatrix} -0.443 \\ -0.619 \\ 0.552 \\ 0.34 \end{bmatrix}. \tag{7.58}$$

d) Write **U**.

$$\boldsymbol{U} = \begin{bmatrix} -0.071 & 0.377 & -0.443 \\ -0.671 & 0.021 & -0.619 \\ -0.448 & 0.7 & 0.552 \\ -0.586 & -0.605 & 0.34 \end{bmatrix} \tag{7.59}$$

Step 3: Find the singular matrix $\sum = \mathbf{U^H\ AV}$.
(a) Find $\mathbf{U}^H$.
From Equation (7.53), we find

$$\boldsymbol{U}^H = \begin{bmatrix} -0.071 & -0.671 & -0.448 & -0.586 \\ 0.377 & 0.021 & 0.7 & -0.605 \\ -0.443 & -0.619 & 0.552 & 0.34 \end{bmatrix}. \tag{7.60}$$

(b) Form the product $\mathbf{U}^H\mathbf{A}$.

$$\boldsymbol{U}^H\boldsymbol{A} = \begin{bmatrix} -0.717 & -5.581 & -6.579 \\ -0.975 & 3.927 & -3.225 \\ -2.971 & -0.278 & 0.56 \end{bmatrix}. \tag{7.61}$$

(c) Form the product $\mathbf{U}^H\mathbf{A}\mathbf{V} = \sum$.

$$\boldsymbol{U}^H\boldsymbol{A}\boldsymbol{V} = \begin{bmatrix} -0.717 & -5.581 & -6.579 \\ -0.975 & 3.927 & -3.225 \\ -2.971 & -0.278 & 0.56 \end{bmatrix} \cdot \begin{bmatrix} -0.083 & -0.188 & -0.979 \\ -0.645 & 0.759 & -0.092 \\ -0.76 & -0.623 & 0.184 \end{bmatrix}$$

$$\Sigma = \begin{bmatrix} 8.657 & 0 & 0 \\ 0 & 5.174 & 0 \\ 0 & 0 & 3.036 \end{bmatrix} \tag{7.62}$$

This is the SVD of A. The singular values are: $8.657 > 5.174 > 3.036 > 0$ and the value of k is 3.

Step 4: Verification. Compare $\mathbf{U} \sum \mathbf{V^H}$ to matrix A.

$$\boldsymbol{U}\Sigma = \begin{bmatrix} -0.071 & 0.377 & -0.443 \\ -0.671 & 0.021 & -0.619 \\ -0.448 & 0.7 & 0.552 \\ -0.586 & -0.605 & 0.34 \end{bmatrix} \cdot \begin{bmatrix} 8.657 & 0 & 0 \\ 0 & 5.174 & 0 \\ 0 & 0 & 3.036 \end{bmatrix}$$

$$\boldsymbol{U}\Sigma = \begin{bmatrix} -0.612 & 1.953 & -1.346 \\ -5.809 & 0.109 & -1.88 \\ -3.882 & 3.624 & 1.675 \\ -5.075 & -3.133 & 1.033 \end{bmatrix} \tag{7.63}$$

Now, form $\mathbf{U} \sum \mathbf{V^H}$

$$\boldsymbol{U}\Sigma \cdot \boldsymbol{V}^H = \begin{bmatrix} -0.612 & 1.953 & -1.346 \\ -5.809 & 0.109 & -1.88 \\ -3.882 & 3.624 & 1.675 \\ -5.075 & -3.133 & 1.033 \end{bmatrix} \cdot \begin{bmatrix} -0.083 & -0.645 & -0.76 \\ -0.188 & 0.759 & -0.623 \\ -0.979 & -0.092 & 0.184 \end{bmatrix}$$

$$\boldsymbol{U\Sigma} \cdot \boldsymbol{V^H} = \begin{bmatrix} 1 & 2 & -1 \\ 2.3 & 4 & 4 \\ -2 & 5.1 & 1 \\ 0 & 0.8 & 6 \end{bmatrix} = \boldsymbol{A} \tag{7.64}$$

7.3.2.2 MIMO interpretation of SVD example

In the above example, we began with the channel matrix,

$$\boldsymbol{A} = \boldsymbol{H} = \begin{bmatrix} 1 & 2 & -1 \\ 2.3 & 4 & 4 \\ -2 & 5.1 & 1 \\ 0 & 0.8 & 6 \end{bmatrix} \tag{7.65}$$

representing an $(n_T, n_R) = (4, 3)$ MIMO system with four transmit and three receive antennas. The SVD result yielded

$$\Sigma = \begin{bmatrix} 8.657 & 0 & 0 \\ 0 & 5.174 & 0 \\ 0 & 0 & 3.036 \end{bmatrix} \tag{7.66}$$

which means that the (4, 3) MIMO system may be represented by an effective three-channel MIMO system of capacity

$$\begin{aligned} \left.\frac{\boldsymbol{C}}{\boldsymbol{B}}\right|_{\text{MMO}} &= \log_2\left(1+\frac{\boldsymbol{P_T/n_T}}{\boldsymbol{N}}\cdot|\sigma_1|^2\right)+ \\ &+\log_2\left(1+\frac{\boldsymbol{P_T/n_T}}{\boldsymbol{N}}\cdot|\sigma_2|^2\right) \quad \text{bits per second per Hz.} \\ &+\log_2\left(1+\frac{\boldsymbol{P_T/n_T}}{\boldsymbol{N}}\cdot|\sigma_3|^2\right) \end{aligned} \tag{7.67}$$

As indicated previously, this equation evokes a scenario in which the system acts as having three *parallel* conduits. Paulraj *et al.* [96] have called this capacity proportionality to linear gain as *spatial multiplexing (SM) gain* because the MIMO system realizes it by the transmission of data signals in the individual antennas that are independent. In particular, if the channel environment exhibits much scattering, then it is possible for the receiver to discern the different streams, resulting in a linear increase in capacity at no additional power or bandwidth cost.

7.4 MIMO Transmit Antenna Input Power Optimization

In the previous sections, we have assumed that the power transmitted by each of the transmitting antennas is identical, namely, equal to the total power divided by the number of transmitting antennas, P_T/n_T. In practice, however, due to the fact that the channel gains h_{ij} may vary drastically, it is necessary, since power is limited, to determine the input power applied to every transmitting antenna in an optimum way. The method utilized for this optimization is the so-called method of Lagrange multipliers, which is presented next.

To efficiently allocate the input power amongst the transmit antennas, the following problem is formulated: Maximize the MIMO capacity equation so that the total input power is less or equal to the maximum available power which can be expressed as,

$$\left.\frac{\boldsymbol{C}}{\boldsymbol{B}}\right|_{\text{MMO}} = \max_{\{P_i\}} \sum_{i=1}^{k} \log_2\left(1 + \frac{\boldsymbol{P}_i}{\boldsymbol{N}} \cdot |\sigma_i|^2\right) \text{ bits per second per Hz} \quad (7.68)$$

subject to

$$\sum_{i=1}^{k} \boldsymbol{P}_i \leq \boldsymbol{P}_T \text{ and } \boldsymbol{P}_i \geq 0. \quad (7.69)$$

In the Lagrange multiplier scheme, an extreme value is found for a function *f(x,y)* subject to a subsidiary condition $\phi(x,y) = 0$ [65]. In particular, the extreme value of **F** = **f** + $\boldsymbol{\lambda\phi}$ is found by solving

$$\frac{\partial \boldsymbol{f}}{\partial \boldsymbol{x}} + \lambda \frac{\partial \phi}{\partial \boldsymbol{x}} = 0 \quad (7.70)$$

$$\frac{\partial \boldsymbol{f}}{\partial \boldsymbol{y}} + \lambda \frac{\partial \phi}{\partial \boldsymbol{y}} = 0. \quad (7.71)$$

In our case,

$$\boldsymbol{f} = \sum_{i=1}^{k} \log_2\left(1 + \frac{\boldsymbol{P}_i}{\boldsymbol{N}} \cdot |\sigma_i|^2\right) \quad (7.72)$$

and

$$\phi = \sum_{i=1}^{k} \boldsymbol{P}_i. \quad (7.73)$$

So

$$\boldsymbol{F} = \sum_{i=1}^{k} \log_2\left(1 + \frac{\boldsymbol{P}_i}{\boldsymbol{N}} \cdot |\sigma_i|^2\right) + \lambda \cdot \left(\boldsymbol{P}_T - \sum_{i=1}^{k} \boldsymbol{P}_i\right) \cdot \quad (7.74)$$

Differentiating this with respect to $\boldsymbol{P}_i$, we have

$$\frac{\partial}{\partial \boldsymbol{P}_i}\left(\sum_{i=1}^{k} \log_2\left(1 + \frac{\boldsymbol{P}_i}{\boldsymbol{N}} \cdot |\sigma_i|^2\right) + \lambda \cdot \left(\boldsymbol{P}_T - \sum_{i=1}^{k} \boldsymbol{P}_i\right)\right) = 0 \quad (7.75)$$

$$\frac{|\sigma_i|^2 / N}{1 + \frac{\boldsymbol{P}_i}{\boldsymbol{N}} \cdot |\sigma_i|^2} + \lambda(0 - 1) = 0 \quad (7.76)$$

$$\lambda = \frac{|\sigma_i|^2 / \boldsymbol{N}}{1 + \frac{\boldsymbol{P_i}}{\boldsymbol{N}} \cdot |\sigma_i|^2} \tag{7.77}$$

$$\frac{1}{\lambda} = \frac{1 + \frac{\boldsymbol{P_i}}{\boldsymbol{N}} \cdot |\sigma_i|^2}{|\sigma_i|^2 / \boldsymbol{N}} \tag{7.78}$$

$$\frac{1}{\lambda} = \frac{1}{|\sigma_i|^2 / N} + \boldsymbol{P_i} \tag{7.79}$$

$$\boldsymbol{P_n} = \frac{1}{\lambda} - \frac{1}{|\sigma_i|^2 / \boldsymbol{N}} \geq 0. \tag{7.80}$$

Now, the value of λ is found by substituting Equation (7.79) in Equation (7.69), i.e.,

$$\sum_{i=1}^{k} \frac{1}{\lambda} - \frac{1}{|\sigma_i|^2 / N} \leq \boldsymbol{P_T}. \tag{7.81}$$

For a given number of antennas *k* and maximum power available $\boldsymbol{P_T}$, λ is found such that Equations (7.81) and (7.80) are satisfied. The algorithm that accomplishes this is called the *water-filling* algorithm [89] and produces a power allocation per channel characteristic typified by that in Figure 7.5. The figure shows that the amount of power allocated to a given channel depends on the noise present in that channel. For example, the power allocated to Channel 1 is lower than that allocated to Channel 2 because the noise in Channel 1 is greater than that in Channel 2.

While power allocation is important to efficiently distribute the total input power amongst the various transmit channels, on the receiver side, it is useful

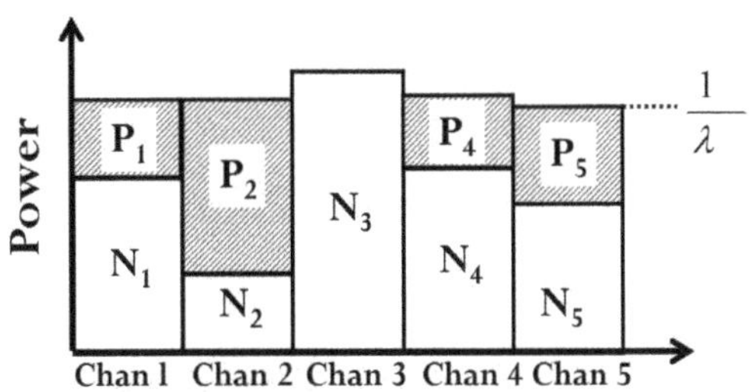

Figure 7.5 Result of power allocation per channel algorithm via the water-filling algorithm. The noise derives from the channel matrix, i.e., $\sigma_1^2 = N_1$, $\sigma_2^2 = N_2$, and so on. *Source:* Ref. [57].

to determine how to best combine the various received signals in light of the differences in channel gains, which include differences in both amplitude and phase. We turn to these next.

7.5 MIMO Receive Antenna Signal Processing

Other key advantages of the MIMO technique over SISO include array gain, diversity gain, and interference reduction [68].

7.5.1 MIMO Array Gain

Array gain refers to the increase in the average SNR at the receiver as a result of signal processing effected at the transmitter and the receiver that produces a coherent combining effect of the signals from the multiple antennas. The magnitude of this gain is, in particular, dependent on the number of transmit and receive antennas, and on knowledge at the receiver, of the channel [68].

7.5.2 MIMO Diversity Gain

Diversity gain [68] refers to the fact that the phenomenon of having the transmitted signals propagate through multiple independent paths tends to nullify the random channel fluctuations introduced. In particular, antenna diversity (spatial), which is not accompanied by extra consumption of transmission time or bandwidth, is preferred. When the $n_T \times n_R$ fluctuation paths comprising the MIMO channel are independent, then the transmitted signal may be reconstructed by proper combination at the receiver. The superposition of the signals results in a large reduction in amplitude fluctuations as compared to a SISO system and it is possible to achieve a diversity of order $n_T \times n_R$. Channel identification techniques [58] enable obtaining the requisite channel knowledge for extracting the gain due to spatial diversity.

7.6 Massive MIMO Detection and Transmission

In a traditional MIMO system, the typical number of transmit, n_T, and receive, n_R, antennas does not normally exceed 10. Massive MIMO, on the other hand, is defined as containing one order of magnitude or more greater number of antennas than traditional MIMO systems, that is, more than 100–1000 antennas [68–73].

7.6.1 Massive MIMO Detection: MRC, ZFBF, and MMSE

The motivation/justification for massive MIMO systems may be surmised from an examination of the reliability in a point-to-point MIMO link. In particular, analysis has shown that, in the context of a quasi-static channel with code words confined to one time and frequency coherent interval, the probability of outage is proportional to $\approx \boldsymbol{SNR}^{-n_T n_R}$, where SNR is the signal-to-noise ratio [73]. This extraordinary potential for highly reliable performance is attractive for enabling the ambitious 5G paradigm and beyond, namely, including future broadband networks, both fixed and mobile, and to enable robust infrastructure for connecting the Internet and the Internet of Things (IoT) to the cloud and other applications.

On the technical implementation side, the very large number of antennas envisioned for massive MIMO systems leads one's attention to systems architectures that are configured to drive very large arrays of antennas in such a way that capacity is maximized and power consumption is minimized. In practice, the arrays of antennas may be operated in two ways, namely, the spatial multiplexing (SM) mode and the *beamforming* (BF) mode [73, 74].

In SM mode, the channel capacity is increased by dividing the signal input to the transmit side into multiple signals applied to the transmit antennas such that each signal copy is transmitted in parallel into the same radio frequency (RF) channel but through n_T individual/separate antennas. Then, due to the separation amongst the locations of the antennas in the array, each transmitted signal experiences different propagation paths, called streams, thus undergoing different frequency-selective time-dispersive channels [74]. On the receive side, n_R multiple antennas superpose all the transmitted signals and feed them to either an individual receiver (single-user MIMO or SU-MIMO) or to multiple independent users (multi-user MIMO or MU-MIMO) so that each of the receivers can reconstitute the originally transmitted signal by exploiting knowledge of the *channel state information* (CSI), i.e., knowledge of the channel matrix. By feeding back to the transmitter the CSI, it is possible for it to effect the technique of *precoding*, whereby the signal streams are sent along the most beneficial directions in the channel, yielding both higher capacity and simpler receiver architectures; this is the so-called *close-loop* SM method. It is possible, in some circumstances, e.g., in time-division duplex (TDD) single-frequency systems, to avoid having to obtain the CSI from the receiver, giving rise to *open-loop* SM [75].

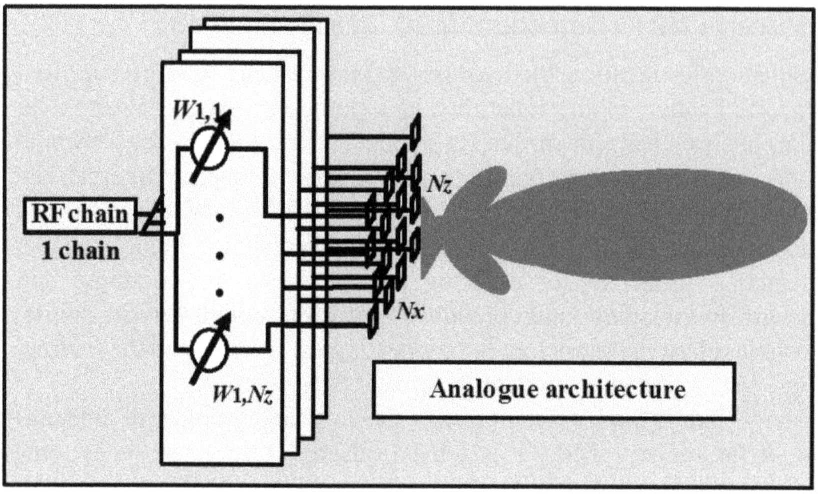

Figure 7.6 Possible RF architectures for mmWave systems, each embodying various tradeoffs in terms of hardware implementation, power, and complexity. (a) Analogue architecture. *Source:* Ref. [73].

In the traditional BF mode [75], the n_T transmit antennas, on the one hand, and the n_R receive antennas, on the other, are adaptively excited/driven to produce a sharp radiation beam pattern concentrated in a specific direction for either transmission or reception of the signal. The shaping of the patterns provide high antenna array gains which, in turn, yield increased SNR, thus mitigating path loss (PL) and even nullifying interference from certain directions (on the receive side). In addition, the PL may be further compensated by smartly processing the signals received by the n_R antennas [76, 77]. Figure 7.6 shows examples of this traditional way of implementing BF which has been considered for MIMO systems at mmWave frequencies [75].

There are many combining strategies [75], but, in the context of mmWave radiation pattern synthesis, the most popular, perhaps, is the MRC [75]. In the MRC BF approach (Figure 7.7), the signal received in each antenna is weighted by its instantaneous carrier-to-noise ratio (CNR). Then, these weighted signals are added to produce

$$s_{n_R} = \sum_{m=1}^{n_R} g_m s_m. \tag{7.82}$$

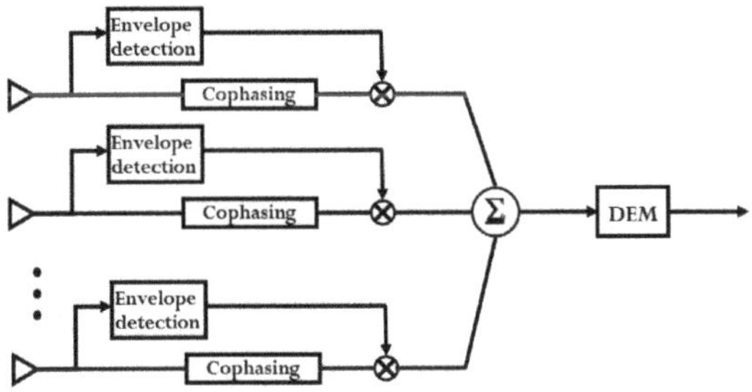

Figure 7.7 Maximum ratio combining. *Source:* Ref. [75].

If the total noise at the output of each antenna is σ_n^2, then the total noise at the output of combiner is

$$\sigma_{n_R}^2 = \sigma_n^2 \sum_{m=1}^{n_R} g_m^2 \tag{7.83}$$

so that the SNR at the combiner output is

$$\gamma_{n_R} = \frac{s_{n_R}^2}{2\sigma_{n_R}^2} \tag{7.84}$$

Now, it may be shown that λ_{n_R} is a maximum when the combiner weights are given by $\mathbf{g}_m = s_m^2/\sigma_n^2$, the SNRs in each input combiner branch. From this, substituting in Equation (8.80, one obtains [75]

$$\gamma_{n_{\bar{R}}} = \frac{1}{2}\frac{\left(\sum_{m=1}^{n_R} \frac{\sigma_m^2}{\sigma_n^2} s_m\right)^2}{\sigma_n^2 \sum_{m=1}^{n_R}\left(\frac{s_m^2}{\sigma_n^2}\right)^2} = \frac{1}{2}\sum_{m=1}^{n_R} \frac{s_m^2}{\sigma_n^2} = \sum_{m=1}^{n_R} \gamma_m. \tag{7.85}$$

Since Equation (7.81) follows a chi-square distribution [83, 45, 103], whose probability density function is

$$p\left(\gamma_{n_R}\right) = \frac{\gamma_{n_R}^{n_R-1} e^{-\frac{2n_R}{\Gamma}}}{\Gamma^{n_R}\left(n_R-1\right)} \tag{7.86}$$

where Γ denotes the mean SNR in each combiner branch, the probability that the SNR is not under a certain threshold is

$$p\left(\gamma_{n_R} < \gamma\right) = \int_0^{\gamma} p\left(\gamma_{n_R}\right) d\gamma_{n_R} = \left(1 - e^{-\frac{\gamma}{\Gamma}}\right)_{m=1}^{n_R} \frac{\left(\frac{\gamma}{\Gamma}\right)^{m-1}}{(m-1)!}. \tag{7.87}$$

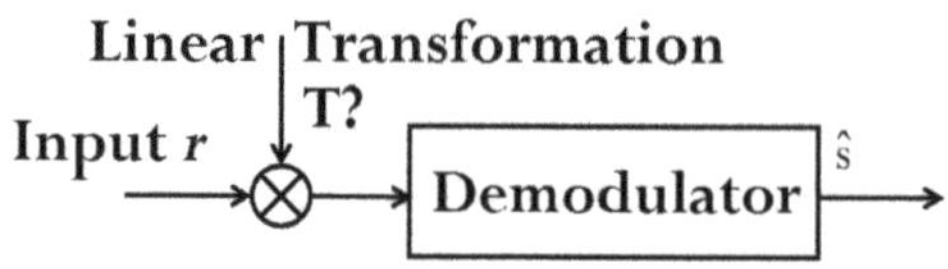

Figure 7.8 Prototypical MIMO model for ZF beamforming. *Source:* Ref. [79].

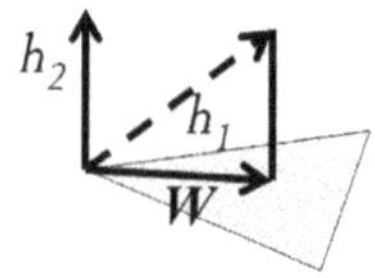

Figure 7.9 ZFBF with one interferer. *Source:* Ref. [76].

Then the average SNR at the combiner output is

$$\bar{\gamma}_{n_R} = \sum_{m=1}^{n_R} \bar{\Gamma} = \boldsymbol{n}_R \Gamma \tag{7.88}$$

The MRC BF approach ignores the presence of interference among users.
A second approach that pertains to extracting the received signal in the context of MIMO systems, particularly at baseband, is the so-called ZFBF approach [73]. In this approach, rather than adjusting the weights and phases of the signals at the outputs of each of the (phased) n_R-antenna array elements to synthesize a desired reception beam pattern (maximum and nulls), the *inverse of the channel matrix* is applied to the received signal to recover the signal radiated into the channel on the transmit side [77–80]. In this process, the noise is ignored; hence, it assumes zero noise is present at the receiver input and, thus, the "zero-forcing" name. In this model (Figure 7.8) [76], there are n_T signal inputs and n_R signal outputs, and the noise is assumed AWGN with zero-mean independent identically distributed (i.i.d.) Gaussian. The goal of the ZFBF is to receive one of the n_T inputs while zeroing out the other n_T-1 inputs, which are considered interference signals, and to maximize the SNR of the desired signal at each of the n_R outputs [76]. The mechanism as to how ZFBF works has been illustrated by Winters [76] for the case of $n_T = 2$ (Figure 7.9). Here, the interfering signal, represented as corresponding to channel matrix coefficient h_2, is projected onto the dimension orthogonal to the desired signal, represented by channel matrix coefficient, h_1. The projection into the dimension orthogonal to h_1 means multiplying h_2 by the weight $\mathbf{W} = \mathbf{h}_1^*$.

In the general (n_T, n_R) MIMO case, the received vector is obtained as [76, 77]

$$r_{ZF} = \left(\boldsymbol{H}^{\boldsymbol{H}}\boldsymbol{H}\right)^{-1}\boldsymbol{H}^{\boldsymbol{H}}\boldsymbol{r} \tag{7.89}$$

where (Figure 7.8) $\mathbf{T} = (\mathbf{H}^{\mathbf{H}}\mathbf{H})^{1}\,\mathbf{H}^{\mathbf{H}}$ is the left pseudo-inverse of $\boldsymbol{H}$ which, for full rank, i.e., $n_T = n_R$, is equal to $\mathbf{H}^{-1}$. Then, multiplying the complete equation, we get

$$r_{ZF} = \left(\boldsymbol{H}^{\boldsymbol{H}}\boldsymbol{H}\right)^{-1}\boldsymbol{H}^{H}(\boldsymbol{H}\boldsymbol{s}+\boldsymbol{v}) = \boldsymbol{s} + \left(\boldsymbol{H}^{\boldsymbol{H}}\boldsymbol{H}\right)^{-1}\boldsymbol{H}^{\boldsymbol{H}}\boldsymbol{v} \tag{7.90}$$

which indicates that the interference is totally removed, but the noise term may be increased [78].

A third approach to MIMO detection, which aims at reducing the impact of channel noise, is the linear MMSE [79]. The MMSE approach aims at minimizing the mean-square error between the transmitted data and the received data disturbed by the effects of the channel, by applying to it the linear transformation matrix, **T**, in Figure 7.7. Mathematically, **T** is expressed as

$$\boldsymbol{T} = \arg\min\ \boldsymbol{E_T}\left(\|\boldsymbol{s} - \boldsymbol{T}\boldsymbol{r}\|_2^2\right). \tag{7.91}$$

This is solved by finding the matrix **T** that makes the error vector, i.e., the difference between the sent and received signal perpendicular (orthogonal) to the conjugate transpose of the received signal, or

$$\boldsymbol{E}\left[(\boldsymbol{s} - \boldsymbol{T}\boldsymbol{r})\boldsymbol{r}^{H}\right]^{T} = 0 \tag{7.92}$$

It can be shown that the resulting matrix is

$$\boldsymbol{T_{MMSE}} = \left(\boldsymbol{H}^{\boldsymbol{H}}\boldsymbol{H} + 2\sigma^2\boldsymbol{I}\right)^{-1}\boldsymbol{H}^{H} \tag{7.93}$$

where E(**s**) = 1 and σ^2 is the noise power per real dimension.

7.6.2 Massive MIMO Transmission: Precoding

By feeding back to the transmitter the CSI, it is possible for it to effect the technique of *precoding*, whereby the signal streams are sent along the most beneficial directions in the channel, yielding both higher capacity and simpler receiver architectures; this is the so-called *closed-loop* SM method.

In this technique, instead of the channel model,

$$\boldsymbol{r} = \boldsymbol{H}\boldsymbol{s} + \boldsymbol{v} \tag{7.94}$$

the substitution, $\mathbf{s} = \mathbf{Gu}$ is made, where

$$\boldsymbol{G} = \boldsymbol{H}^H \left(\boldsymbol{H}\boldsymbol{H}^H\right)^{-1} \tag{7.95}$$

is the Moore–Penrose pseudo-inverse of $\boldsymbol{H}$ and $\mathbf{u} \in \mathbf{C}^{\mathbf{n_R}}$. In this way, n_R noninterfering channels are created into which transmission is effected [77].

In the elegant *eigen-beamforming* approach, Meitzner [78] posits the following, perhaps, more pedagogically transparent formulation.

For a MIMO system with $n_T = n_R > 1$, if the channel matrix is known at both the transmitter and the receiver, then it may be decomposed as

$$\boldsymbol{H} = \boldsymbol{U}\Lambda\boldsymbol{U}^H \tag{7.96}$$

where $\boldsymbol{U}$ is a unitary square matrix such that $\boldsymbol{U}^{\mathbf{H}}\boldsymbol{U} = I$, Λ is a square diagonal matrix containing the eigenvalues $\lambda_1, \lambda_2, \ldots, \lambda_{n_R}$ of $\boldsymbol{H}$. Then, instead of transmitting the signal vector, $\boldsymbol{s}$, the transmitter sends the signal, $\boldsymbol{s}' = \boldsymbol{Us}$. At the receiver, to recover the signal vector $\boldsymbol{s}$, the vector $\boldsymbol{r}'$ is multiplied by $\boldsymbol{U}^H$. In other words

$$r = U^H\boldsymbol{r}' = \boldsymbol{U}^H\left(\boldsymbol{H}\boldsymbol{s}' + v\right) = \boldsymbol{U}^0\boldsymbol{H}\boldsymbol{U}\boldsymbol{s} + \boldsymbol{U}^{\boldsymbol{H}}\boldsymbol{v} \tag{7.97}$$

which, using the decomposition of $\boldsymbol{H}$ in Equation (7.92), yields

$$\boldsymbol{r} = \boldsymbol{U}^H\boldsymbol{U}\Lambda\boldsymbol{U}^H\boldsymbol{U}\boldsymbol{s} + \boldsymbol{U}^H\boldsymbol{v} = \Lambda\boldsymbol{s} + \overline{\boldsymbol{v}} \tag{7.98}$$

which results in

$$\boldsymbol{r} = \Lambda\boldsymbol{s} + \overline{\boldsymbol{v}} \tag{7.99}$$

so that if none of the eigenvalues are zero, the precoding produces n_R parallel channels devoid of interference, thus increasing the capacity by a factor of n_R, while the noise, $\bar{\mathbf{v}} = \mathbf{U^H v}$, remains i.i.d. with zero mean and variance σ_n^2.

The MMSE has been pointed out to be superior to the ZF approach in that it results in a better balance between the elimination of multi-user interference and channel noise because it minimizes simultaneously the total error due to both. MMSE performs better than ZF at low SNRs [80].

7.7 Massive MIMO Systems Architectures

Massive MIMO systems architectures are predicated on the fact that they have to transmit and receive/discern a multitude of simultaneous signals which usually interfere with one another and are distorted by random noise.

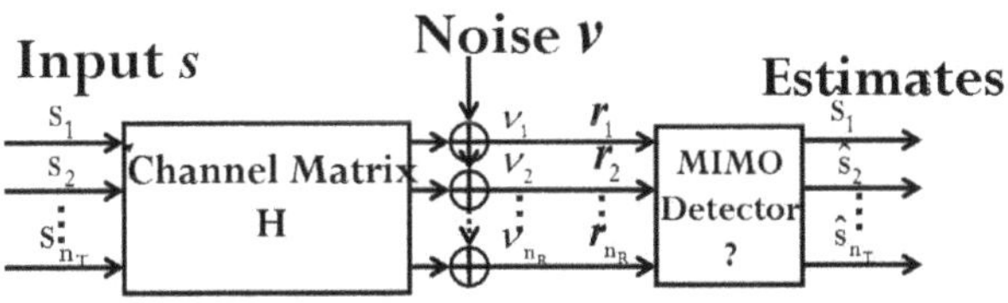

Figure 7.10 Conceptual illustration of the MIMO detection problem. *Source:* Ref. [80].

Because of this, the problem of signal detection has been a rather important one that has been dealt extensively. An extensive review of MIMO detection covering 50 years (1965 – 2015) has been recently published by Yang and Hanzo [80]. In this section, we focus on a perusal of a few prototypical architectures. The general architecture of the prototypical MIMO system is shown in Figure 7.10. As stipulated previously, the system's goal is to solve the equation

$$\boldsymbol{r} = \boldsymbol{H}\boldsymbol{s} + \boldsymbol{v} \tag{7.100}$$

The first prototypical massive MIMO architecture to be discussed is shown in Figure 7.11. It consists of a transmitter, the channel matrix, and a receiver [81]. The transmitter includes 1) a precoder with closed-loop SM, and 2) the ability to do BF. The receiver includes 1) a decoder, 2) a data combining section, and 3) the ability to send CSI to the transmitter, which allows a reduction in the receiver complexity. In general, the availability of a process to obtain and disseminate CSI varies. In the first place, there may be no CSI, in which case the transmitter would lack any knowledge pertaining to the channel or interferences at the receiver. In the second place, there may be perfect CSI, in which case the transmitter would have complete knowledge of the instantaneous channel matrix, allowing the mitigation of inter-user interference and improvement of the SNR at the receiver. In the third place, there may be imperfect CSI, in which case the transmitted will be supplied with inaccurate knowledge about channel matrix.

The second prototypical MIMO architecture is shown in Figure 7.12 [82–84]. This MIMO architecture addresses the issue of limited feedback precoding. The architecture is motivated by the fact that, as the number of antennas increases and the channel environment is rapidly changing, the amount of information and bandwidth needed to update the state of the channel at the transmitter becomes exorbitant. Love *et al.* [82] point out, for example, that whereas in a single antenna link, only one parameter is needed for fast power control, a complex (4, 4) MIMO system requires that

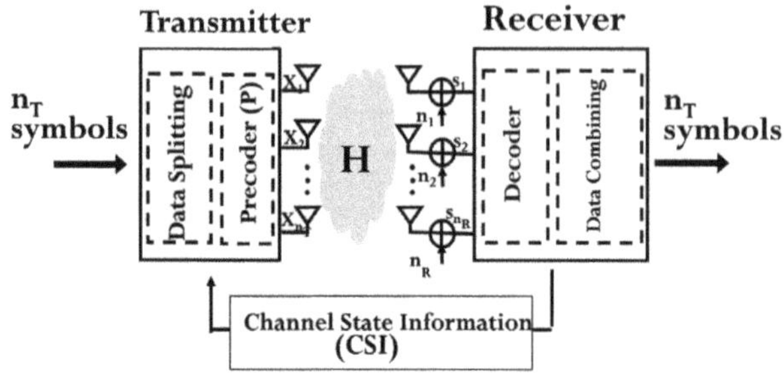

Figure 7.11 Massive MIMO architecture. *Source:* Ref. [81].

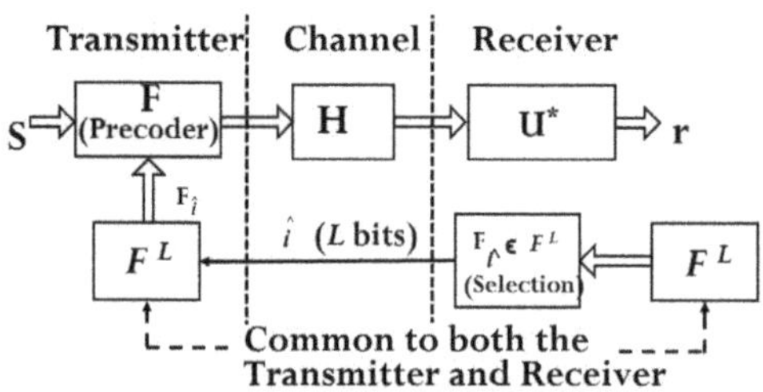

Figure 7.12 Block diagram of a limited feedback unitary precoding MIMO system. *Source:* Ref. [83].

32 parameters be specified every time the channel changes. Evidently, this situation would be untenable for large-scale/massive MIMO.

One approach adopted to overcome this limited feedback hurdle was the concept of the *codebook* [83]. The "cookbook" is a library of indices to a set of precoding matrices, known *a priori* at both the transmitter and the receiver so that, based on its knowledge of the current state of the channel, the receiver can send to the transmitter the index pertinent to the optimal channel precoding matrix that would result in the best capacity. The indexes of the codebook are a few bits that capture/select the required channel containing the required degrees of freedom of the channel. This is shown in Figure 7.11, where a set of 2^L matrices, $F_i \in F^L = \{F_0, F_1, \ldots, F_{2^L-1}\}$ is stored at the receiver and transmitter, and instead of sending the full such matrixes to the transmitter, the receiver only sends an *L-bit* word corresponding to index ***i***. As suggested above, in massive MIMO, the required amount of resources, e.g., the feedback overhead, the storage requirement, and the computational complexity accompanying codeword searching will increase tremendously for growing antenna array sizes. Therefore, further schemes have been pursued to expedite these functions.

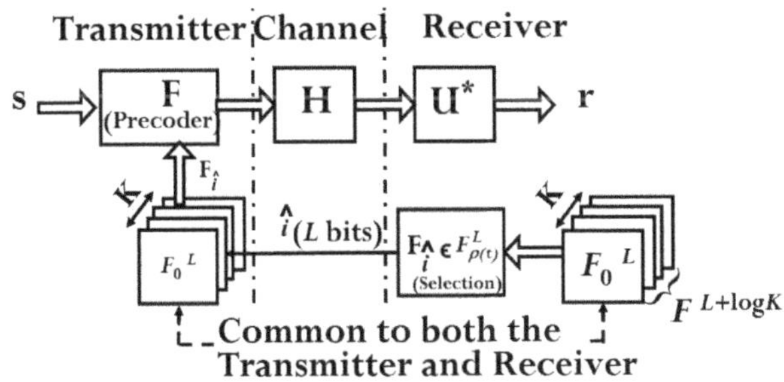

Figure 7.13 Schematic block diagram of the rotating codebook precoding system where the codebook used at feedback time t is $F^L_{\rho(t)}$. *Source*: Ref. [83].

Another such scheme is the *rotating codebook precoding* approach (Figure 7.13) [83]. This scheme extends the size of the cookbook but without increasing the required number of feedback bits employed in the unitary precoding system. In particular, this scheme involves a two-step process. In the first place, multiple codebooks, say K, are stored at the receiver and transmitter, which enable a local iterative selection of one of them. Once one of the K local codebooks is selected, the same number of bits as for the unitary scheme L is required for sending its index to the transmitter. At the transmitter, the precoding matrix chosen may be either a previous precoding matrix or a combination of a previous precoding matrix with the matrix whose newly sent index has been received. The rotating codebook approach is limited by the maximum number of precoding matrixes, K, whose indexes can be sent in a coherence time, the time within which the channel may be construed as being constant [83].

The third prototypical MIMO architecture to be discussed is shown in Figure 7.14 [85–188]. This architecture aims at providing selectable numbers of transmit and receive antennas. There are two reasons for pursuing this approach. First is the prohibitive increase in cost accompanying a large number of antennas, where on the transmit side, each one is driven by an RF chain that usually includes a low noise amplifier (LNA), frequency up- and down-converters, analog-to-digital and digital-to-analog converters, and

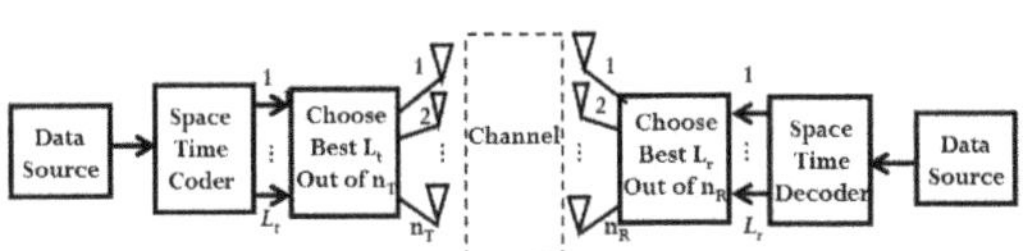

Figure 7.14 MIMO system architecture with antenna selection. *Source:* Ref. [86].

several filters. And second is the reality that there are performance advantages that are gained by judiciously discarding, amongst the antennas in a given array, the ones resulting in a channel matrix with maximized norm, $||\boldsymbol{H}^2||$, giving rise to the highest capacity. No exact solution exists to arrive at this optimal channel matrix; therefore, *ad hoc* search algorithms are required. The same algorithms may be applied to both the transmit and receive sides, but the transmit side requires that CSI be fed back from the receiver. The antenna selection approach is not always useful. In particular, when the assumption of the channel frequency response being flat is not met (i.e., the channel being frequency-selective), this would dictate having a different channel response at different bands leading to a different antenna selection for optimality which would be impractical.

7.8 Massive MIMO Limiting Factors

7.8.1 Pilot Contamination

In the practical implementation of MIMO systems, special (overhead) signals are required to organize and synchronize the interactions amongst transmitters and receivers; such signals are called *pilots* [88]. They are also employed in the determination of CSI at the transmitter and receiver, the generation of the MIMO channel matrix, measuring the channel over the entire frequency band, and other functions [88].

In a massive MIMO system, every terminal is assigned an orthogonal uplink pilot signal; however, the maximum number of orthogonal pilot signals is limited by the duration, the time within which the channel may be construed as constant, the coherence time interval, divided by the channel delay spread [71]. When neighboring cells are sufficiently close and they are using the same frequencies (frequency sharing), there exists the possibility of one cell's pilot being mistaken by that of the adjacent one; this is called *pilot contamination*. This manifests negatively in that the obtained channel estimate will be wrong. In addition, when operating with a contaminated pilot, downlink BF will result in interference being directed at those terminals that employ the same pilot together with interference directed at the uplink transmissions.

7.8.2 Radio Propagation

The radio propagation limitation refers to the fact that, due to the large number of antennas in massive MIMO arrays, the experienced channel behavior may be different from that perceived by smaller antenna arrays [71]. This, in turn, may result in there being large-scale fading throughout the array, or the small-scale statistics may change over the extent of the array. As a result, the assumptions regarding channel matrix statistics may be wrong.

7.9 Summary

In this chapter, several aspects of the concept of MIMO systems have been addressed. In particular, we began our study of MIMO by first discussing the *channel capacity* of the simpler SISO in terms of which that of a MIMO system was interpreted. Then, we addressed the topics of MIMO channel models and propagation models. This was followed by a discussion of the SVD approach and its application to the channel matrix, and the interpretation of MIMO in terms of SVD. Next, the water-filling approach to MIMO transmit antenna input power optimization and MIMO receive antenna signal processing were presented. MIMO detection and transmission techniques, in particular, MRC, ZFBF, and MMSE were then addressed. Lastly, we addressed the topic of precoding, massive MIMO architectures, and massive MIMO limitations.

7.10 Problems

1. Explain the concept of *codebooks*. Why are they useful?
2. Discuss the advantages and disadvantages of MIMO.

8

Aerospace/Electronic Warfare RADAR

8.1 Introduction

The term *electronic warfare* (EW) refers to a number of electronic technologies employed for the gathering of enemy intelligence or for interfering with enemy operations [90–92]. In the context of RADARs, EW deals with the latter, i.e., diminishing the capabilities of enemy RADAR to detect the target threat; this area of operation is referred to as *electronic countermeasures* (ECM) [91]. In order to successfully operate while carrying out its ECM mission, however, a RADAR must be able to countermeasure the enemy's own EW operations against it; thus, it must possess means to protect itself from enemy's electronic attacks, i.e., be equipped with *electronic counter countermeasures* (ECCM).

As indicated in Chapter 1, the basic function of a RADAR is to determine the distance (range) or the speed of a target in its field of view (FOV) [14]. This is accomplished by a system possessing a transmitter, which sends an electromagnetic (EM) wave whose line of propagation is intercepted by the target, reflects from it, and travels back to a receiver where it is processed to extract the target's position and/or speed.

In this chapter, we begin by addressing the principles of RADARs, in particular, the types of RADAR, radio detection and ranging, the RADAR-target geometry/coordinate system, based on which the target location can be discerned in a display. Then, we discuss the topic of RADAR pulses and their relation to range ambiguities, range resolution, and range gates. This is followed by the topic of RADAR sensitivity and Doppler shift, and we explain the tracking versus search modes of operation. Next, we discuss RADAR cross section (RCS). We focus then on the systems aspects of RADARs such as their architectures for various types of RADAR, in particular, continuous wave (CW) Doppler, frequency modulation CW (FM-CW), and pulse Doppler RADARs. We then go on to discuss how RADARs effect ECMs, including searching for signal sources, noise jamming, deception jamming, and then ECCM techniques such as pulse compression, frequency hopping, side lobe blanking, polarization, and artificial intelligence (AI) based jammer-nulling.

8.2 Principles of RADARs [92–95]

8.2.1 Types of RADAR

While the fundamental function of a RADAR involves determining the distance (range) and/or the speed of a target, there are various ways of designing/implementing a RADAR system to accomplish these.

In a *bistatic* RADAR (Figure 8.1), the transmit and receive antennas are at different locations with respect to the target; for instance, one can have a transmitter on the ground and a receiver on an airplane.

In a *monostatic* RADAR, the transmitter and the receiver are located on the same platform, with respect to the target, so that they share the same antenna.

In a *quasi-monostatic* RADAR, the transmitter and the receiver have separate antennas, but they are on the same platform separated by a small distance so that, as perceived by the target, they appear to be at the same location, for instance, as may be the case of the transmit and receive antennas on an aircraft.

The regular functions of a RADAR include, determining 1) the range of the target, which is based on the round trip delay of a transmitted pulse; 2) the velocity of the target, which is based on the Doppler frequency [92] shift incurred by the transmitted pulse; and 3) the angular direction of the target relative to the transmitting (pointing) antenna.

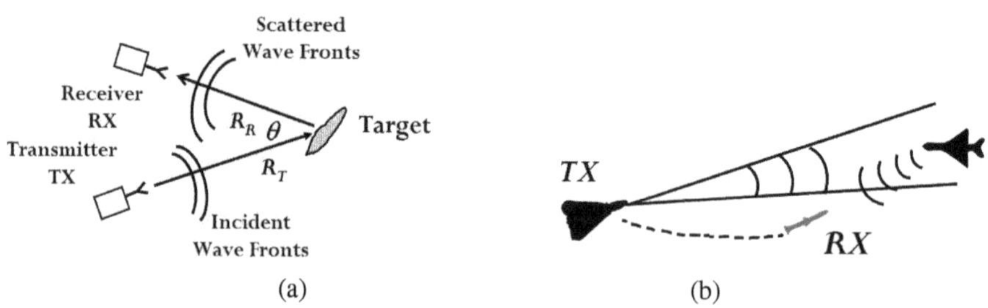

Figure 8.1 (a) Sketch of bistatic RADAR. (b) Example of bistatic RADAR situation: Fighter plane transmits signal and launches missile, and missile receives return echo from target enemy airplane. *Source:* Refs. [92, 93].

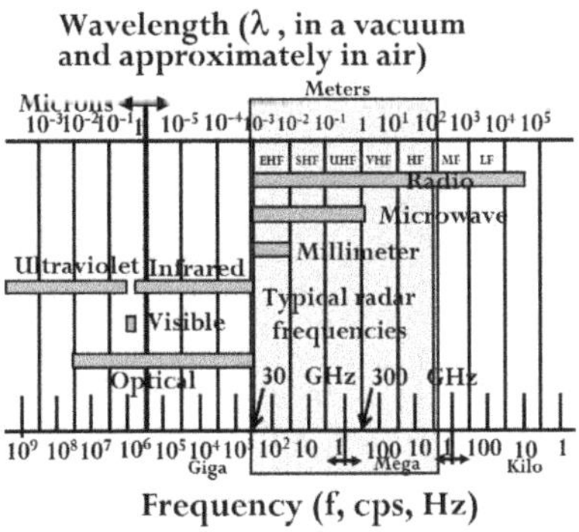

Figure 8.2 Frequencies and respective wavelengths at which RADARs operate. *Source:* Ref. [92].

In addition, based on performing analyses of the return signal, it may determine other features of the target, including signature analysis, to identify another RADAR, and inverse scattering. The latter includes obtaining 1) the target size, which is assessed based on the magnitude of the return; 2) the shape of the target and its components, which is based on determining the return as a function of direction; 3) the existence of moving parts in the target, which is based on the detection of modulation in the return; and 4) the material composition of the target.

It is apparent that the dimension of the target must be such that it is commensurate with the wavelength of the RADAR transmission. For this reason, RADARs operate at a variety of frequencies depending on their intended target. Figure 8.2 shows the range of frequencies at which RADARs operate.

The frequencies at which RADARs operate are also designated by letters; these are given in Table 8.1.

Clearly, RADAR systems find application in both civilian and military markets.

8.2.2 Radio Detection and Ranging [93]

It could be said that, perhaps, the most fundamental parameter embodying the workings of a RADAR is the RCS of the target. The RCS gives a description of the strength of the EM wave reflected from the target threat and, thus, captures the detectability of the target. The RCS of a target is defined as

$$\sigma = \lim_{R\to\infty} 4\pi \boldsymbol{R}^2 \frac{|\boldsymbol{E}_s|^2}{|\boldsymbol{E}_i|^2} \tag{8.1}$$

Table 8.1 Letter designations of RADAR frequency bands. *Source:* Ref. [93].

Band Designation	*Frequency Range*	*Application*
HF	3–30 MHz	Over the horizon (OTH) surveillance
VHF	30–300 MHz	Very-long-range surveillance
UHF	300–1000 MHz	Very-long-range surveillance
L	1–2 GHz	Long-range surveillance En-route traffic control
S	2–4 GHz	Moderate-range surveillance Terminal traffic control Long-range weather
C	4–8 GHz	Long-range tracking Airborne weather detection
X	8–12 GHz	Short-range tracking Missile guidance Mapping, marine RADAR Airborne intercept
K_u	12–18 GHz	High-resolution mapping Satellite altimetry
K	18–27 GHz	Little use (water vapor)
Ka	27–40 GHz	Very-high-resolution mapping Airport surveillance
Millimeter	$40–100^+$ GHz	Experimental

where $\boldsymbol{E}_s$ represents the magnitude of the electric field scattered from the target, $\boldsymbol{E}_i$ represents the magnitude of the electric field incident onto the target, and R is the distance between the transmitter antenna and the target.

Obtaining the target range entails the determination of the time, T_R, it takes the signal pulse transmitted by the RADAR antenna to make a round trip to the target, calculated as follows for a bistatic RADAR:

$$\text{Bistatic} = \boldsymbol{R}_T + \boldsymbol{R}_R = c\boldsymbol{T}_R \tag{8.2}$$

and for a monostatic RADAR

$$\text{Monostatic} = \boldsymbol{R} = \frac{cT_R}{2} \quad (\boldsymbol{R}_T = \boldsymbol{R}_R = \boldsymbol{R}) \tag{8.3}$$

In other words, for the bistatic RADAR, the product of the speed of light in free space, $c = 3 \times 10^8 m/s$, and the time taken for the signal to travel from the transmitter in one platform to the receiver in the other far away platform equals the total distance traveled by the pulse, the range. The proper speed of light must be used according to the deviation of atmospheric conditions from

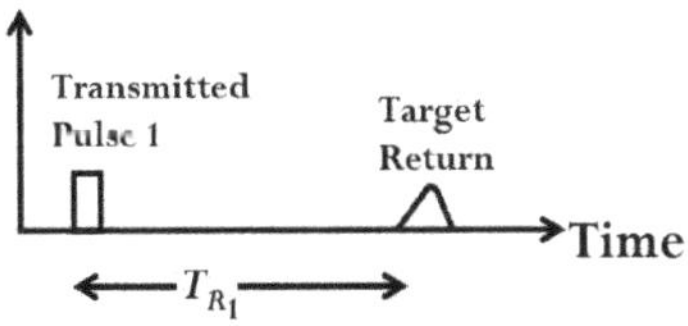

Figure 8.3 RADAR round trip time T_R for calculating range. *Source:* Ref. [93].

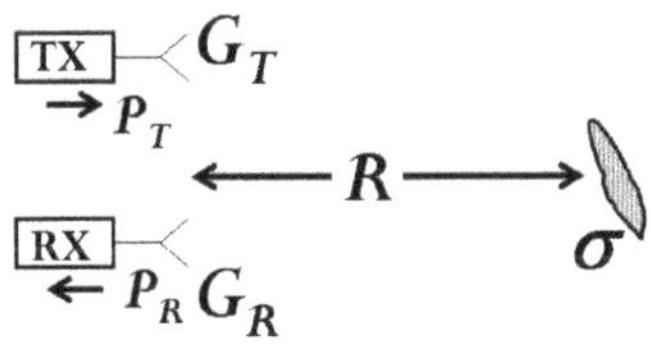

Figure 8.4 Sketch of quasi-monostatic RADAR. P_T is the transmit power in W. P_R is the received power in W. G_T is the transmit antenna gain. G_R is the receive antenna gain. *R* is the range. σ is the target RCS (m^2). *Source:* Ref. [93].

strictly free space, i.e., to account for water vapor [94]. On the other hand, for the monostatic RADAR, in which the transmitter and receiver share the same platform, the distance from transmitter to target, R_T, is the same as that from target to receiver, R_R; so the range equals one-half the product of the speed of light and T_R (Figure 8.3).

The key RADAR equation gives the range or distance *R* to the target (Figure 8.4). This is given in terms of the power reflected from the target and received at the antenna, P_R, which, in turn, is given in terms of the power radiated by the transmit antenna, $P_T G_T$, where G_T is the transmit antenna gain and is subsequently attenuated en-route to the target after traveling a distance *R* by a factor $1/4\pi R_T^2$, from where it gets reflected by the target's RCS, σ, and is further attenuated as it travels en-route back to the receiver a distance *R* by another factor $1/4\pi R_R^2$ and then gets captured by the effective area of the receiver antenna A_{eR}. Mathematically, we can express this sequence as of events as

$$P_R = P_T G_T \cdot \frac{1}{4\pi R_T^2} \cdot \sigma \cdot \frac{1}{4\pi R_R^2} \cdot A_{\varepsilon R} \tag{8.4}$$

If $\boldsymbol{R}_\mathbf{T} = \boldsymbol{R}_\mathbf{R} = \boldsymbol{R}$, then

$$\boldsymbol{P}_R = \frac{\boldsymbol{P}_T \boldsymbol{G}_T \sigma \boldsymbol{A}_{eR}}{(4\pi \boldsymbol{R}^2)^2} \tag{8.5}$$

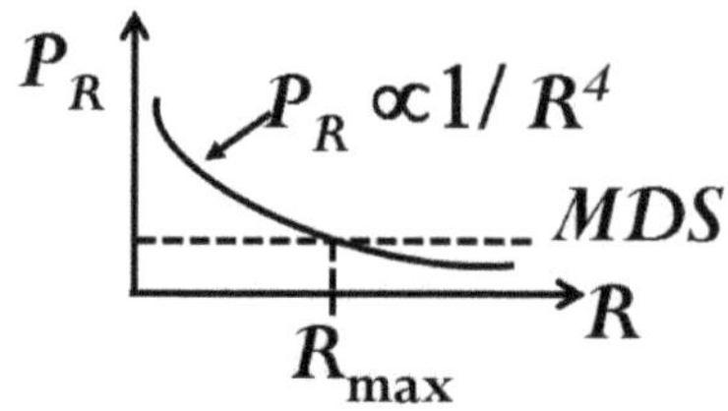

Figure 8.5 Sketch of variation in received power versus range. *Source:* Ref. [93].

Now, using $\mathbf{A_{eR}} = \mathbf{G_R}\lambda^2/4\pi$ (see Chapter 2), Equation (8.5) may be expressed as

$$\boldsymbol{P_R} = \frac{\boldsymbol{P_T G_T G_R}\sigma\hat{\lambda}^2/4\pi}{(4\pi\boldsymbol{R}^2)^2}$$

$$\boldsymbol{P_R} = \boldsymbol{P_T} \cdot \frac{\boldsymbol{G_T G_R}\lambda^2}{(4\pi)^3\boldsymbol{R}^4} \cdot \sigma \tag{8.6}$$

On the other hand, Equation (8.5) may also be expressed as

$$\boldsymbol{P_R} = \frac{\boldsymbol{P_T G_T}\sigma\boldsymbol{A_{eR}}}{(4\pi\boldsymbol{R}^2)^2} = \boldsymbol{P_T} \cdot \frac{\boldsymbol{A_{eT}A_{eR}}}{4\pi\lambda^2\boldsymbol{R}^4} \cdot \sigma \tag{8.7}$$

The expression (8.7) indicates that the received power varies inversely proportional to the fourth root of the range. A sketch of this variation is shown in Figure 8.5.

The maximum range the radar is capable of detecting will depend on the smallest signal it can detect, namely, the *minimum detectable signal* (*MDS*). If we insert this signal, P_R=*MDS*, into the range equation, then, solving for *R*, we arrive at the maximum range as

$$P_R = MDS = \frac{P_T G_T G_R \sigma\lambda^2}{(4\pi)^3 R^4} \Rightarrow R_{\text{max}} = \left(\frac{P_T G_T G_R \sigma\lambda^2}{(4\pi)^3 MDS}\right)^{\frac{1}{4}} \tag{8.7}$$

Examination of the above equations and sketches indicates that the geometry of the RADAR with respect to the target is important. We will discuss this next.

8.2.3 RADAR-Target Geometry/Coordinate System

The general geometrical situation describing the relative positions of the RADAR and the target are given, in a spherical coordinate system, in Figure 8.6.

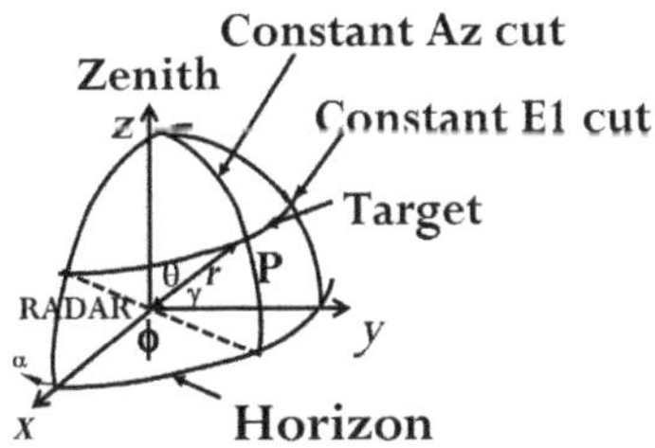

Figure 8.6 Coordinate system with respect to which a number of RADAR parameters is defined. *Source:* Ref. [93].

With the RADAR at the origin, the target position is described by the spherical polar coordinate systems, namely, its coordinates, (r, θ, φ), and its azimuth/elevation, (Az, El), or $(\alpha\gamma)$. Alternatively, the RADAR position may be taken as being at the origin of the Cartesian $x-y$ plane. The azimuth (α) is measured clockwise with respect to a reference axis, e.g., the x-axis in Figure 8.6, whereas the azimuth angle (α) is measured counterclockwise from the x-axis. Accordingly, we have $\gamma = 90 - \theta, \alpha = 360 - \phi$.

8.2.4 RADAR Pulses

As mentioned previously, the RADAR functions are predicated upon the transmission and reception of EM pulses. These functions, in turn, are varied and tailored by the properties of the pulses. For instance, a multitude of pulses may be designed for 1) conducting search patterns in a region of space; 2) tracking moving targets; and 3) increasing target detectability by dwelling on the same target to accumulate (add) multiple returns from it.

The transmitted RADAR pulses are periodic, i.e., they are a *pulse train* (Figure 8.7). These are characterized by the following parameters: 1) the instantaneous power, $\mathbf{P}_0$, in W; 2) pulse width (PW), τ, in sec; 3) pulse repetition frequency (PRF, in Hz), $\mathbf{f_p} = \mathbf{1/T}$; 4) inter-pulse period, $\mathbf{T}_p$, in sec; and 5) number of pulses, N.

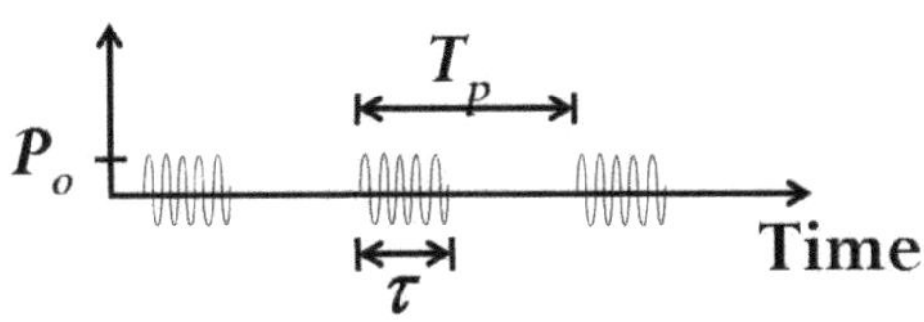

Figure 8.7 RADAR pulse characteristics. *Source:* Ref. [93].

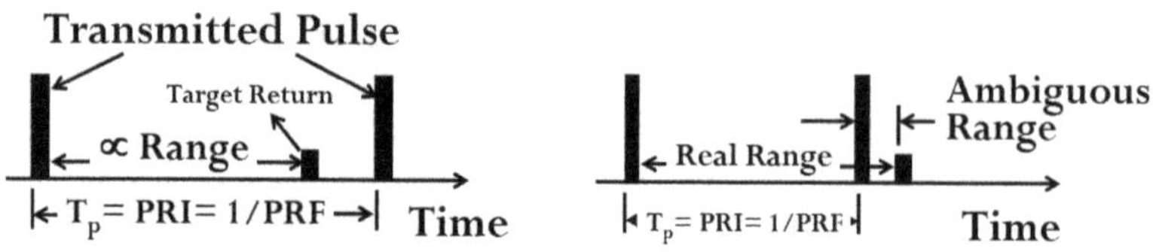

(a) (b)

Figure 8.8 RADAR pulse ambiguity. *Source:* Ref. [96].

8.2.5 Range Ambiguities

In the context of a RADAR transmitting repetitive pulses with a repetition frequency PRF, it may occur that the receiver may detect reflections that arrive from distances that are greater than the distance between transmit pulses corresponding to the PRF [96] (Figure 8.8).

This situation is said to introduce a *range ambiguity*. To resolve the range ambiguity and obtain the true range, it is necessary that [95, 96]

$$\text{Range} = \text{Distance} > \frac{c}{2 \times PRF} \tag{8.8}$$

or

$$\mathrm{T_p} \geq \frac{2\mathrm{R}}{\mathrm{c}} \tag{8.9}$$

In other words, we must have

$$\mathrm{R} = \frac{\mathrm{cT}}{2} = \frac{c}{2f_p} \tag{8.10}$$

A number of sophisticated techniques may be found in the literature to resolve range ambiguity issues [95, 96].

8.2.6 Range Resolution

Range resolution refers to what is the minimum distance between two targets so that their returns may be detected as separate by the receiver, i.e., as coming from two distinct targets. Figure 8.9 illustrates the situation. Two targets may be resolved when their returns do not overlap. The range resolution corresponding to a PW τ is

$$\Delta\mathrm{R} = \boldsymbol{R}_2 - \boldsymbol{R}_1 = \frac{c\tau}{2}. \tag{8.11}$$

Physically, the range resolution equals the distance traveled during half the duration τ of the pulse. Any separation closer than that will not permit the return pulses to be separated/discerned as coming from two different targets.

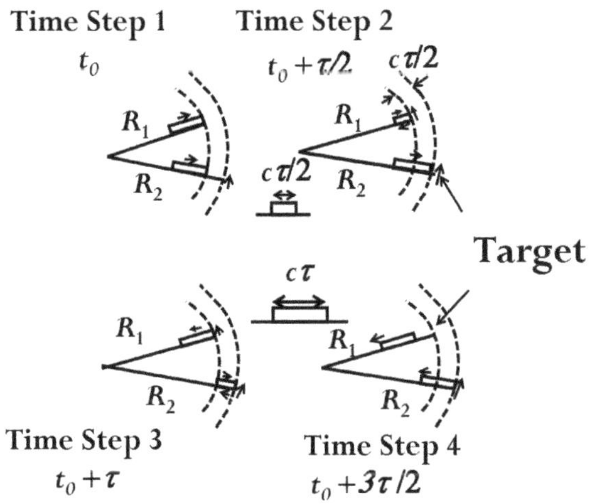

Figure 8.9 RADAR range resolution. *Source:* Ref. [93].

Range ambiguity resolution is required to obtain the true range when the measurements are made using a system where the inequality, $T_p \geq 2R/C$ is true.

8.2.7 Range Gates

Since range and propagation round trip time are directly related, a technique to "hardwire" this relationship has been adopted in some RADAR designs. The *range gate*, an electronic circuit that selects signals within a given time period, accomplishes this task (Figure 8.10). In particular, this circuit permits the passing of signal pulses only within a preset time interval [97]. In the context of RADARs, range gates are employed to select specific targets for further examination. In practice, the gates are activated in sequence so that every time each gate is closed corresponds to a range increment. The gates must block the complete period of time between pulses according to the distances/ranges of interest. In the case of target tracking, a single gate remains closed until the target moves out of the stipulated range.

One instance in which range gates are very useful is in the context in which the RADAR FOV is populated by many objects besides the desired targets (Figure 8.11). It is said, in those circumstances, that the FOV is *cluttered.* Clutter gives rise to interference, as the transmitted pulse may experience multi-path propagation, resulting in simultaneous return pulses from different targets.

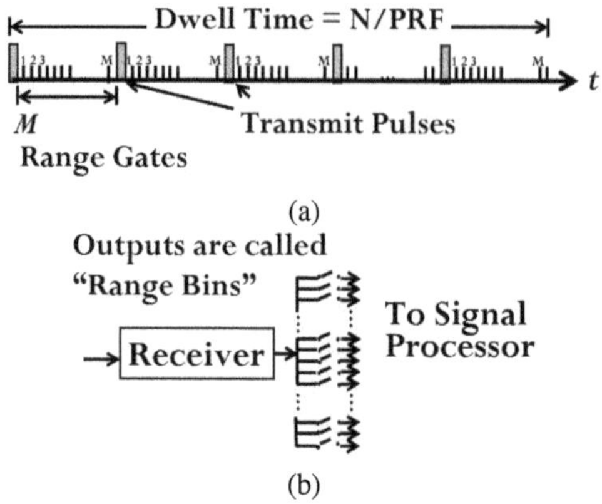

Figure 8.10 (a) Range gates. (b) Analog implementation. *Source:* Ref. [93].

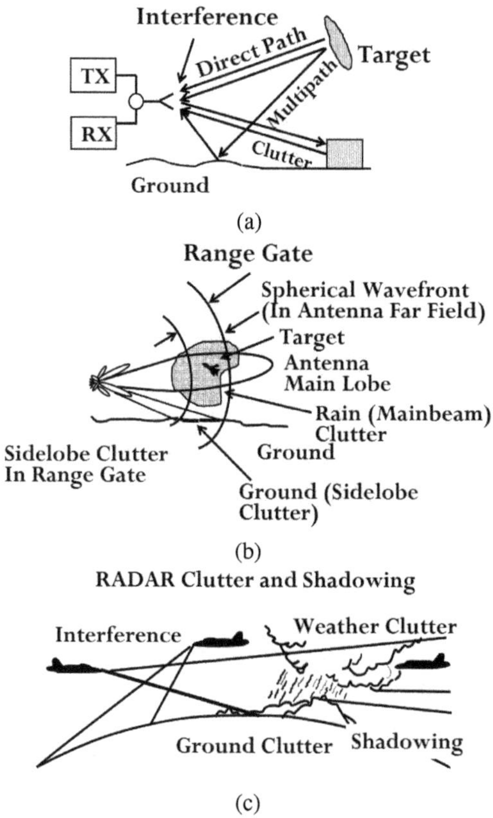

Figure 8.11 (a), (b), (c) Clutter, interference, and shadowing scenarios. *Source*: Refs. [92, 93].

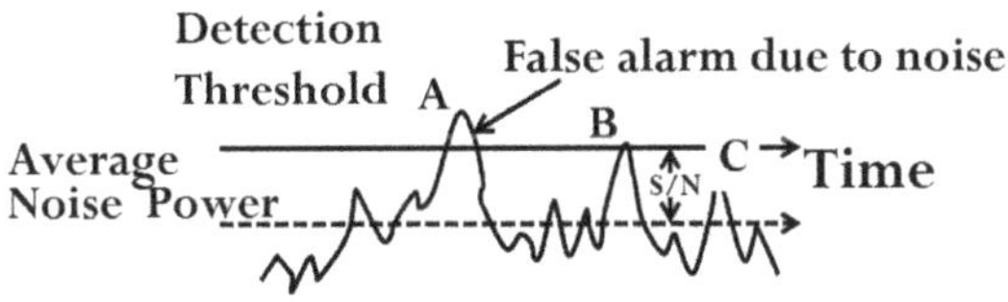

Figure 8.12 Typical RADAR return display. *Source:* Refs. [92, 94].

8.2.8 RADAR Sensitivity

As should be expected by now, the received return pulses at a RADAR receiver are contaminated by noise present in the environment. This noise includes thermal noise, clutter, and interference and manifests itself in that it "competes" with the amplitude of the real return signal (Figure 8.12).

An approach which is often employed to select the return signal from the background noise is the so-called *threshold detection*. According to this technique, whenever the received power is greater than a certain threshold, a target is declared as having been captured. In Figure 8.14, for example, signal **A** is a false alarm since the noise is greater than the threshold level; yet, there is no target. In the case of **B**, while it represents a target, due to its power being less than the threshold, it is missed, i.e., not detected. As we saw in an earlier chapter, the *MDS* is related to the detection threshold.

To evaluate the detection threshold for a RADAR receiver, we need to consider all the noise sources with which the return signal will be competing. These include the noise arising from the RADAR hardware itself such as antenna noise, mixer noise, cable attenuation, and so on. For a received signal to be detectable, it must be greater than the system noise. Thus, we must ensure than the signal-to-noise ratio (SNR) at the receiver input is greater than unity. Now, in analogy with the noise power due to thermal noise,

$$N_0 = k_B T_0 B \tag{8.12}$$

we can define the noise power due to the RADAR system's own noise as

$$N_0 = k_B T_S B \tag{8.13}$$

and, with it, define the SNR

$$\boldsymbol{SNR} = \frac{\boldsymbol{P_R}}{\boldsymbol{N_0}} = \frac{\boldsymbol{P_T G_T G_R} \sigma \lambda^2 \boldsymbol{G_P L}}{(4\pi)^3 \boldsymbol{R^4 k_B T_S B}} \tag{8.14}$$

where G_P is the *processing gain*, which may be introduced by, e.g., integration of multiple returns, and L is an implementation loss, e.g., cable

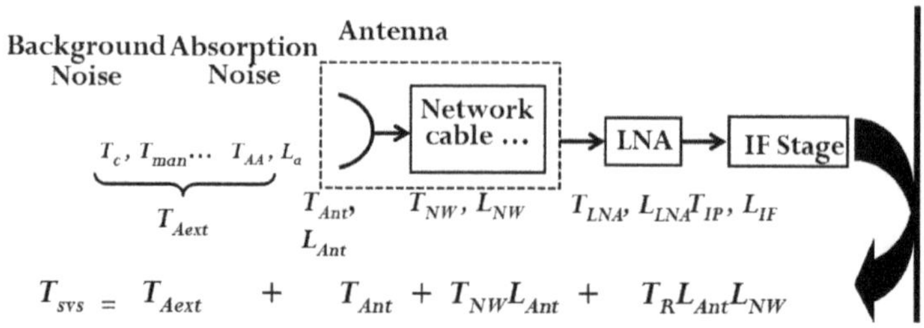

Figure 8.13 System noise components. *Source:* Ref. [94].

attenuation. In Equation (9.15), the bandwidth $\boldsymbol{B}$ is usually equal to the inverse PW, $\mathbf{B} = 1/\tau$.

The RADAR system noise temperature consists of three sources, namely (Figure 8.13) [94]: 1) cosmic noise and atmospheric absorption noise collected by the antenna (T_A); 2) the increase of the noise figure (NF) resulting from the losses and attenuation in the antenna, interconnects, switches, circulator, and so on; 3) the effective noise temperature of the receiver building blocks such as amplifiers, mixers, filters, and so on.

As suggested by Figure 8.13, the overall system may be analyzed by the techniques discussed in chapter 3, with the antenna noise, T_A, and the effective system noise temperature, T_e, which results from combining the gains and temperatures of the system building blocks determining the total system temperature as given in Equation (8.15).

$$\boldsymbol{T_S} = \boldsymbol{T_A} + \boldsymbol{T_e} \tag{8.15}$$

A prototypical diagram of a *coherent RADAR* system, that is, one in which the transmitter and the receiver share the same oscillator, is shown in Figure 8.14.

A number of techniques have been pursued to increase the *SNR*. In *noncoherent integration (NCI)* (postdetection integration), the magnitudes of the returns from all the received pulses are added; this increases the *SNR* in proportion to $\sqrt{N}$, where N is the number of pulses summed.

In *coherent integration (CI)*, the pulse addition is performed prior to integration, while the phase information is still available; it causes the *SNR* to increase in proportion to N.

Another scheme to increase the return signal magnitude and, thus, the *SNR* is increasing the *dwell time* (DT) of the beam on the target (Figure 8.15). Since the beam of an antenna has a certain extent within which its gain is constant, namely, the half-power beam width, characterized by the

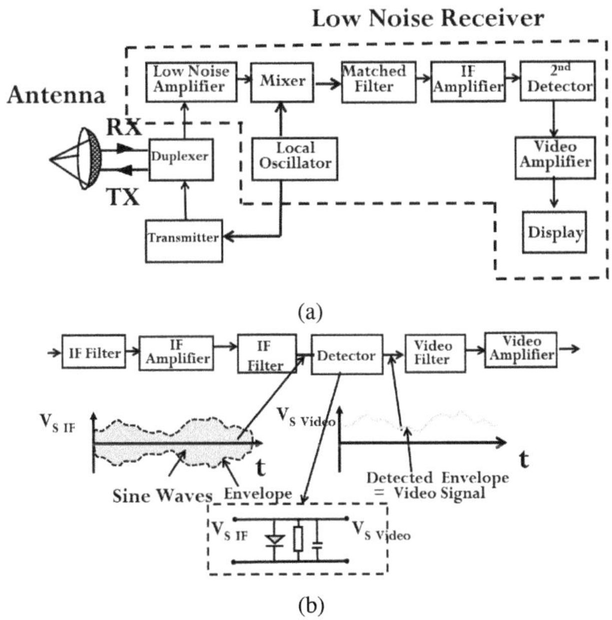

Figure 8.14 (a) Schematic of coherent RADAR system. (b) Schematic of detector. *Source:* Ref. [93]. (b) Schematic of envelope detector. *Source:* Ref. [94].

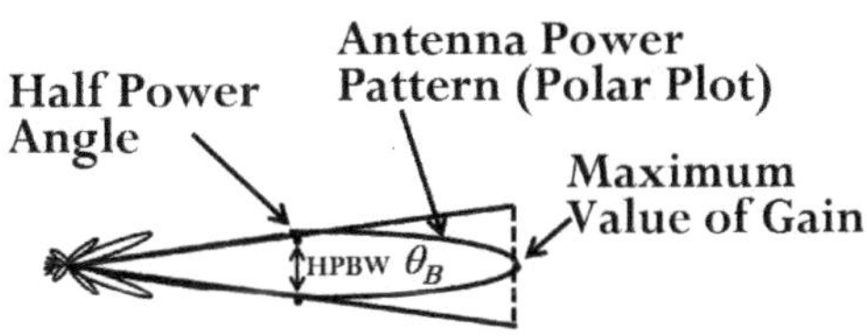

Figure 8.15 Schematic of envelope detector. *Source:* Ref. [93].

angle $\theta_B \approx \lambda/\mathbf{D}$, where *D* is the dimension of the antenna aperture, it is clear that it will take some time for the beam to traverse the target. This time is called variously *DT*, *look time*, or *time on target*, and it is given by

$$\boldsymbol{t}_{ot} = \frac{\theta_B}{\dot{\theta}_S} \tag{8.16}$$

where $\dot{\theta}_\mathbf{B} = \mathbf{d}\theta_B/\mathbf{dt}$ in degrees per second, which derives from the beam scan rate ω_S in revolutions per minute. In an interval of time, t_{ot}, the number of pulses hitting the target is $n_B = t_{ot} f_p$.

8.2.9 Doppler Shift

As previously indicated, one of the RADAR functions is to measure the velocity of the target. This is accomplished by measuring the Doppler shift, which is the change in return signal frequency, relative to the transmitted signal frequency, due to the relative motion between the RADAR and the target (Figure 8.16). Figure 8.16(c) shows the two-way (transmitter and reflector, e.g., airplane, are both in motion) Doppler frequency [92].

The frequency shift occurs when the relative velocity vector has a radial component, $\mathbf{V_r}$. Then, the return frequency will be modified according to

$$f_D = -2\frac{v_r}{\lambda}. \tag{8.17}$$

In particular, we have two situations. When the distance R between RADAR and target is decreasing,

$$\frac{\mathrm{dR}}{\mathrm{dt}} < 0 \Rightarrow f_d > 0 \text{ (RADAR and target approching)} \tag{8.18}$$

and when R is increasing,

$$\frac{\mathrm{dR}}{\mathrm{dt}} > 0 \Rightarrow f_d < 0 \text{ (RADAR and target receeding)} \tag{8.19}$$

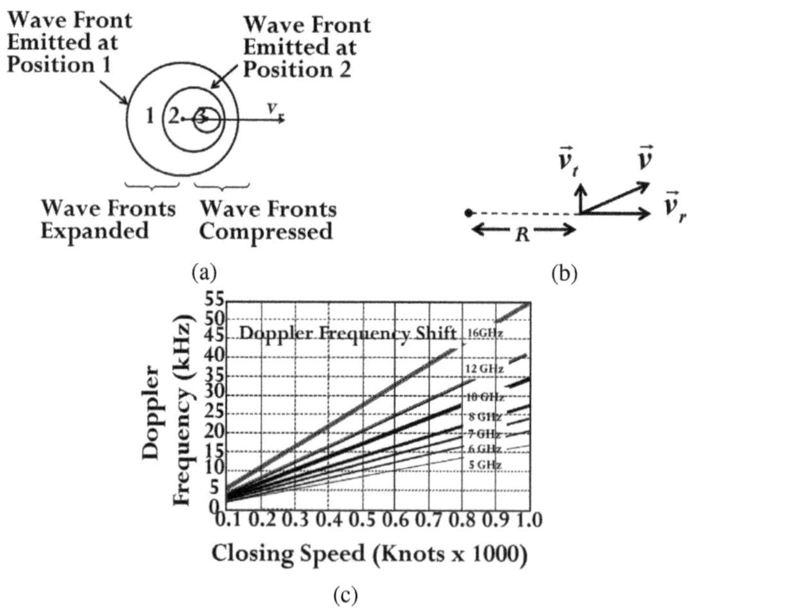

Figure 8.16 (a) Schematic illustration of Doppler shift. (b) Relation between velocities. (c) Two-way Doppler frequency shift. *Source:* Refs. [92, 94].

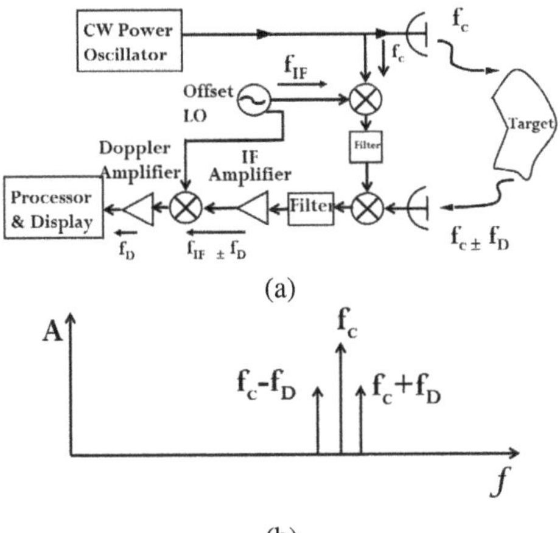

Figure 8.17 (a) Block circuit diagram of radar with IF. (b) Typical Doppler signal of CW radar. f_c+f_D denotes *R* decreasing and f_c-f_D denotes *R* increasing. *Source:* Ref. [94].

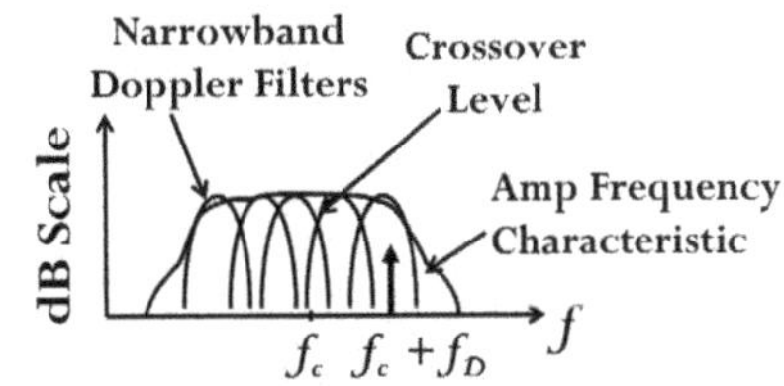

Figure 8.18 Doppler filter banks. *Source:* Ref. [93].

The Doppler shift is detected by examining the spectrum of the return signal (Figure 8.17).

To facilitate the detection of the return frequency, the RADAR's band of operation is divided into narrow sub-bands by Doppler filters (Figure 8.18) whose bandwidth is narrow relative to the total bandwidth of the RADAR, thus reducing the noise bandwidth for detection within each sub-band and increasing the *SNR*. The velocity is estimated by monitoring the power at the output of each filter so that, if a signal is present, the range of velocity possessed by the target is known.

The relative velocity between RADAR and target introduces ambiguity in determining the target velocity (Figure 8.19). This manifests itself in modifications to the return spectrum. In particular, if f_D is increased, then the true target Doppler-shifted return moves out of the Doppler filter passband and a lower sideband lobe enters. Therefore, the Doppler measurement becomes ambiguous.

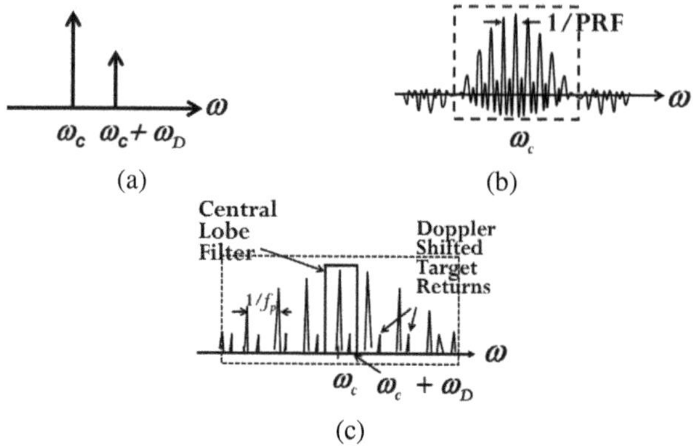

Figure 8.19 (a) Spectrum of Doppler shifted CW signal. (b) Coherent pulse train spectrum (fixed target – no Doppler). (c) Expanded central lobe region with target Doppler shift. *Source:* Ref. [93].

Table 8.2 PRF determined Doppler and range ambiguities. *Source:* Ref. [93].

PRF	*Range*	*Doppler*
High	Ambiguous	Unambiguous
Medium	Ambiguous	Ambiguous
Low	Unambiguous	Ambiguous

The maximum Doppler frequency delimiting velocity ambiguity is

$$f_{D\_\max} = \pm\frac{f_p}{2} \tag{8.20}$$

which gives an unambiguous velocity

$$v_u = \lambda\frac{f_{D\_\max}}{2} = \pm\lambda\frac{f_p}{4} \tag{8.21}$$

and

$$\Delta v_u = \lambda\frac{f_p}{2} \tag{8.22}$$

The degree of velocity ambiguity is a function of the PRF (Table 8.2).

8.2.10 Track Versus Search

The design of RADARs differs according to whether their intended function is to search for a target within an ample field or to track a moving target.

Search RADARs usually seek targets in the 20–200 km range. They illuminate the target with a high power density beam, which exhibits high

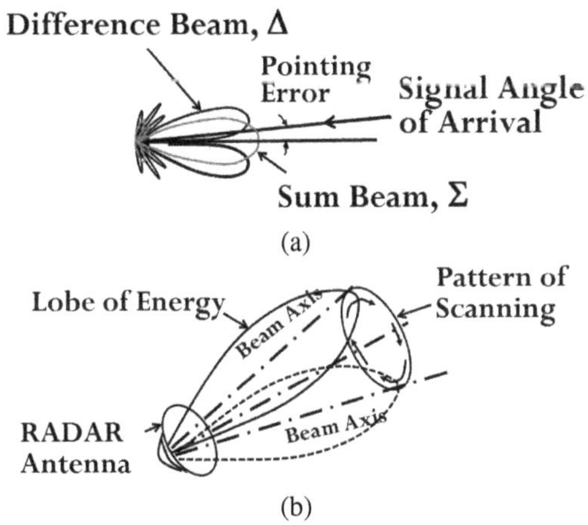

Figure 8.20 (a) Monopulse RADAR technique. In the monopulse radar technique, the pulse transmitted is especially encoded so as to facilitate extracting accurate range and direction information from a single signal pulse. (b) Conical scan. This antenna, used in automatic-tracking radar systems, consists of a parabolic reflector combined with a radiating element which is caused to move in a small circular orbit about the focus of the antenna with or without change of polarization. The radiation pattern is in the form of a beam that traces out a cone centered on the reflector axis. The process is also known as nutating conical scanning. *Source*: Refs. [92, 93, 95].

peak power, long duration pulses, long pulse trains with low PRFs, and large frequency bins. In addition, they utilize high-gain antennas. They may be operated in two ways, namely, at a fast search rate with narrow beams or at a slower search rate with wide beams.

Tracking RADARs measure the precise location of a target and provide information to determine its path and predict its subsequent position. Its design must minimize the DT on the target to achieve rapid processing time and may use special tracking techniques such as monopulse (Figure 8.20(a)), conical scan (Figure 8.20(b)), or beam switching. In this mode, the accurate measurement of the beam angle and of the target range is required.

8.2.11 RADAR Cross Section

For an electrically large, perfectly reflecting surface of Area A viewed directly by the RADAR, the RCS may be defined as

$$\sigma \approx \frac{4\pi A^2}{\lambda^2}. \tag{8.23}$$

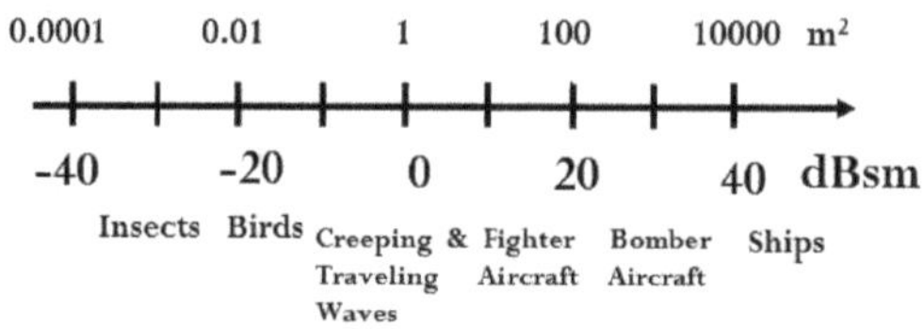

Figure 8.21 Typical values of RCS. *Source:* Ref. [93].

The RCS is usually expressed in decibels relative to a square meter (dBsm) as

$$\sigma_{dBsm} = 10\log_{10}(\sigma). \tag{8.24}$$

Figure 8.21 shows the relative RCS of various objects.

In practice, it is found that certain patterns correspond to given scatterers. The scattered field pattern of a large aircraft is found in [93, 94].

8.3 RADAR Architectures

Thus far, our discussions on RADAR systems have centered on transmitting a pulse train and relating the time delay between the transmitted and received pulse trains to the range; this is the pulse RADAR. In addition to using pulses to determine range, there are other approaches that are employed to determine target range and velocity (Figure 8.22).

In this section, we briefly summarize the principles upon which these RADARs are designed (Figure 8.23) and then present their prototypical architectures.

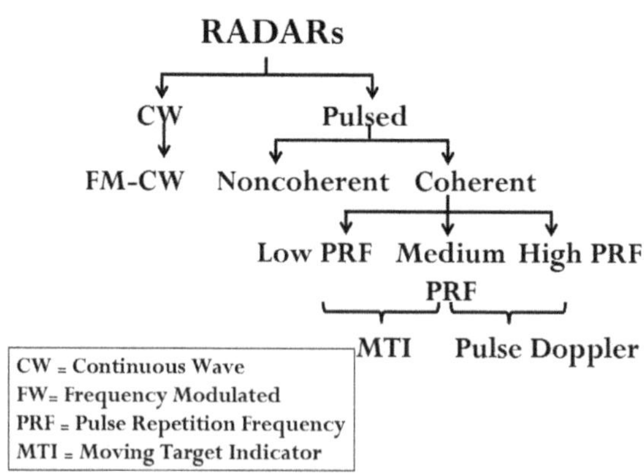

Figure 8.22 Types of RADAR systems. *Source:* Ref. [93].

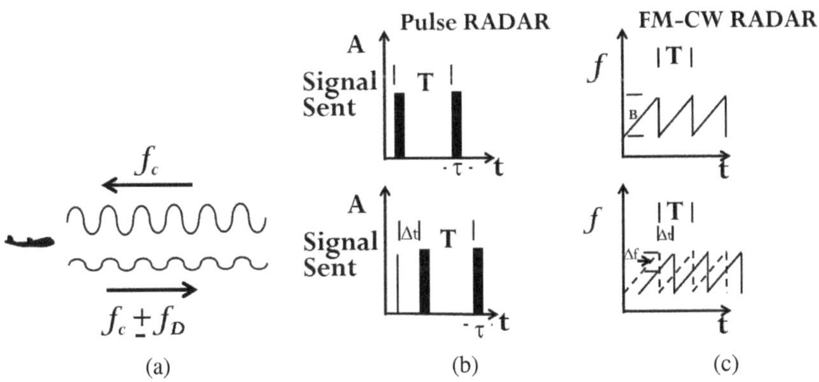

Figure 8.23 Schemes for measuring range. (a) **CW RADAR**: A continuous wave of frequency f_c is transmitted, and the returned echo signal, modified by the Doppler shift, $f_c \pm f_p$ is detected and related to the target velocity via $f_D = -d\frac{V_r}{\lambda}$. (b) **Pulse RADAR**: A periodic pulse of period T_p and duration τ is transmitted, and the time delay from transmission to echo reception, $\triangle$t, is related to the range. (c) **FM-CW RADAR**: A signal, whose frequency is linearly modulated periodically over a frequency bandwidth $\boldsymbol{B}$ is transmitted, and the time delay $\triangle$t and corresponding frequency shift $\triangle$f between the transmitted sawtooth and received sawtooth signals, namely, $\triangle t_t = \triangle f \frac{T}{B}$, is related to the range by, $R = \triangle f \frac{T}{2B} c$. *Source*: Refs. [94, 95].

8.3.1 CW Doppler RADAR Architecture

The CW Doppler RADAR architecture is shown in Figure 8.24. A CW sine wave of frequency f_c is transmitted. Then the received echo of frequency $f_c \pm f_D$ is mixed with f_c, resulting in the extraction of the Doppler frequency shift, f_D.

8.3.2 FM-CW RADAR Architecture

The FM-CW RADAR architecture is shown in Figure 8.25. An modulated FM transmitter transmits a periodic sawtooth signal. The transmitted signal is mixed with the returned echo from the target, it is then amplified, limited, and fed to a frequency counter (calibrated in distance), whose output is fed to a display.

8.3.3 Pulse Doppler RADAR Architecture

The pulse Doppler RADAR architecture is shown in Figure 8.26. In this architecture, the Doppler frequency shift is detected with respect to a transmitted pulse echo, rather than with respect to a transmitted CW echo. It

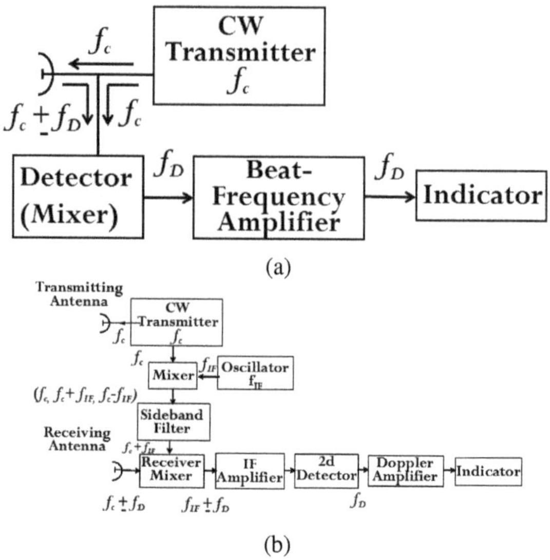

Figure 8.24 (a) Sketch of CW Doppler RADAR. (b) Block diagram of CW Doppler RADAR with nonzero IF receiver, sometimes called *sideband super-heterodyne. Source:* Ref. [95].

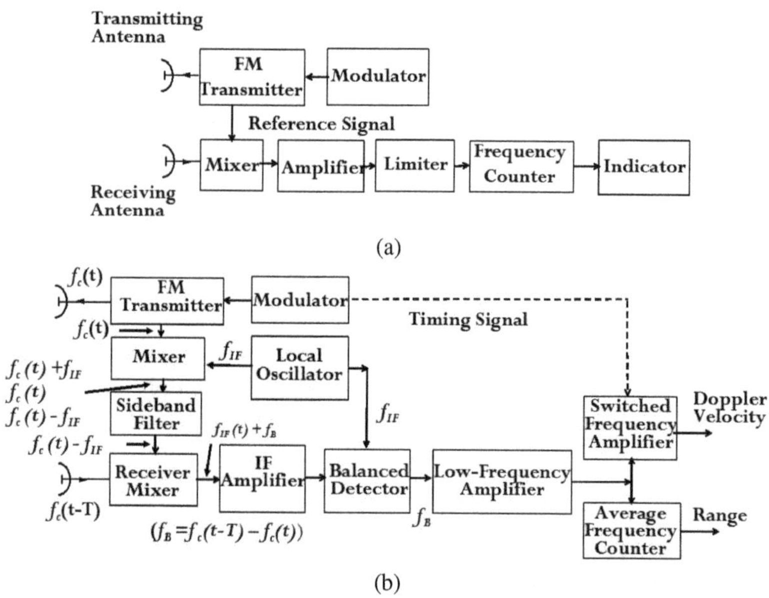

Figure 8.25 (a) Sketch of FM-CW RADAR. (b) Block diagram of FM-CW RADAR using sideband super-heterodyne receiver. *Source:* Ref. [95].

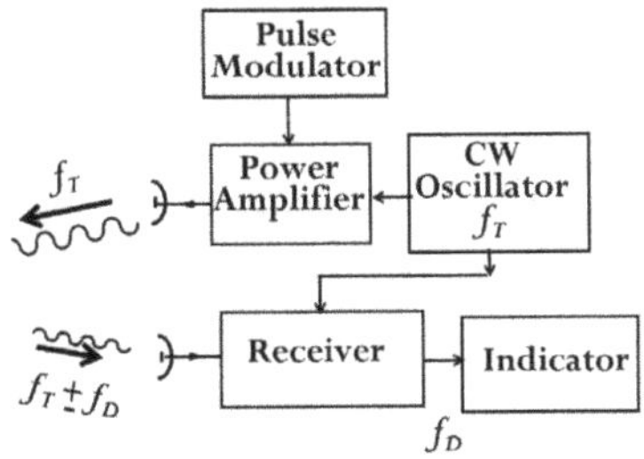

Figure 8.26 Sketch of pulse radar using Doppler information. *Source:* Ref. [95].

is employed to determine the relative velocity with respect to a single target or to discern moving targets that are of interest out of a clutter of undesired stationary targets. This RADAR is also called moving target indicator (MTI).

Having discussed the general principles of RADARs, we will next go on to discuss them in the context of EW.

8.4 ECM Capabilities of an EW RADAR

The ECM capabilities of a RADAR operating in an EW environment focus on target detection and monitoring (or tracking) and on precluding the target from having access to the electronic spectrum. This is effected by producing interference (jamming), making their information ambiguous, or by generating unreal information (spoofing) [90]. To effect target detection, the RADAR must be equipped to obtain knowledge of the nature of the signals being used by the target in order to jam it or deceive it [91]. These capabilities have been specified as follows [90, 91]: 1) searching for signal sources, in particular, in what pertains to their frequency, their azimuth, and their elevation; 2) detecting and identifying characteristic RADAR signals; 3) establishing the importance of a detected signal; and 4) initiating pertinent countermeasures.

8.4.1 Searching for Signal Sources

The search and capture of RADAR signal sources is effected by a system consisting of [92, 93] 1) a receiver, which effects, upon a short acquisition time, the detection of radiation emitted far away and measures its key parameters, namely, (i) *radio (carrier) frequency (RF)*, (ii) *PW or pulse duration (PD)*, (iii) *pulse repetition interval (PRI)*, (iv) *scan pattern and rate*, (v) *beam width and side lobe levels*, (vi) *angle of signal arrival* (AoA), (vii) *signal amplitude* (A), and (viii) *electric field polarization*; 2) a signal processor, which discerns and performs the accurate analysis of complex RADAR signals immersed in a congested EM environment and, for each of them, provides a pulse. Once these signals are analyzed, the *signature* of the parameters from the RADAR that emitted them may be identified and incorporated into a database; 3) a signal recognition system, which extracts the features of the incoming data and feeds it to a classifier where the data is compared with a library of known emitter characteristics to arrive at the possible identities/source of the signals detected. To associate the extracted data, the recognition system organizes it by mapping it to pulse chains that have been associated with the known RADARs. Further processing by the recognition system determines a number of derived RADAR parameters, in particular, PRF and PRI modulation (i.e., agility, stagger, dwell and switch, jitter, etc.). Together, these parameters embody a radar "fingerprint" for future use.

8.4.2 ECM Techniques: Jamming

The jamming physical concept is depicted in Figure 8.27. Next, we discuss various approaches to jamming.

8.4.2.1 Noise jamming

The *noise jamming* approach aimed at a RADAR involves directing EM energy to it in order to disable its normal operation [91]. There are several ways of accomplishing this.

In the first place, there is *spot jamming*, in which all of the EM energy of the RADAR may be focused at a particular frequency (Figure 8.28(a)). In the second place, there is *sweep jamming* (Figure 8.28(b)), in which the RADAR aims its EM energy at multiple frequencies sequentially, one at a time. And in the third place, there is barrage jamming (Figure 8.28(c)), in

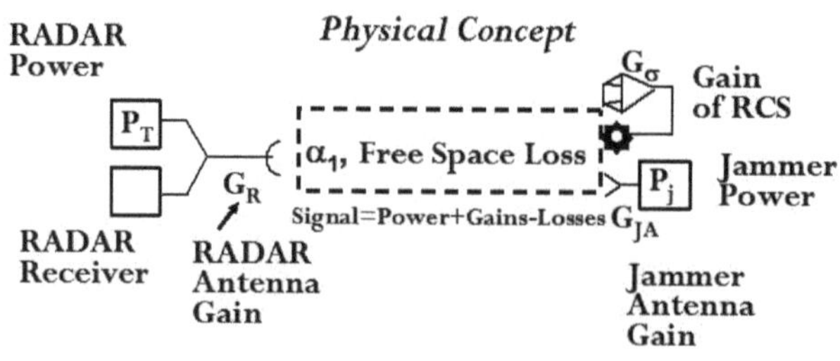

Figure 8.27 RADAR jamming visualized. *Source*: Ref. [92].

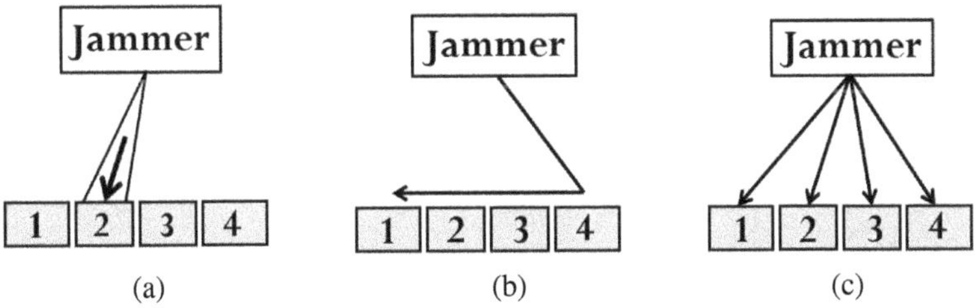

Figure 8.28 (a) Spot jamming, the ECM system targets one frequency. (b) Sweep jamming, one frequency at a time is sequentially targeted. (c) Barrage jamming, a band of frequencies are targeted simultaneously. *Source:* Ref. [91].

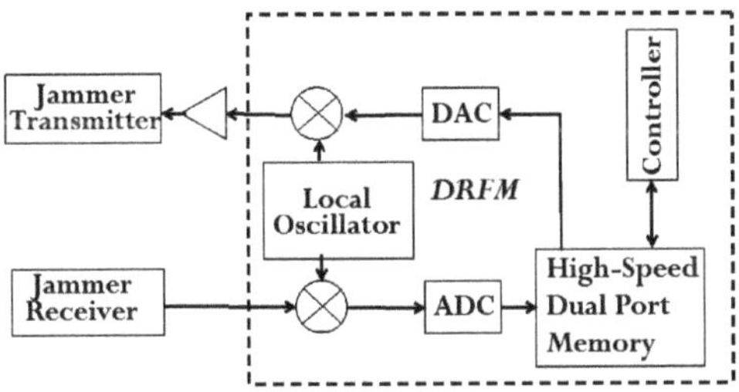

Figure 8.29 Block diagram of DRFM. *Source:* Ref. [91].

which the RADAR aims its EM energy concurrently at multiple frequencies, thus precluding the enemy RADAR from operating at any of the unjammed frequencies.

8.4.2.2 Deception jamming

The deception *jamming* approach aimed at a RADAR involves sending false signals to it in order to deceive it so that it misinterprets the information received and the target information detected is false [91]. The technique of digital radio frequency memory (DRFM) (Figure 8.29) is the main tool employed in deception jamming.

In this device, the incoming signal is first sampled at high speed and stored in a digital memory. Then, the stored signal is mixed or corrupted, and reconstructed, and sent back to the target RADAR. The majority of fighter aircrafts possesses DRFM systems. Next, we discuss several ways of accomplishing deception jamming.

In the first place, there is *range deception* (Figure 8.30), in which the jammer first locks on/tracks the target, uses pulses that are generated in the jammer, and uses time delays to confuse the enemy RADAR by providing

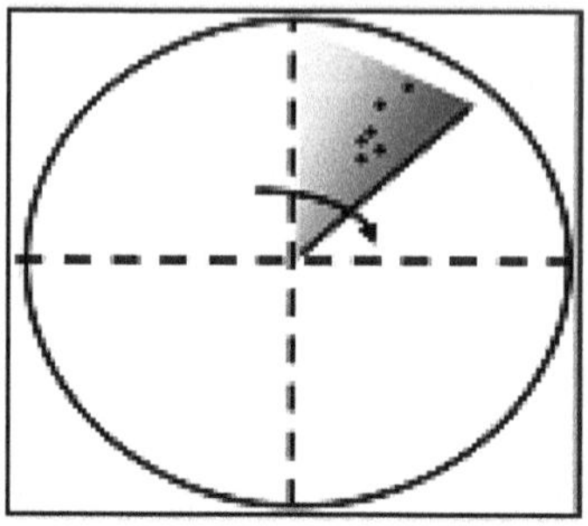

Figure 8.30 Deception jamming, multiple false targets appearing on the Plan Position Indicator (PPI) of the radar. (b) *Source:* Ref. [91].

false information about the original location of the RADAR; multiple targets at different ranges/locations and velocities are seen on the enemy RADAR display. In the second place, there is *velocity deception*, in which the jammer aims at altering the frequency or phase of the signal sent to the enemy RADAR so that it cannot identify the correct location of its target.

The mechanism by which a jammer disables the operation of a target RADAR is by altering its SNR. In particular, in the presence of the jammer, the SNR at the RADAR input is given by [90]

$$\boldsymbol{SJNR} = \frac{\boldsymbol{S}}{\boldsymbol{J}+\boldsymbol{N}} = \frac{\left(\frac{\boldsymbol{P_T G_T \sigma A_{eR} \tau}}{(4\pi)^2 \boldsymbol{R^4 L}}\right)}{\left(\frac{(\boldsymbol{ERP})\boldsymbol{A_{eR}}}{4\pi \boldsymbol{R^2 B_J}} + \boldsymbol{k_B T_0}\right)} \tag{8.25}$$

where ERP is the effective radiated power of the jammer, given by $P_J G_J / L_r$, where PJ, GJ, and LJ are the jammer power, antenna gain, and power loss, respectively. The impact of the jammer power on the enemy RADAR's input SNR (SNJR) is shown in Figure 8.32. It is apparent that, when the jammer power is not present, the SNR is greater than about 25 dBm over a range up to 50 km. However, when a jammer power of 500 kW is present, the SNR decreases to approximately between −50 and −75 dB, in which case it is impossible for the enemy RADAR to operate under the jamming power.

In order for the jamming to be effective, it must transmit a power large enough that it can overcome the power of the enemy's RADAR. This tradeoff is shown in Figure 8.32, where the jammer's and RADAR's peak powers are shown versus range at which they cross over range for a RADAR operating at 6 GHz. It is seen that, in order to be effective, the jammer must come closer to the RADAR. This vulnerability is exploited by ECCM systems to overcome jamming systems. The cross over range is given by [91]

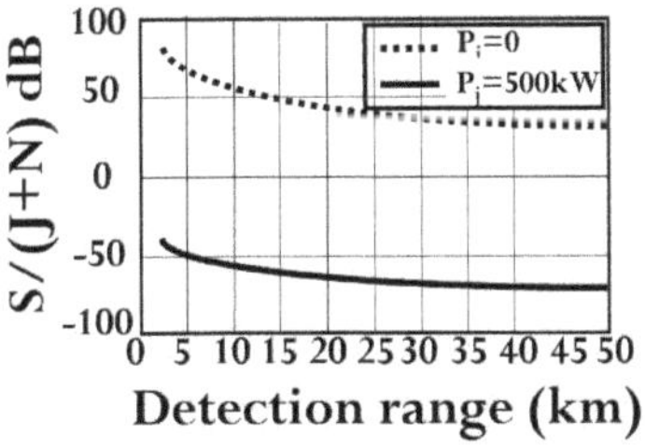

Figure 8.31 Simulation comparing the signal to noise plus jamming ratio (S/(J+N) versus detection range for a RADAR operating at 5.6 GHz. *Source:* Ref. [90].

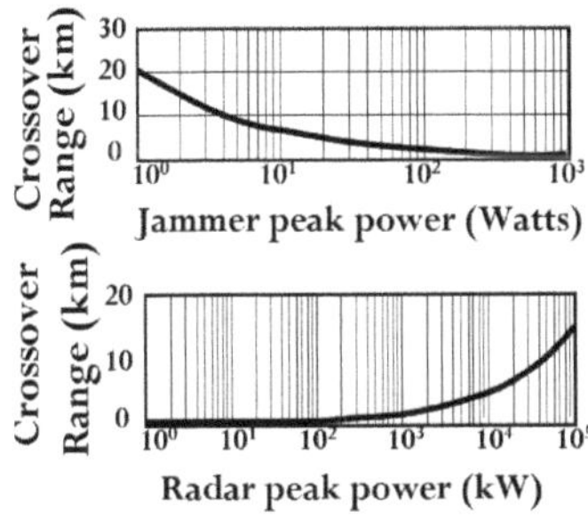

Figure 8.32 Comparison of jammer and radar's peak power vs. cross over range. *Source:* Ref. [91].

$$R_{co} = \sqrt{\frac{P_T G_T \sigma B_J}{4\pi B_R L(ERP)}} \tag{8.26}$$

Next, we address counter–counter measures techniques.

8.4.3 ECCM Techniques

The purpose of ECCM is to neutralize the various ECM techniques. Therefore, it is first necessary to identify the type of ECM encountered so that it can be appropriately countered. Once the decision has been made as to how to counter the ECM, a number of techniques may be employed, see next.

8.4.3.1 Pulse compression

In the *pulse compression* technique, the RADAR return echo is processed by the receiver to add a delay as a function of frequency so that it appears stronger. In particular, a signal strength greater than that employed for noise jamming is produced. In addition, the produced noise signal's frequency variation (chirp) will usually be different. As a byproduct of pulse compression, it is seen that, since the range resolution ΔR is given by

$$\Delta R = \frac{c\tau}{2} = \frac{c}{2B} \tag{8.27}$$

where $\boldsymbol{B}$ is the bandwidth, and τ is the PW, high resolution, while keeping the average transmitter power reasonable, is attainable [91].

8.4.3.2 Frequency hopping

In the *frequency hopping* technique, the carrier frequency is switched at a high rate so that it becomes difficult for the jammer to jam a particular frequency within a certain time interval. The carrier frequency is, in addition, changed at random so that the sequence of transmission frequencies cannot be anticipated. The technique is mainly employed in the context of barrage jamming because it makes the jammer distribute its finite power among multiple frequencies, reducing its magnitude. Furthermore, since the RCS of large objects such as aircraft is a strong function of frequency, it will drastically vary, i.e., at some frequencies being small and at others being large. In this context, a RADAR operating at a single frequency may lose detection if it coincides with a small RCS frequency, but, on average, the overall probability of detection will be larger if multiple frequencies are employed.

8.4.3.3 Side lobe blanking

Normally, a RADAR antenna is pointed in the direction of interest as established by its main radiation pattern lobe. The antenna beam pattern, however, does contain side lobes of lower gain that point in directions different from that of the main beam. In *side lobe blanking*, the EM energy aimed at jamming a RADAR through the signal received by its main lobe may be so strong that it is also jammed by the EM energy received via its side lobes. The approach to overcome this side lobe jamming is to use an omni-directional antenna so that the signals received via the main lobe and that received via the omni-directional antenna are compared. If they are equal, no side lobe jamming is present as the main lobe signal is the desired signal [98].

8.4.3.4 Polarization

In this ECCM technique, the polarization of the received signal is employed to filter out the jamming signals. In particular, when the RADAR's receiver does not have the same polarization as the jamming signal, it cannot be detected by the receiver, thus becoming ineffective. The protection of the receiver may be extended by using multiple antennas with different polarizations. Figure 8.33 illustrates the concept of polarization.

8.4.3.5 Artificial-Intelligence-Based Jammer-Nulling

This ECCM technique finds application in modern cognitive RADARs (CRr) (Figure 8.34) [99].

A CRr is a closed-loop knowledge-aided dynamic RADAR that is architected to achieve high performance in the context of spectrally congested EW environments. The fundamental idea of this ECCM concept is to

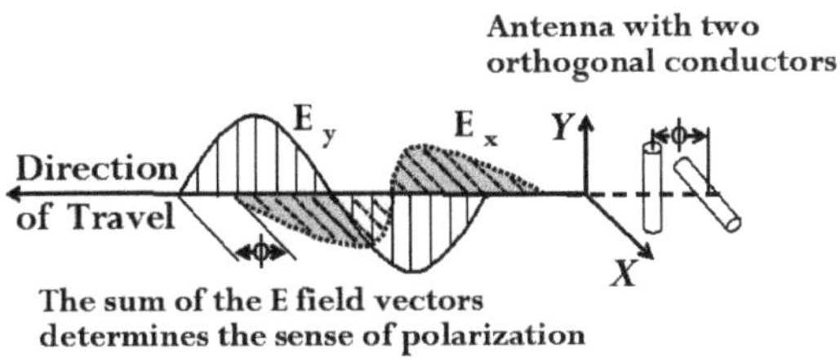

Figure 8.33 Block diagram of signal model and cognitive radar platform for target recognition using jammer-nulling adaptive matched waveform design. *Source:* Ref. [99].

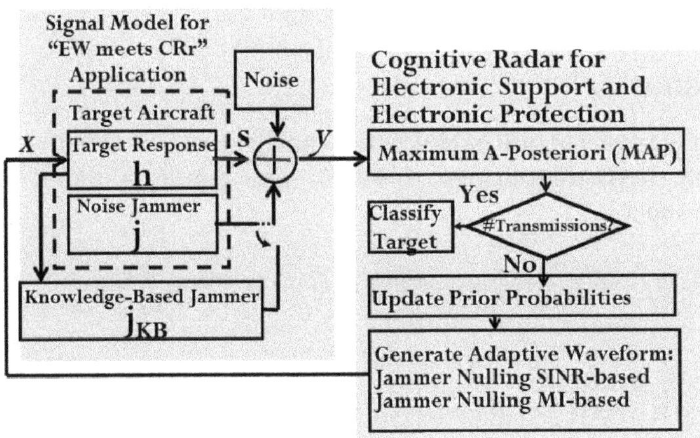

Figure 8.34 Block diagram of signal model and cognitive radar platform for target recognition using jammer-nulling adaptive matched waveform design. The received signal plus noise jammer interference and AWGN vector is y = s+j+w where the received target echo *s* is the convolution of the transmit waveform and target response (s= x*h) j_{KB} represents the noise jammer. *Source:* Ref. [100].

exploit target recognition, including the target RCS response, its narrowband, frequency sweep, and basic noise jamming waveforms to synthesize and transmit adaptive jamming-nulling waveforms to counter them while continuing operation.

8.5 Summary

This chapter has dealt with the topic of aerospace/EW RADAR. We began by addressing the principles of RADARs, in particular, the types of RADAR, radio detection and ranging, the RADAR-target geometry/coordinate system, based on which the target location can be understood in the display. Then, we discussed the topic of RADAR pulses and their relation to range ambiguities, range resolution, and range gates. This was followed by the topic of RADAR sensitivity and Doppler shift, and we explained the tracking versus search modes of operation. Next, we discussed RCS. We focused then on the systems aspects of RADARs such as their architectures for various types, in particular, CW Doppler, FM-CW, and pulse Doppler. We then went on to discuss how RADARs effect ECM, including, searching for signal sources, noise jamming, and deception jamming, and then ECCM techniques such as pulse compression, frequency hopping, side lobe blanking, polarization, and AI-based jammer-nulling.

8.6 Problems

1. Explain the concept of RADAR pulse ambiguity.
2. Two-way RADAR equation (bistatic). The power at the RADAR receiver input is

$$\boldsymbol{P_R} = \frac{\boldsymbol{P_T G_R \lambda^2 \sigma}}{(4\pi)^3 \boldsymbol{R_{TX^2} R_{RX^2}}} = \boldsymbol{P_T G_T G_R} \left[\frac{\sigma c^2}{(4\pi)^3 \boldsymbol{f^2 R_{TX^2} R_{RX^2}}} \right] \quad \text{(P8.1)}$$

   Express p_T in dB.
3. Two-way RADAR equation (bistatic).
   Calculate and plot p_T in dB for all the objects with RCS given in Figure 8.24, for $R_{TX} = R_{RX} = 5000$ m, for frequencies f from 1 to 20 GHz.
4. Can a RADAR system be used also for communications?

9

Tutorials

9.1 Introduction

The tutorials presented in this book are conducted with the Keysight PathWave System Design (*SystemVue*) electronic system level (ESL) design software. Should you need a SystemVue license, you may contact Keysight (please follow this link: **https://www.keysight.com/find/delossantos-book-eval**) to obtain a free time-limited evaluation license. In the tutorials, you will gather practical experience with slightly simplified real-world design examples of 5G, MIMO, and aerospace/EW RADAR systems. The tutorials cover state-of-the-art system applications and consider the characteristics of typical radio frequency (RF) hardware components. The tutorials will help you in the following:

- enhancing your comprehension of the theory discussed by immersing yourself in practical examples;
- familiarizing yourself with the performance of real RF hardware components and how they impact system limitations;
- start learning to use a state-of-the-art software tool for communications system-level design.

Before starting a tutorial, or while working through them, you may consult the pertinent theoretical chapters of the book. The chapters accompanying each tutorial are indicated. The presented screenshots and results have been taken with the **Version 2020 Update 1 of SystemVue**. Familiarization of the student with the current SystemVue version will be preparation for understanding the "look and feel" of subsequent versions of it.

In this chapter, we begin by addressing the principles of beamforming antennas, in particular, their prototypical architecture, geometry/coordinate system, and resulting beam visualization.

9.2 Tutorial 1: Introduction to SystemVue and Basic Phased Array (Beamforming) Analysis

Learning objective: To enable the reader to construct a simulation model in SystemVue, parameterize the components, run the simulation, and read the important quantities from the simulation results.

Prerequisites: Basic knowledge of properties of amplifiers, attenuators, phase shifters, and antenna performance (Chapters 2–5).

Design task: A basic transmitter beamforming antenna amenable to 5G/MIMO applications is implemented considering the characteristics of real RF hardware.

9.2.1 Preliminaries

To run an **RF Phased Array** simulation, you need to:

1. Define a phased array design in a *schematic*:
   For a design to be considered a phased array design, it must contain parts from the **RF Phased Array category** of the **RF Design Library**. Other parts from the **RF Design Library** as well as other **RF libraries** (Analog Devices Inc. library, Mini Circuits library, or any X-MW library) can also be used in a phased array design. Furthermore, additional requirements need to be satisfied in order for the design to be considered a valid phased array design amenable to simulation. These include the following.
   (i) The phased array design must have exactly one component representing the antenna array, the **ArrayAnt** model, and one representing the source signal, the **ArrayPort** model.
   (ii) The RF Phased Array design must be a **single cascade of models**. The **ArrayAnt** model and one **ArrayPort** model must be located **at the two end points of the cascade.** No branches are allowed unless they converge back to the single cascade. As an example, switches can be used to create two branches to model a transmit/receive module.
   (iii) In order to achieve practical simulation speeds for large arrays, the parts from the **RF Design Library** are replicated N times **internally** for an N element array. **This means that users are unable to customize individual parameters for the replicated**

parts. For instance, if the user wants to customize the gain for each amplifier, this can be done using an RF Design Library RF amplifier that has constant gain for every element followed by a phased array attenuator (ArrayAttn) using a Custom Window to specify the attenuation of each of the N elements.

(iv) Smart Components:

Although a phased array system can have hundreds or thousands of parts and paths, SystemVue's special "smart" **RF Phased Array models** allows the user to create the system as a simple **single chain**. Under a phased array simulation, the **RF Phased Array models** as well as other **RF models** used in the design **know how to replicate themselves to match the number of paths at any point in the system based on the splitter/combiner stages before and after them**. This allows a very easy and compact representation of the system regardless of the number of elements the antenna has. This is a facility that saves time and avoids errors.

2. Invoke **Phased Array Analysis**:

   (i) The **Phased Array Analysis** specifies the simulation mode (whether the design will be simulated as a receiver or transmitter).
   (ii) The measurements to be performed (most measurements are common for Rx and Tx, but there are a few that are specific to the simulation mode).
   (iii) Any additional simulation settings.
   (iv) Transmitter (Tx) phased array simulation:

   - The **ArrayPort** model is considered the input and the **ArrayAnt** model is considered the output.
   - The splitter (ArraySplit) stages between ArrayPort and ArrayAnt must split the single signal coming out of ArrayPort to a number of signals equal to the antenna elements.
   - For a **Tx** phase array, the ArraySplit model cannot be used as a combiner.
   - Simulation frequency: The phased array simulation is a frequency-domain simulation performed at a **single frequency**. In the Tx mode, the phased array simulation frequency is specified in the **Freq** parameter of the ArrayPort model. Frequency sweeps can also be used to obtain the system performance over a frequency range.

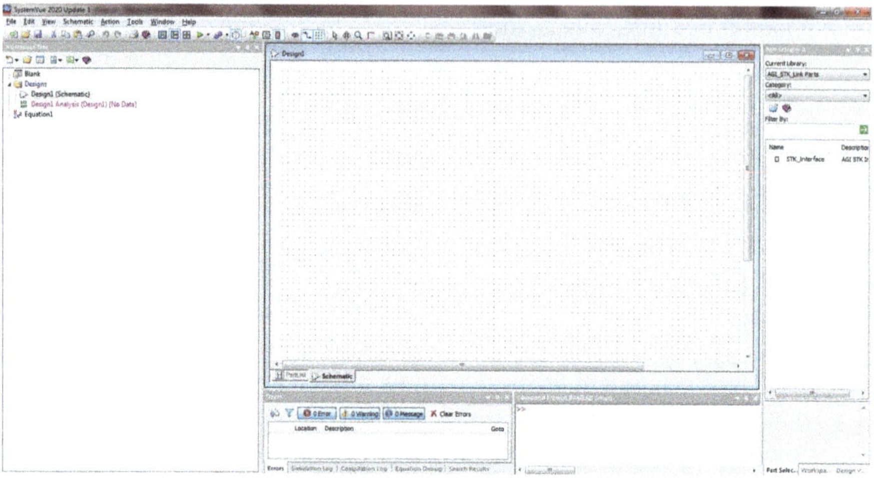

Figure 9.1 New empty workspace in SystemVue.

9.2.2 Getting Started and Schematic Window

After the startup of SystemVue, a window with a welcome message appears. Click on the *Close* button to get to another dialog window (*Getting Started with SystemVue – Please Select an Action*). In this window go to the section *Create a NEW workspace from a template* and select *Blank* to create a new empty design, then press *OK*. Maximize the *Design1* window in the center. Your screen should now look similar to Figure 9.1.

In the *Workspace Tree* column on the left side, new windows can be created, which will become important later to display the simulation results. In the *Part Selector A* column on the right side, the available component models are displayed. Under *Current Library*, select *RF Design* to see the available RF component models. To get an overview of all available models, you may browse through the complete list.

9.2.2.1 Implementation of Basic Phased Array (Beamforming) Antenna Model

To get started with SystemVue, we implement a basic transmit (Tx) phased array (Beamforming) antenna model consisting of a transmitter (Tx) to generate the signal to be radiated, a power splitter to provide a path toward each of the antenna elements in the array, a variable attenuator to adjust the amplitude of the signal exciting each antenna element, a phase shifter to adjust the phase of the signal exciting each antenna element, an amplifier to compensate for the loss in the phase shifter, and an array of antennas. To set up the design in SystemVue, the required components simply need to be dragged from the *RF Design* Library into the *Schematic* window and connected.

We now, in principle, should proceed to define a **Tx** phased array design. However, Keysight has already developed an RF phased array template, which we will depart from. This RF phased uniform linear array (ULA) (Figure 9.2) is contained in workspace Tutorial **5G_MIMO_Beamforming_ULA_1 x 4**.wsv.

The ULA has an N=4 element antenna array (see Figure 9.2) and has the following components, from left to right.

Phased Array RxPort/TxSource (ArrayPort): This model generally represents the Rx output port or Tx source port of a phased array. By default, upon dragging this component into the Schematic window and clicking on it, you may find the **RxTx** parameter to be set to **1:Rx** in the ArrayPort Properties window. Click on the **1:Rx** and select **2:Tx**. This model represents the Tx source port and uses the parameters (**TxPwrIn** and **Freq**), which will be the source power and frequency specific to Tx. In the Workspace file shown in Figure 9.2, the source parameters are defined as: RxTx = Tx; Freq = 10 GHz [Fsignal]; TxPwrIn = –10dBm; Phase = 0°; MainCarrierIndex = 1; PORT = 2. All parameter definitions are accessed by clicking on the icon and then clicking on the *Model Help* button.

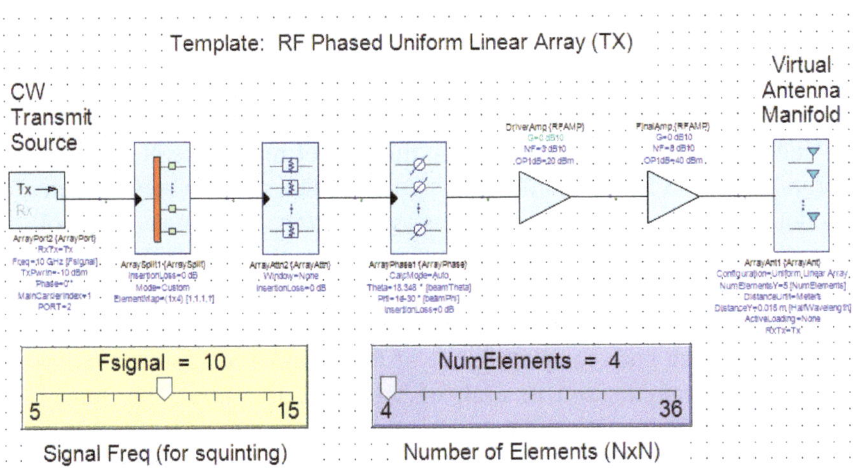

Figure 9.2 Phased uniform linear array template in SystemVue.

Power Splitter (ArraySplit): This model is used as a splitter/combiner in phased array antenna (PAA) designs. When in Tx mode, this model will act as a splitter. Under Tx mode, the signal should travel from the Tx port to the "splitter" and then finally to the antenna array. In this case, the ArraySplit model is used under split mode, and the signal should go from the common port (small triangle) of the multiport side (multiple green boxes). Note that depending on how the user draws the schematic, the signal does not need to flow from right to left for Tx mode. In the Workspace file shown in Figure 9.2, the splitter properties are: InsertionLoss = 0 dB; Mode = Custom; ElementMap = (1 × 4) [1 1 1 1]. The **Mode=Custom** option allows custom splitter/combiner output/input mapping. The parameter **ElementMap** is an integer array that specifies this mapping. This mapping mode can be used with any antenna array **configuration** (*Custom*, *ULA*, *uniform rectangular array (URA)*, *Circular Array*, etc.). All parameter definitions are accessed by clicking on the icon and then clicking on the *Model Help* button.

As a splitter, **ElementMap**(i) specifies the input signal index that the *i*th output signal is split from. The size of the **ElementMap** must be equal to the number of output signals. For example, suppose **ElementMap** = [1 1 2 2 1 1 2 2 2 2]. This map defines 2 splitters to split 2 input signals into 10 output signals. The first splitter is 1-to-4 split, and it splits the first input signal into the first, second, fifth, and sixth output signals. The second splitter is 1-to-6 split, and it splits the second input signal into the third, fourth, seventh, eighth, ninth, and tenth output signals. In our (1 × 4) ULA example, ElementMap = (1 × 4) [1 1 1 1] means that we have one splitter, whose single input is split into four outputs. All parameter definitions are accessed by clicking on the icon and then clicking on the *Model Help* button.

Attenuator (ArrayAttn): This model is used as an amplitude taper for PAAs. Under the Phased Array Analysis, this model represents N different attenuators, where N is the number of signals at the specific stage where this model is connected (for a single-stage system, only one stage of combiners/splitters N is the number of antenna elements). In the Workspace file shown in Figure 9.2, the attenuator properties are: InsertionLoss = 0 dB and Window = None. This means that in (1 × 4) ULA, we have one attenuator with four paths, each path connecting one of the inputs to its corresponding output, where the attenuation in each of the paths is 0 dB. This model represents an attenuator employed to model an amplitude taper (i.e., the signal attenuation in the path to each antenna varies in a prescribed fashion) for PAAs. Under the Phased Array Analysis, this model represents N different attenuators, where N is the number of signals/signal paths being attenuated. In our example, N is the number of antenna elements. This model simulates the behavior of a real attenuator. All relevant parameters (e.g., insertion loss, 1-dB compression point, etc.) can be set by the user. Clicking on the component icon, a window where access to setting its properties opens up.

In general, the Window parameter may be set such that the distribution of attenuations among the paths varies from the first path to the last path in a prescribed way, such as None (our case here), Bartlett, Hann, Hamming, Blackman, GeneralizedCosine, Ready, BlackmanHarris, Taylor, Custom, and Gaussian. All parameter definitions are accessed by clicking on the icon and then clicking on the *Model Help* button.

Phase Shifter (ArrayPhase): This model is utilized to apply a phase shift to the individual signals in a PAA system. Under the Phased Array Analysis, this model represents N different phase shifters, each one driving one of the N signals in the antenna array. In the Workspace file shown in Figure 9.2, the phase shifter properties are: CalcMode = Auto; Theta = 18.348° [bea,Theta]; Phi = 1e-30° [beamPhi]; InsertionLoss = 0 dB, corresponding to each of the N = 4 antenna elements. This model simulates the behavior of a real phase shifter. All relevant parameters (e.g., phase shift, insertion loss, etc.) can be set by the user. The CalcMode parameter specifies how the phase shifts of the individual phase shifters will be computed.

- When **CalcMode** = *Auto*, the phase shifts are calculated so that the resulting beam points in the direction specified by the **Theta** (Boresight zenith angle of the array) and **Phi** (Boresight azimuth angle of the array) parameters (see Figure 9.3 for the definition of **Theta** (θ) and **Phi** (φ) in a spherical coordinate system). In this case, the array **Configuration** (specified in the ArrayAnt model) is also taken into consideration in the calculation of the phase shifts. For this reason, **CalcMode** = *Auto* can **ONLY** be used in a single combining/splitting stage system. In a multi-combing/splitting stage system, phase shifters can be used at different stages; however, they **MUST** use **CalcMode** = *Manual.*

All parameter definitions are accessed by clicking on the icon and then clicking on the *Model Help* button.

DriverAmp (RFAmp): This model simulates the behavior of a real amplifier. All relevant parameters (e.g., gain, noised figure, 1-dB compression point, etc.) can be set by the user. This model is used here as a first stage to boost the power of the signal driving the antenna array. In the Workspace file shown

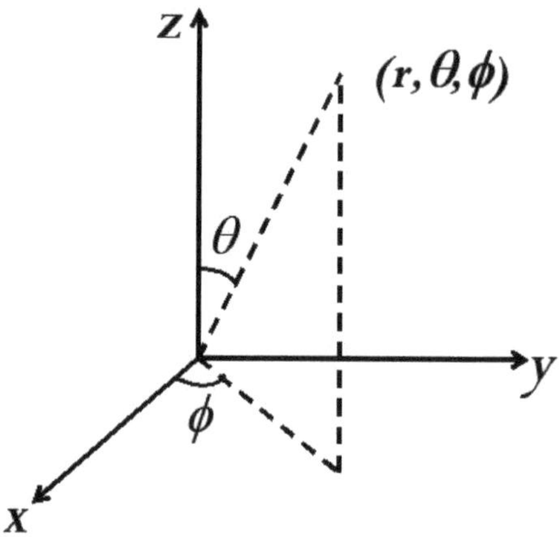

Figure 9.3 Definition of Theta (θ) and Phi (ϕ) in a spherical coordinate system. $\theta = 90°$ and $\phi = 0°$ is the direction along the y-axis.

in Figure 9.2, the amplifier parameters are: Gain G = 0 dB10; NF = 3 dB10; OP1dB = 20dBm. All parameter definitions are accessed by clicking on the icon and then clicking on the *Model Help* button.

OutputAmp (RFAmp): This model simulates the behavior of a real amplifier. All relevant parameters (e.g., gain, noised figure, 1-dB compression point, etc.) can be set by the user. This model is used here as the output power stage driving the antenna array elements. In the Workspace file shown in Figure 9.2, the amplifier parameters are: Gain G = 0 dB10; NF = 5 dB10; OP1dB = 40dBm. All parameter definitions are accessed by clicking on the icon and then clicking on the *Model Help* button.

Antenna (ArrayAnt): This model represents an antenna array. Under the Phased Array Analysis, the Rx/Tx parameter of the model must be consistent with the **Rx/Tx parameter in the General tab of the Phased Array Analysis**. When the Phased Array Analysis is set to Rx mode, this model represents the Rx antenna array and uses the parameters (RxPwrDensity, Freq, SkyTemperature, etc.) specific to Rx. When the Phased Array Analysis is set to Tx mode, this model represents the Tx antenna array. All relevant parameters (e.g., insertion loss, 1-dB compression point, etc.) can be set by the user. Clicking on the component icon, a window where access to setting its properties opens up. In the Workspace file shown in

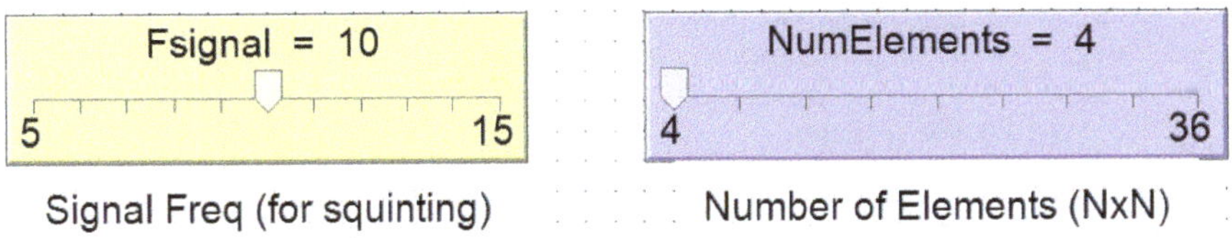

Figure 9.4 Annotations containing tuning parameters.

Figure 9.2, the antenna parameters are: Configuration = Uniform Linear Array; NumElementsY = 4, DistanceUnit = Meters, DistanceY = 0.015m [HalfWavelength], ActiveLoading = None, RxTx = Tx, PORT = 1. All parameter definitions are accessed by clicking on the icon and then clicking on the *Model Help* button.

Tuning: The reader will have noticed that, in Figure 9.2, in addition to the schematic, there are two "Annotation" boxes (Figure 9.4).

These annotation boxes define the *tunable variables* "Fsignal" and "NumElements." As the "handles" are moved between the minimum and maximum values in their respective ranges, simulations are executed and the results, plotted on the same graph, enable one to observe and compare the effects of the parameters varied.

The overall look of the workspace for this Tutorial **5G_MIMO_ Beamforming_ULA_1 x 4**.wsv is shown in Figure 9.5.

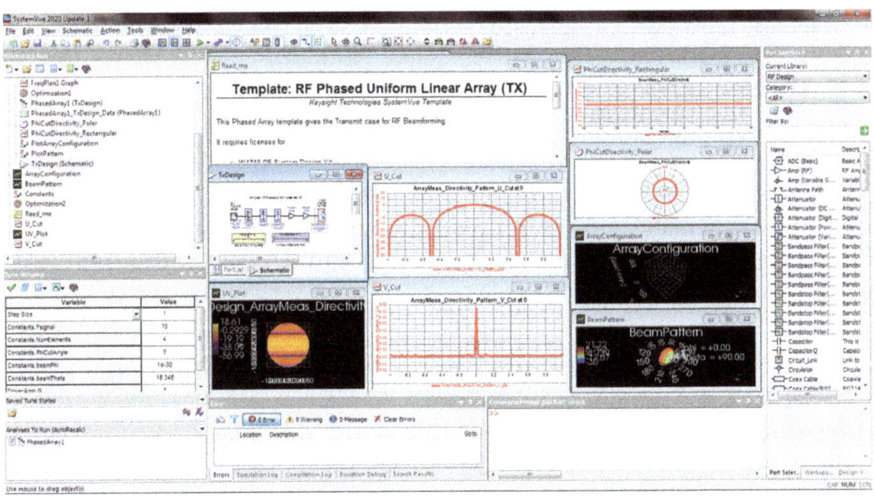

Figure 9.5 SystemVue screen of workspace file 5G_MIMO_Beamforming_ULA_1 x 4.wsv.

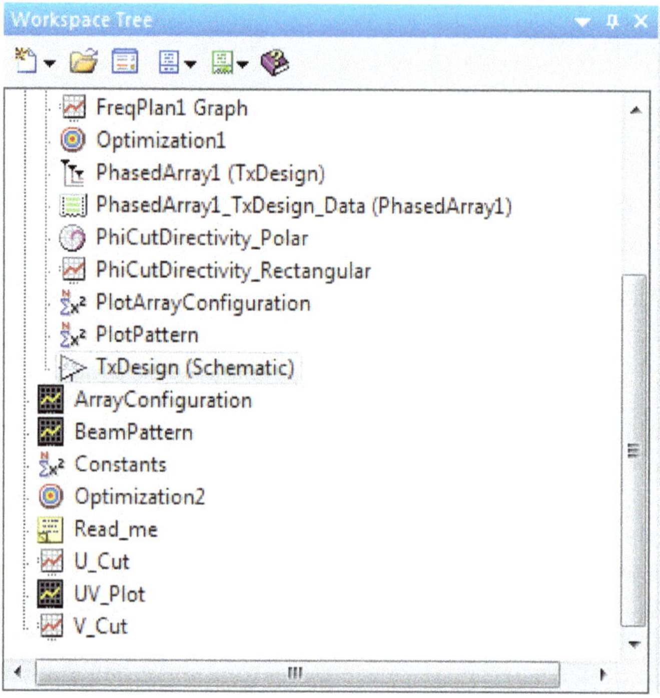

Figure 9.6 Workspace Tree of workspace file 5G_MIMO_Beamforming_ULA_1 x 4.wsv.

On the upper left-hand corner (Figure 9.6), we have the "Workspace Tree," which displays the various simulations performed and plots generated.

Below the Workspace Tree is the "Tune Window," which shows the tunable variables and their values. The simulation results (e.g., plots/graphs) for every set of tunable variable values may be assigned a name in the "Save Tune States" and saved in a file with such a name.

At the center of the screen (Figure 9.8), we have several sub-windows containing the schematic and various plots whose calculation has been set.

On the right-hand side of the screen (Figure 9.9) is the "Parts Selector" pallet, from where the components inserted in the schematic are picked.

9.2.2.2 Running the Workspace file 5G_MIMO_Beamforming_ULA_1 x 4.wsv.

To run a simulation, click on the left tree and hover the cursor over the leftmost icon (Figure 9.10(a)), which will show "New Items" (Figure 9.10(b)).

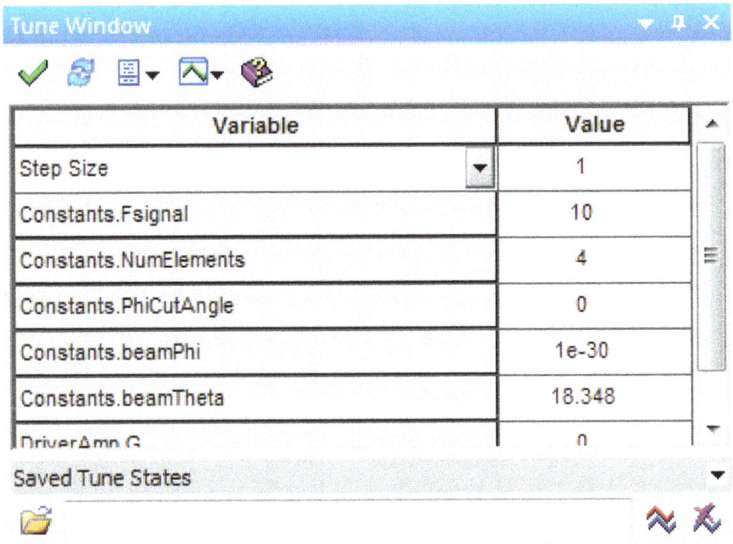

Figure 9.7 Tune Window in workspace file 5G_MIMO_Beamforming_ULA_1 x 4.wsv.

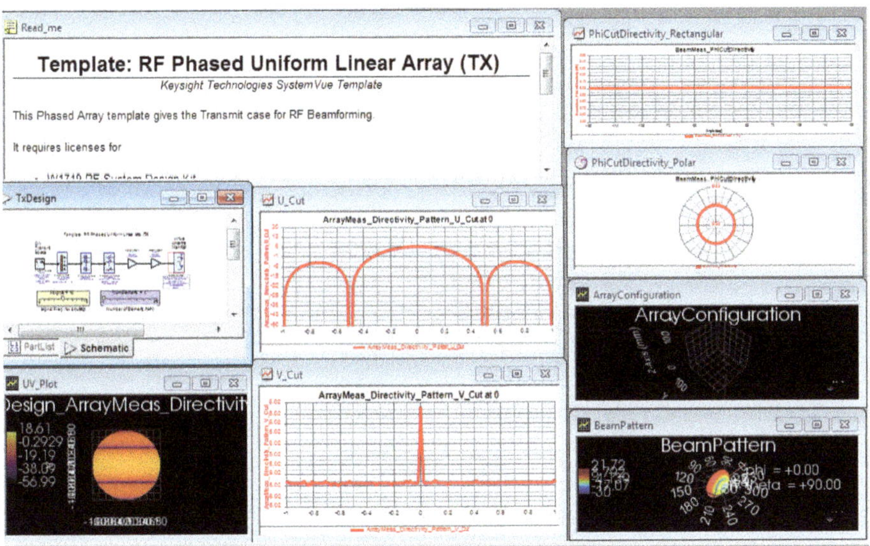

Figure 9.8 Schematic and Results windows in workspace file 5G_MIMO_Beamforming_ULA_1 x 4.wsv.

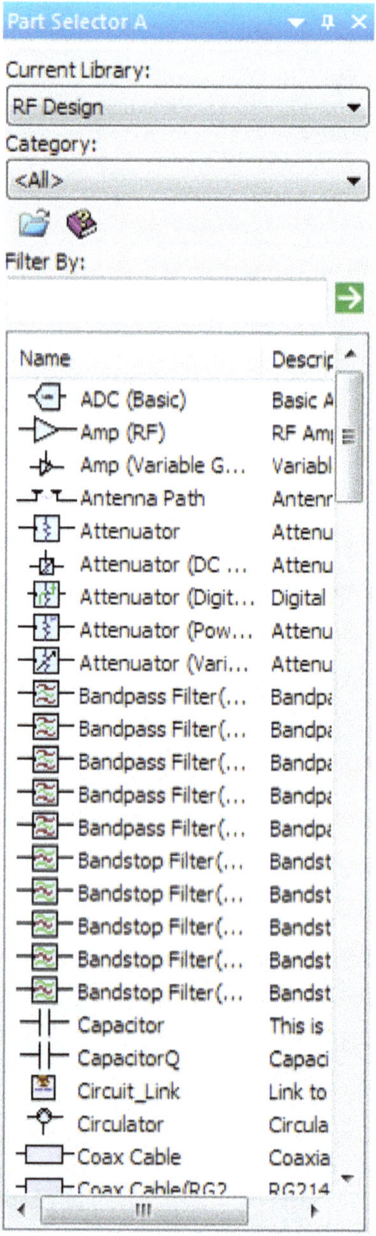

Figure 9.9 Parts selector pallet windows in workspace file 5G_MIMO_Beamforming_ULA_1 x 4.wsv.

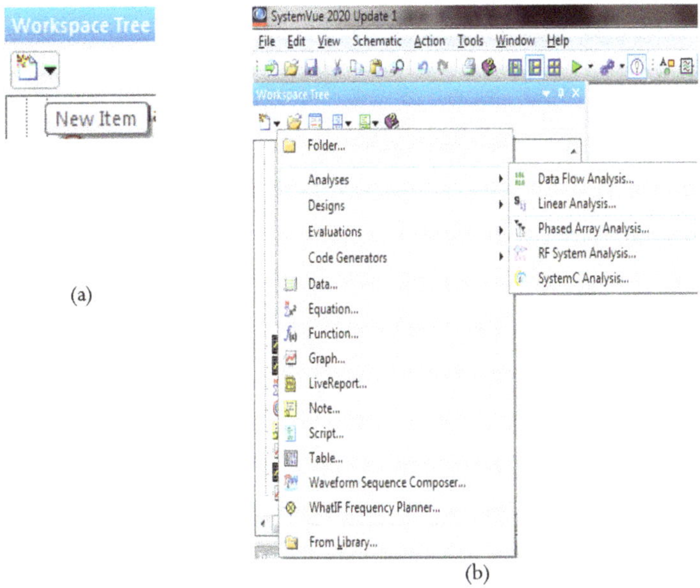

Figure 9.10 Steps to run analysis in workspace file 5G_MIMO_Beamforming_ULA_1 x 4.wsv.

Click on "New Items," and from the Menu that opens up (Figure 9.10(b)), click on "Analysis," and, again, from the Menu that opens up, click on "Phased Array Analysis." Then, the window "Phased Array Analysis" opens up (Figure 9.11).

Make sure the RxTx box is set to 2:Tx; if not, set it to that. Click on the button "Calculate Now." While the simulation is running, the window in Figure 9.12 comes up.

The Array Configuration may also be obtained by double-clicking on the ArrayAnt icon of the schematic, which will cause the following window to open (Figure 9.13).

Then, clicking on the "Display Array" button will open a window with the array configuration. The calculated array directivity, and beam pattern, are found by expanding those minimized windows in the central portion of the screen. They are shown in Figure 9.14.

Now, as we saw in Chapter 2, the directivity of a ULA with N antenna elements is D (dB) = $10\log_{10}(N)$. Thus, for N = 4, D(dB) = 6.02 dB. Compare with the simulation result in Figure 9.14.

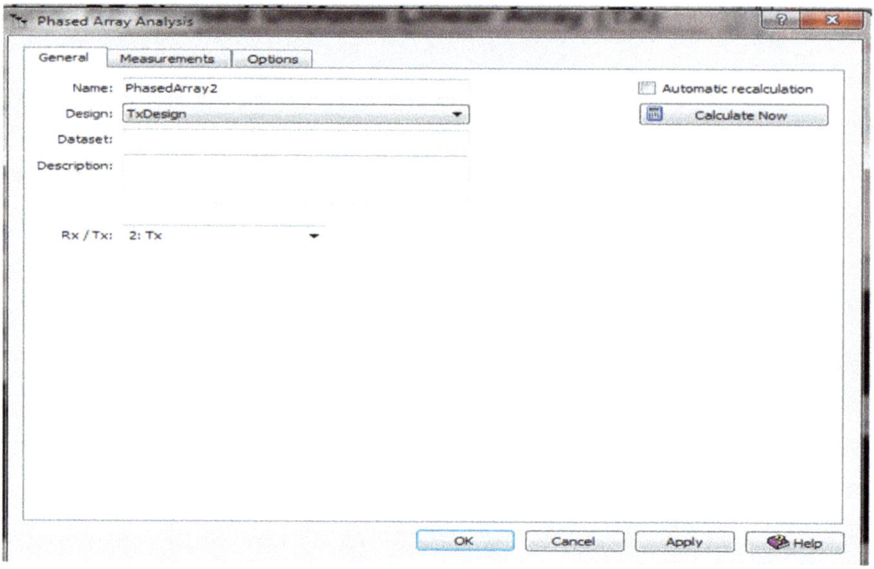

Figure 9.11 Screen that comes up upon clicking on "Phase Array Analysis" in workspace file 5G_MIMO_Beamforming_ULA_1 x 4.wsv.

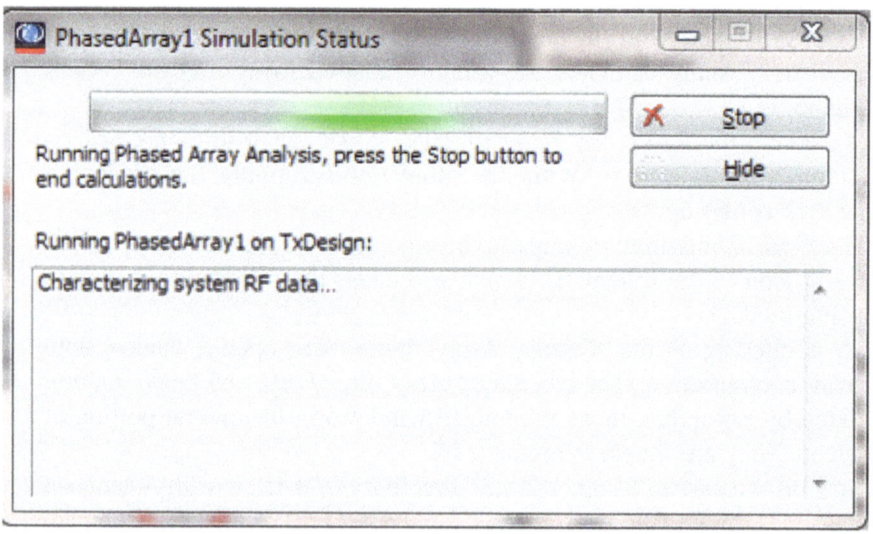

Figure 9.12 Screen that comes up upon clicking on "Calculate Now" in workspace file 5G_MIMO_Beamforming_ULA_1 x 4.wsv.

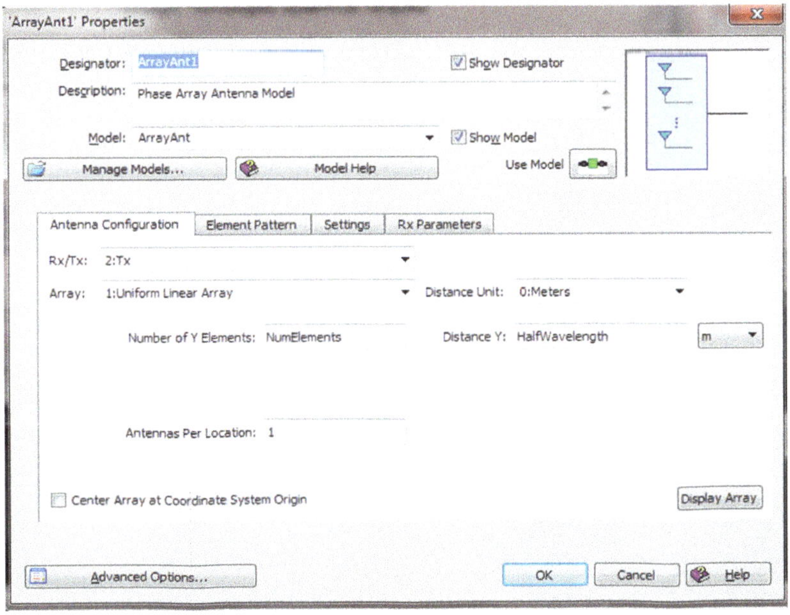

Figure 9.13 Screen showing Array Configuration in workspace file 5G_MIMO_Beamforming_ULA_1 x 4.wsv.

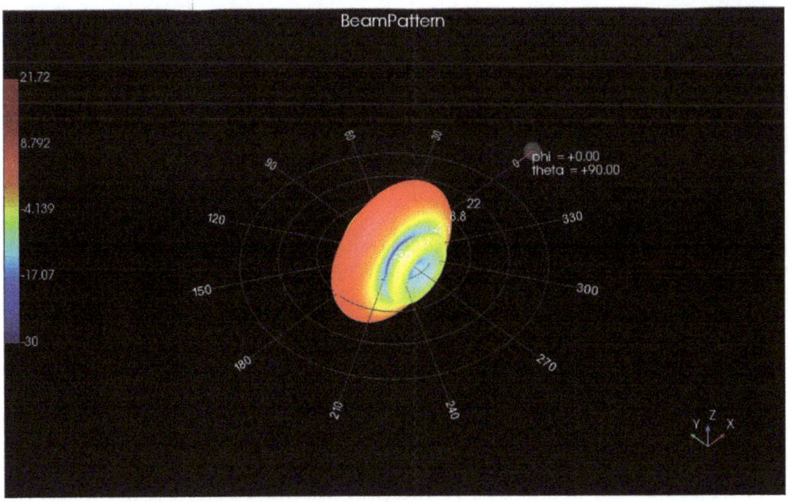

(a)

Figure 9.14 *Continued*

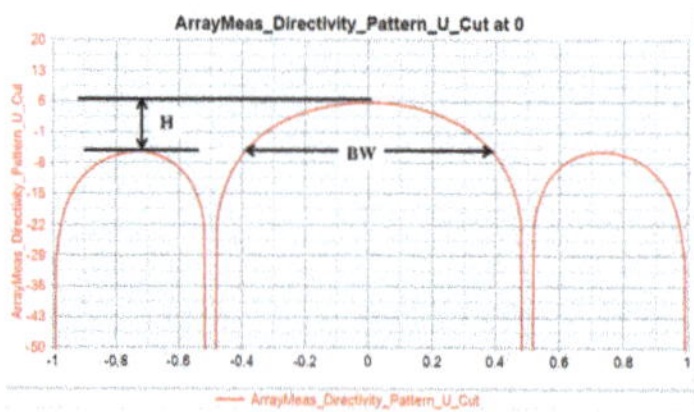

(b)

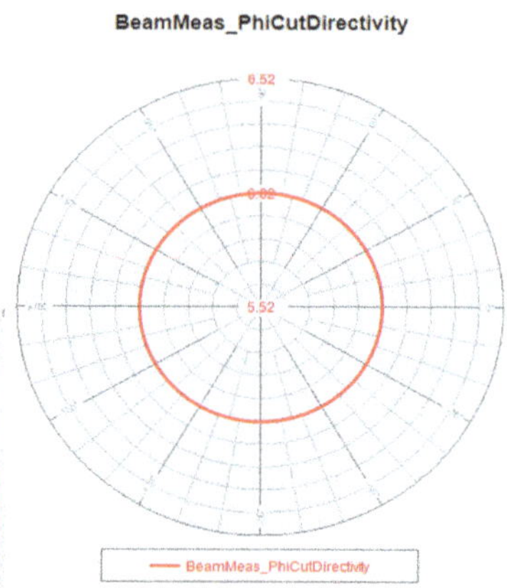

(c)

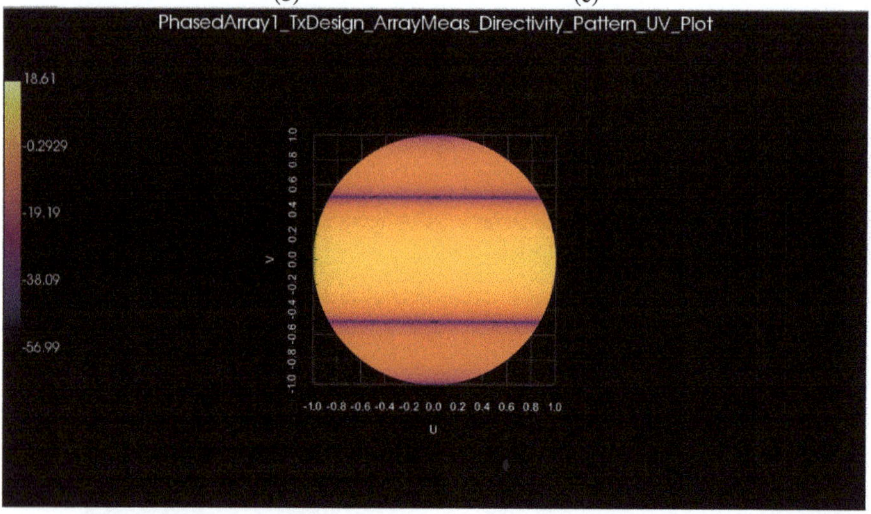

(d)

Figure 9.14 (a) Screen shot of beam pattern of four-element ULA disposed along the y-axis. (b) Directivity along the $90°-270°$ axis cut (U-cut) in (a). (c) Directivity along the $0°-180°$ axis cut (V-cut) in (a). (d) Directivity pattern along direction in (b) vs. that along (c).

9.2.2.3 Effect of Number of Elements on ULA Directivity

To determine how the directivity changes as a function of the number of elements, N, it is useful to employ the "Tune" capability of SystemVue. We will vary N from 4 to 12, in steps of 1, and plot the directivity for the different values of N on the same graph; so we can easily see its effect. We proceed as follows.

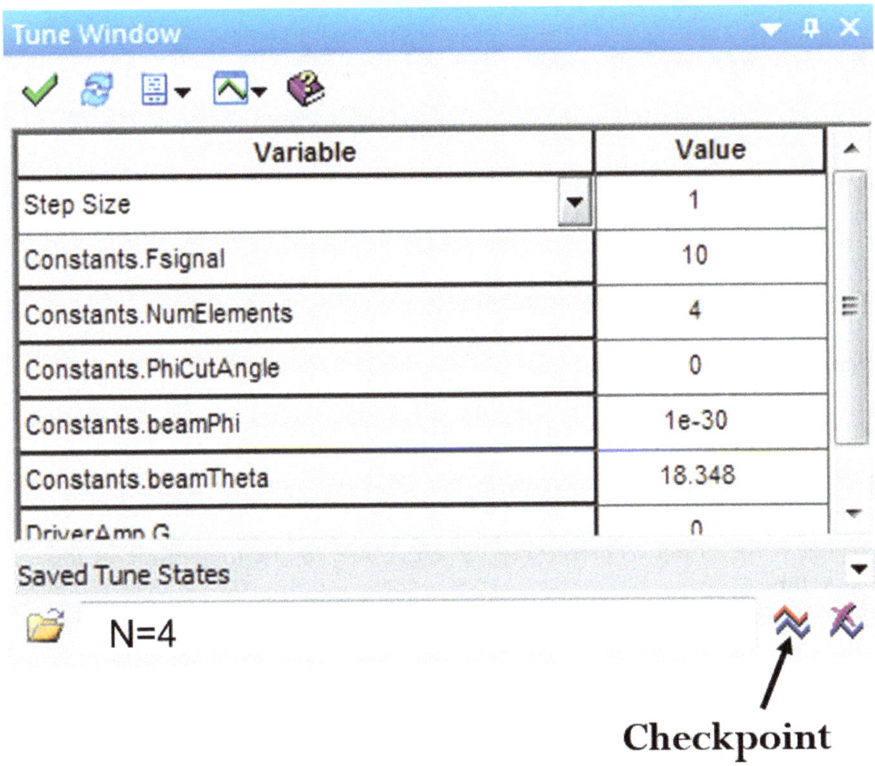

Figure 9.15 "Tune" window.

1) We run the simulation at the initial value, N=4. Then, in the "Tune Window," we enter N=4 to identify those results and click on the "Checkpoint" symbol to save them (Figure 9.15).
2) Next, we go to the schematic window and in the "NumElements" annotation, slide the slider to the position 5; this automatically sets the size of the 1 × N array to 1 × 5.
3) We then, still in the schematic window, double-click power splitter icon and modify the "ElementMap" bracket to contain the number "1" five times (since N=5).
4) Run the simulation by clicking on "Run Analysis" (Figure 9.16).
5) Go back to 1), name the results file, e.g., N=5, click on "Checkpoint," and continue with step 2) defining the next value of N, e.g., N=6, and so on.

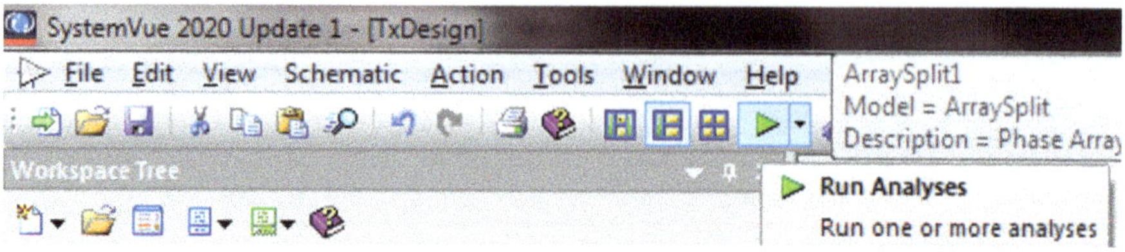

Figure 9.16 "Run Analysis" button.

The results of the above exercise are shown in Figure 9.17.

Exercises

1) What is the effect of increasing the ULA size on the magnitude of the directivity?
2) What is the effect of increasing the ULA size on the beam pattern bandwidth?
3) What is the effect of increasing the ULA size on the magnitude of the side lobes?
4) Upon examining its beam pattern, can you think of any shortcomings of the ULA?
5) Why is the radiated beam pattern surrounding the line of antenna elements as opposed to radiating in just one direction?
6) Change the array from ULA to URA and repeat the ULA simulations for an $N \times N$ URA for the cases N=1 to N=12, and answer questions 1)–5) above.

9.2.2.4 Element Antenna Radiation Pattern

In the previous ULA simulations, SystemVue modeled the antenna elements as isotropic radiators. In practice, antenna elements are supported by a dielectric substrate which has a ground place. Therefore, the radiation of the individual elements is no isotropic but away from the substrate. This is modeled in SystemView clicking on the ArrayAnt icon, and upon clicking on the "ElementPattern" tab, changing the "Type," which, by default, is set to "0:isotropic," to "1:Cosine." Doing this and re-running the simulation for N=12 gives the results in Figure 9.18.

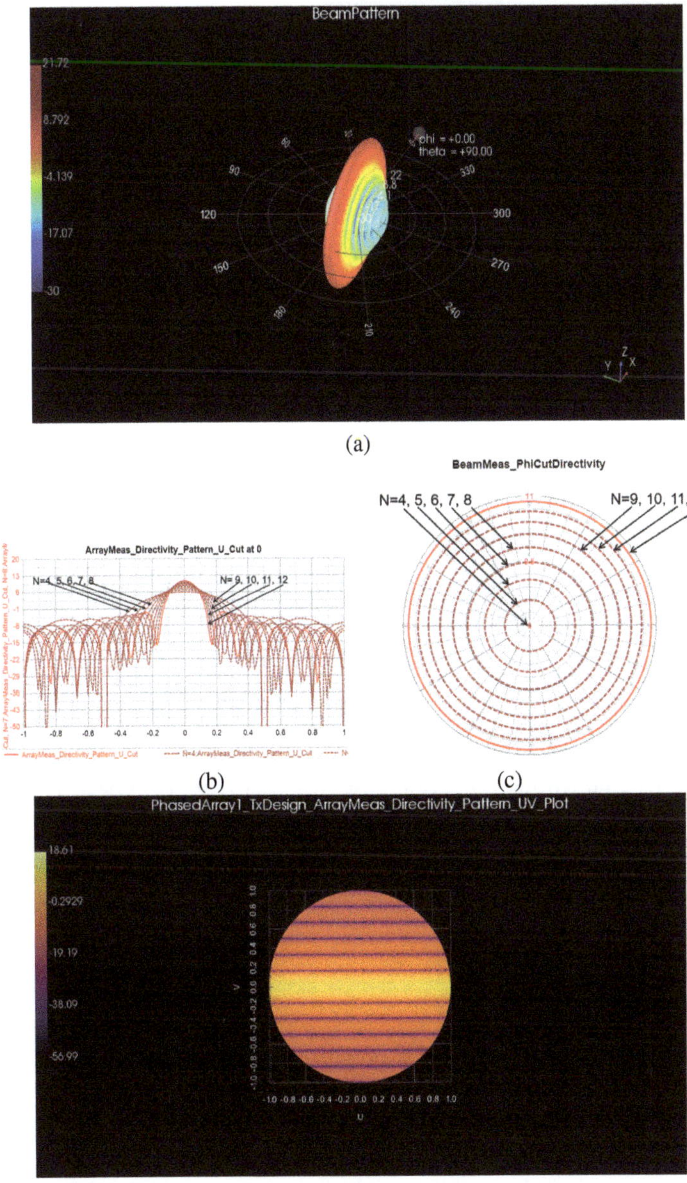

Figure 9.17 (a) Screen shot for N=4 to 12 of beam pattern of four-element ULA disposed along the *y*-axis. (b) Directivity along the $90° - 270°$ axis cut (U-cut) in (a). (c) Directivity along the $0° - 180°$ axis cut (V-cut) in (a). (d) Directivity pattern along direction in (b) vs. that along (c).

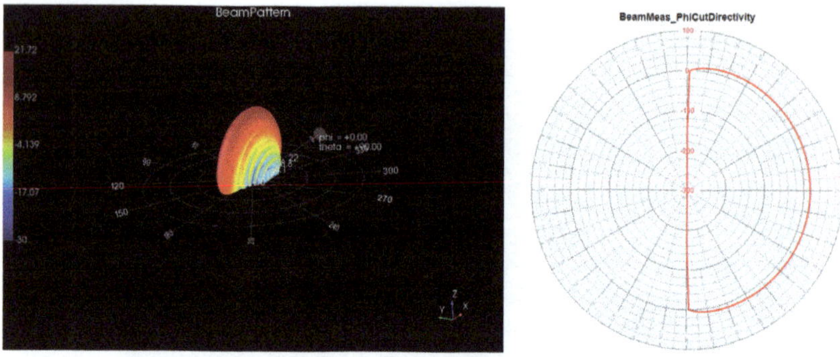

(a) (b)

Figure 9.18 Results of simulating ULA with N=12 and antenna elements modeled as "Cosine" radiators. (a) Radiated beam pattern. (b) Directivity.

9.3 Tutorial 2: Codebook Design for 28GHz 5G/MIMO Antenna Array Transmission

Learning objective: To enable the reader to construct a simulation model in SystemVue, parameterize the components, run the simulation, and read the important quantities from the simulation results.

Prerequisites: Tutorial 1, Chapters 2–7.

Design task: To determine the codebook for a 12 × 12 URA with six angular beam directions (Theta, Phi), namely, (45°, 0°), (−45°, 0°), (−45°, 45°), (−45°, −45°), (45°, 45°), and (0°, 0°). Provide the set of attenuation and phase shifts corresponding to these beam directions, considering the characteristics of real RF hardware.

9.3.1 Preliminaries

In the uniform array, with the inter-element spacing being equal, and the amplitude and phases of the excitations being equal, the resulting beam is symmetric with respect to the center of the array geometry. Assuming all antenna elements to be identical, there are three degrees of freedom one can tune to point the beam in a desired (Theta, Phi) direction, namely, the spacing, and the excitation amplitude and phase to each element. SystemVue contains the "Optimization" facility to achieve this (Figure 9.19).

In particular, by clicking on "Optimization1," the "Optimization Properties" window with four tabs opens up (Figure 9.19(a)).

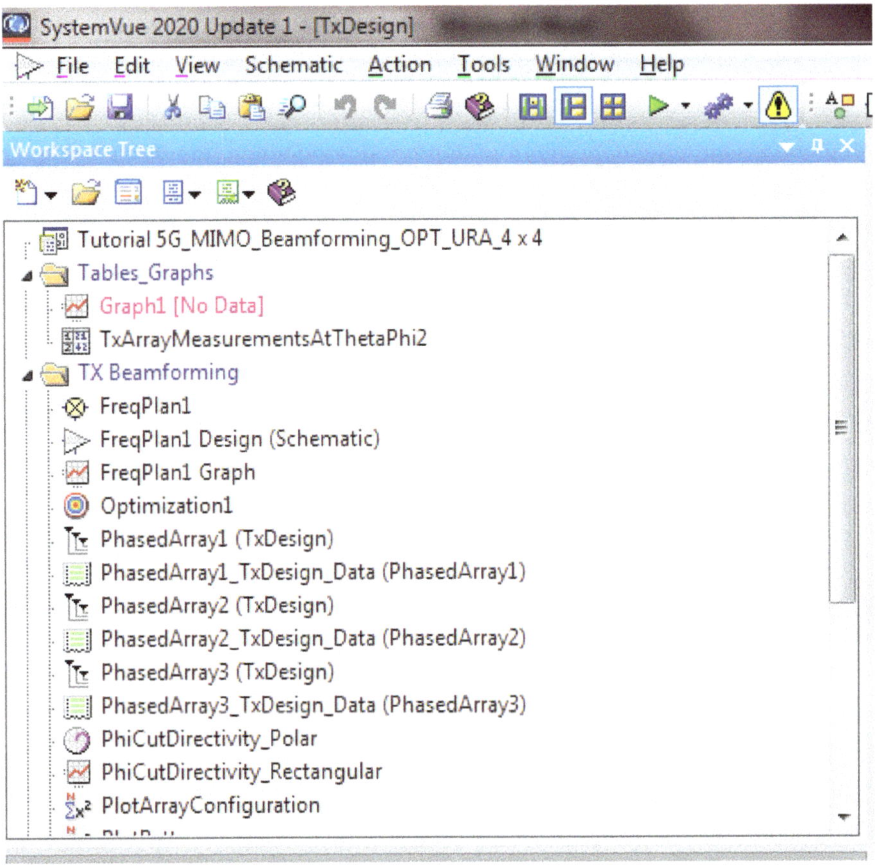

Figure 9.19 Screen shot Workspace Tree showing "Optimization" setup button.

The sub-windows contain entries for the goals of the optimization, e.g., antenna gain (Figure 9.19(b)), variables to be tuned/optimized, e.g., Driver Amplifier Gain (Figure 9.19(c)), and algorithmic method of optimization (Figure 9.19(d)), which is beyond the scope of this book; the default method is "Automatic." Further details may be found by clicking on "Help."

A look at the schematic window (Figure 9.20) shows the variables that may be optimized.

The ArrayAttn has the "Window" variable. This refers to specifying the attenuation for each of the paths leading to the N antenna elements, in particular, to the relative attenuation level for neighboring paths so as to

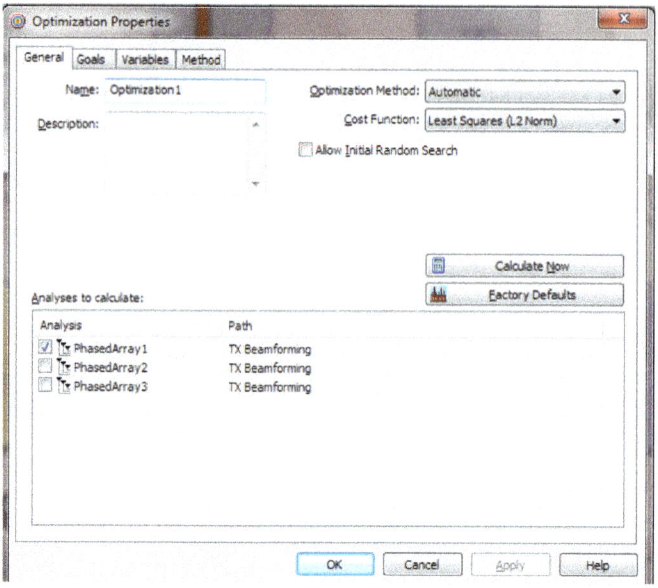

(a)

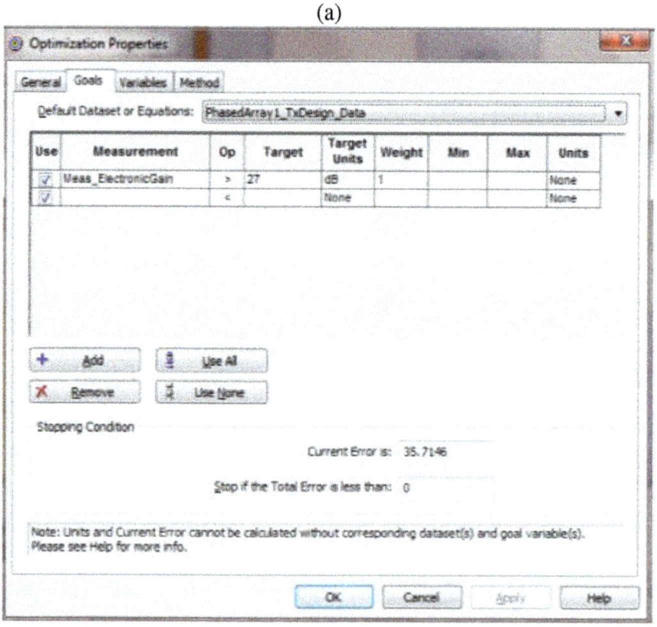

(b)

Figure 9.20 *Continued*

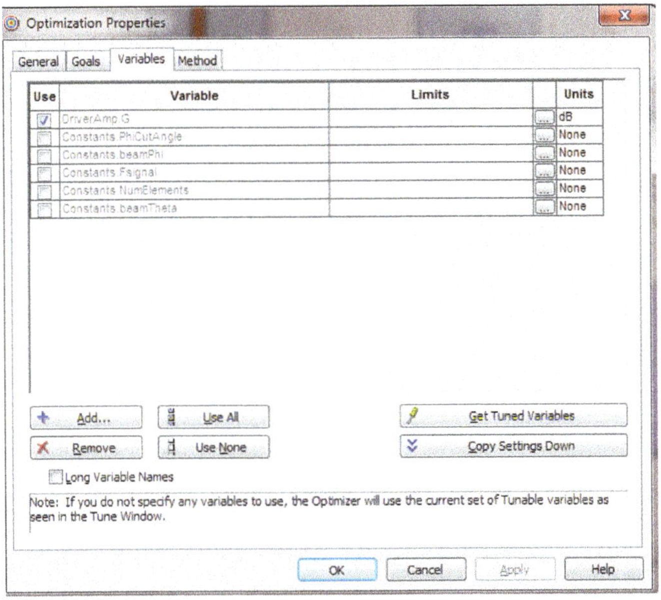

(c)

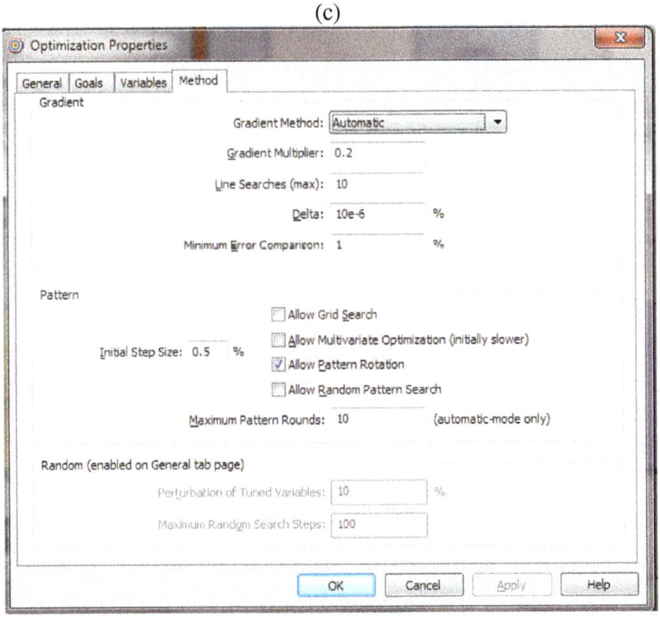

(d)

Figure 9.20 Optimization Parameters windows.

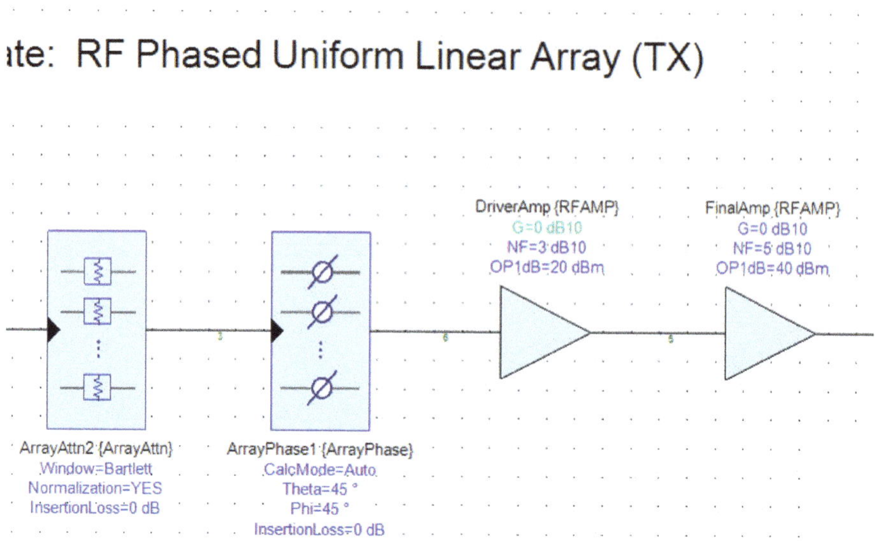

Figure 9.21 Component in schematic and their optimizable variables.

obtain a certain beam width, ratio of beam magnitude to side lobe magnitude, etc. The topic of "Windowing" is beyond the scope of this book.

The ArrayPhase1 component in Figure 9.20 is a little bit strange. There, what we see is the desired beam-pointing direction (Theta=45°, Phi=45°), which will be arrived at upon optimization of the N internal phase shifts in the paths of each of the N antenna elements. The resulting phase shifts may be accessed as follows. In the Workspace Tree window (Figure 9.21(a)), we click on the "PhaseArray1_TxDesign_Data(PhaseArray1," which opens the window in Figure 9.21(b).

Finally, to run the optimization "Optimization," we click on the analysis tab (Figure 9.23), and in the window that opens, click "TX Beamforming\Optimization1."

The optimized results for the beam pointing in the (Theta=45°, Phi=45 °) are shown in Figure 9.24.

9.3.2 Determination of Codebook for 12 × 12 MIMO URA

We begin with the schematic shown in Figure 9.25, from the workspace **Tutorial_2_5G_MIMO_Beamforming_CODEBOOK_OPT_URA_4 x 4. wsv** file. It will be noticed, upon clicking on the ArrayAttn2 and ArrayPhase1

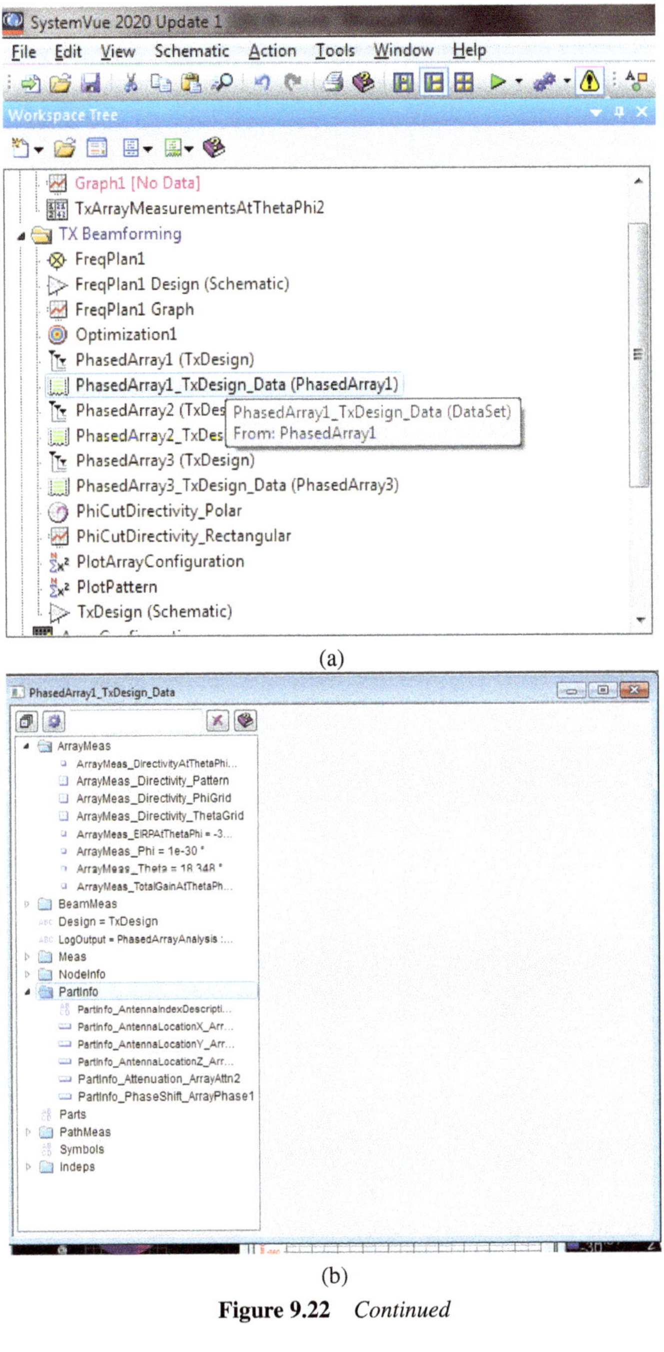

Figure 9.22 *Continued*

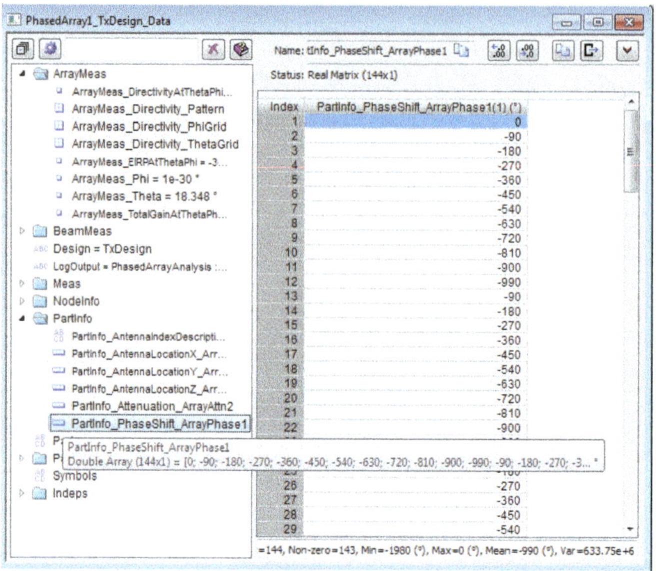

(c)

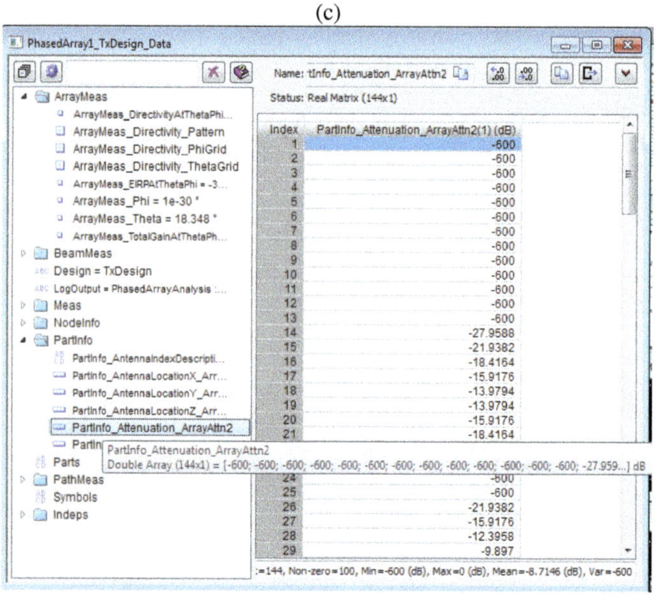

(d)

Figure 9.22 How to access the phase shifter values stored internally. (a) Click on "PhaseArray1_TxDesign_Data." (b) Click on the PartInfo tab. (c) See an array of phase shift values, the result of the optimization. (d) See an array of attenuator values, the result of the optimization.

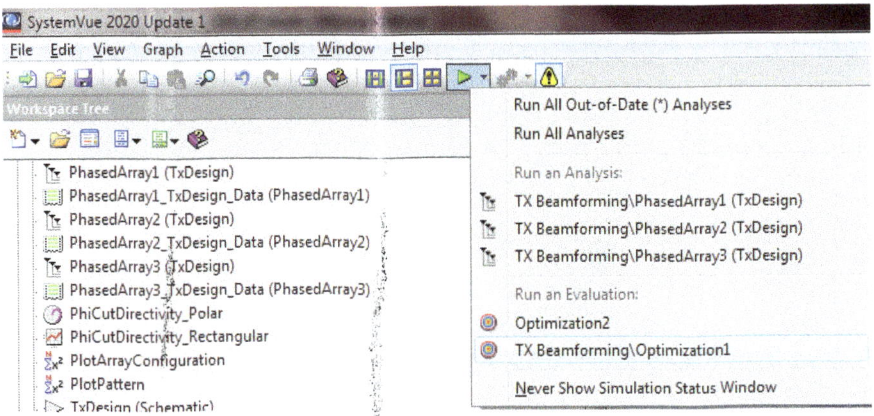

Figure 9.23 Running Optimization1.

components (Figure 9.25), that the properties of **real RF hardware** may be entered, Figure 9.26.

To find real devices we searched for Monolithic Microwave Integrated Circuits (MMICs) that implement 28 GHz amplifiers, attenuators, and phase shifters and found representative devices in [102, 104], respectively. Although S-parameter files of these devices may be incorporated into SystemVue, see "UseSParameters" entry, we will focus on the simple models that capture the typical performance at this frequency such as "Gain," "NF," "OP1dB," "InsertionLoss," "NumBits," and so on. The parameter values in question were inserted in the respective component "Properties" windows (Figure 9.26). The reader may click on the "Model Help" to look at all pertinent parameter definitions.

To proceed, we changed the frequency, "Freq," in the transmit port, "ArrayPort2," to 28 GHz, and the inter-element spacing to 5 mm, which is 0.5 mm less than the half-wavelength at 28 GHz. Then, setting the angle (Theta, Phi) of the ArrayPhase1 component to each of those given earlier, one at a time, we run "Optimization1," until no further changes are observed in the error. The results of these optimizations are given in Table 9.1.

The reader may compare his results with those of the codebook files included in the corresponding workspace.

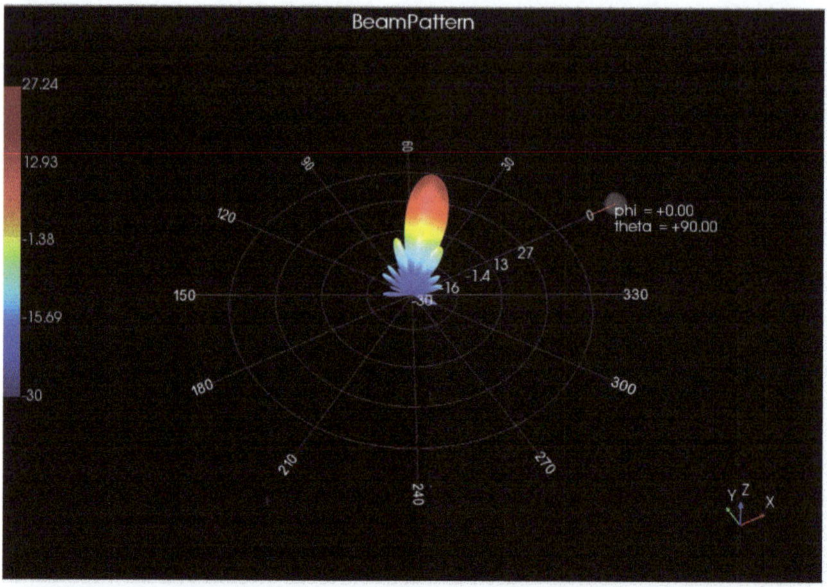

(a)

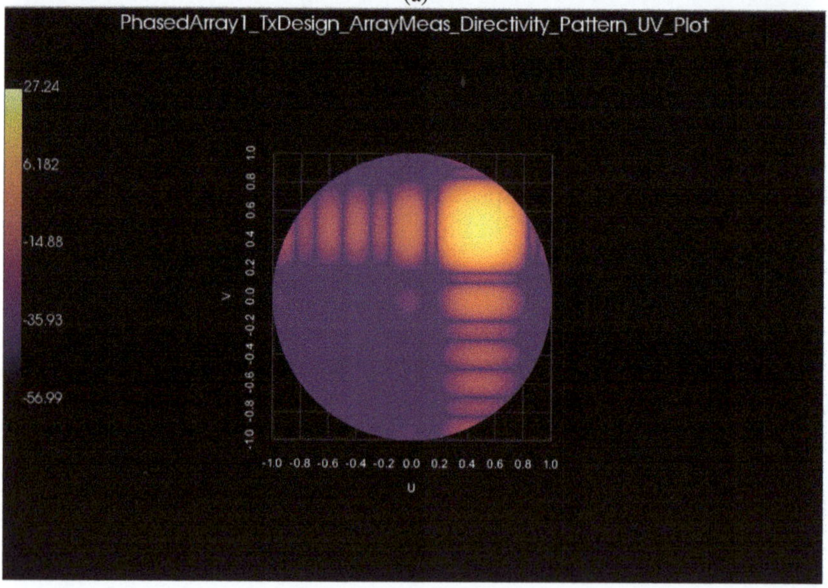

(b)

Figure 9.24 (a) Beam pattern pointing along the Theta=45°, Phi=45° direction. (b) Directivity pattern along the V-cut vs. U-cut.

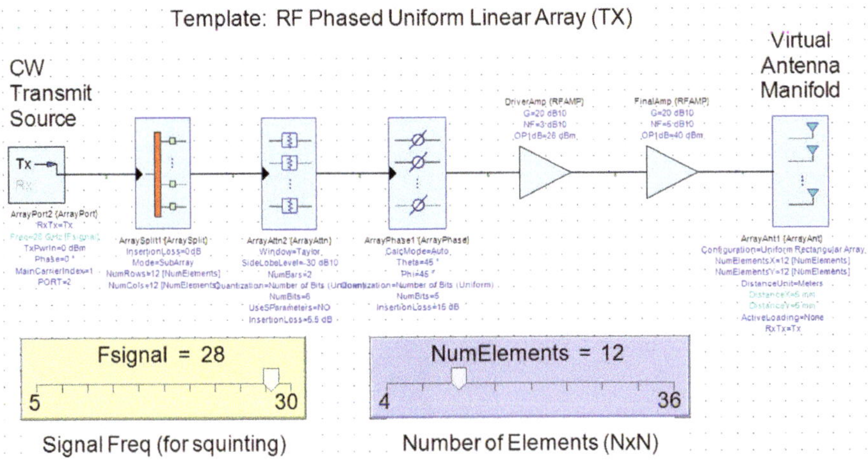

Figure 9.25 Schematic of URA for codebook design simulations.

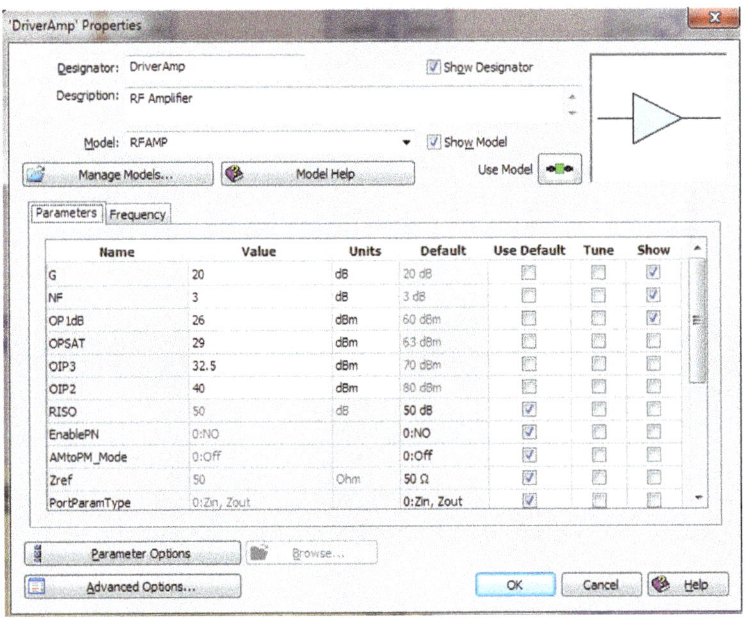

(a)

Figure 9.26 *Continued*

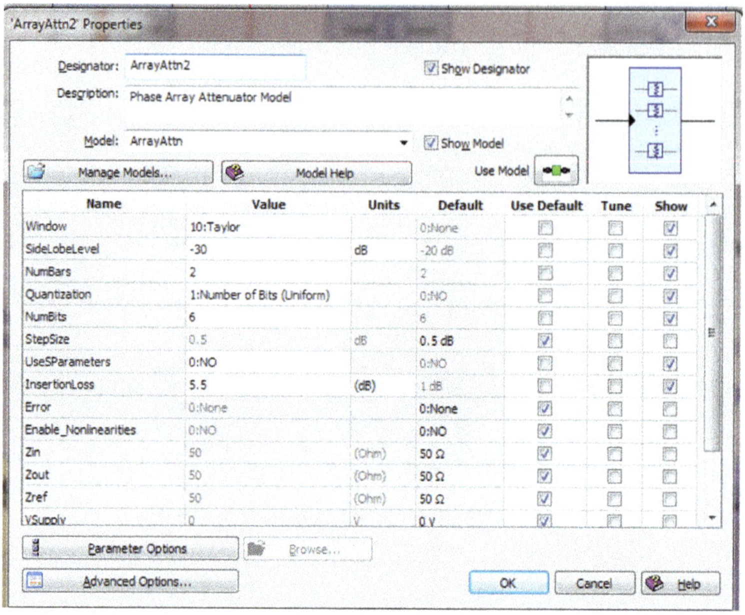

(b)

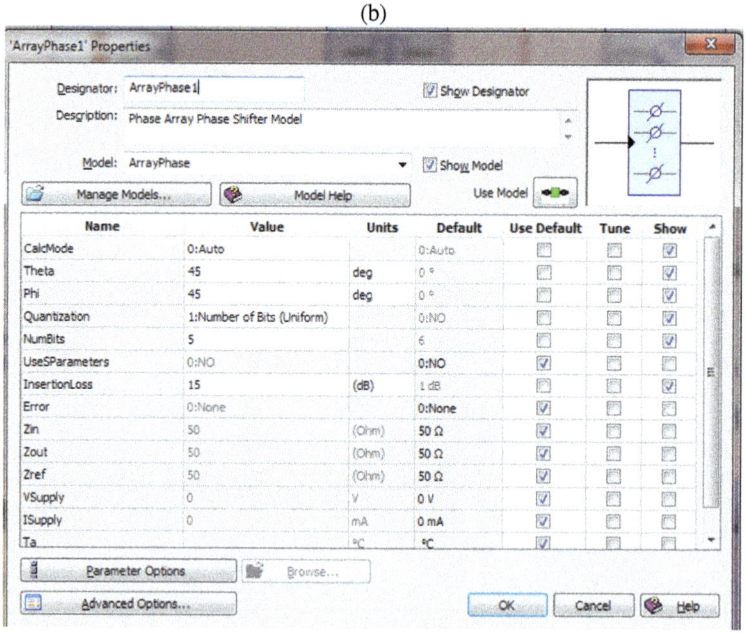

(c)

Figure 9.26 Properties of (a) ArrayAttn2 and (b) ArrayPhase1 components of URA.

Table 9.1 Beam patterns whose codebooks were obtained.

(-45°, -45°)

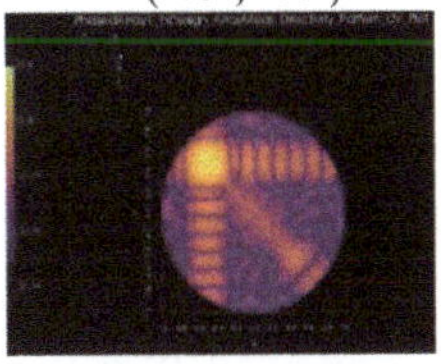

(45°, 45°)

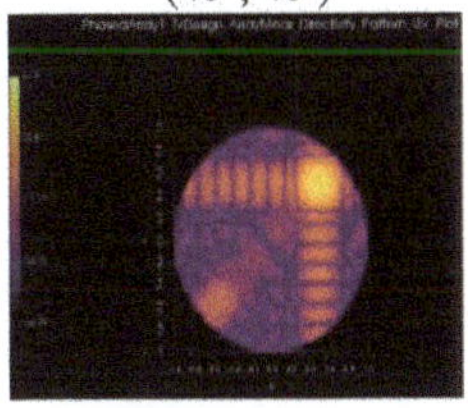

(-45°, 0°)

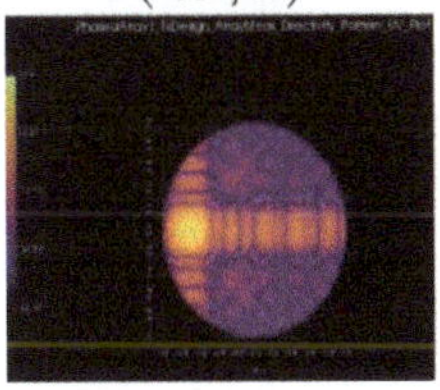

(0°, 0°)

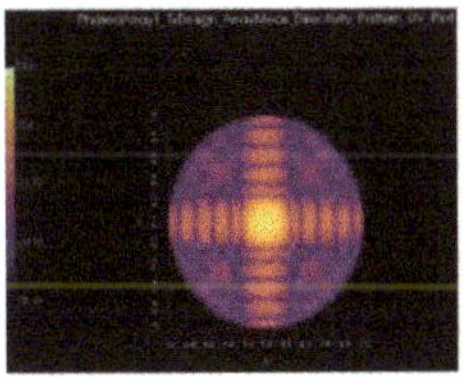

(45°, 0°)

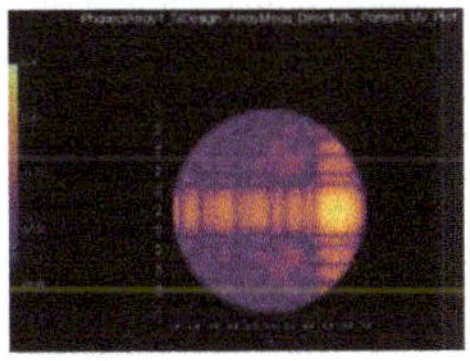

(-45°, 45°)

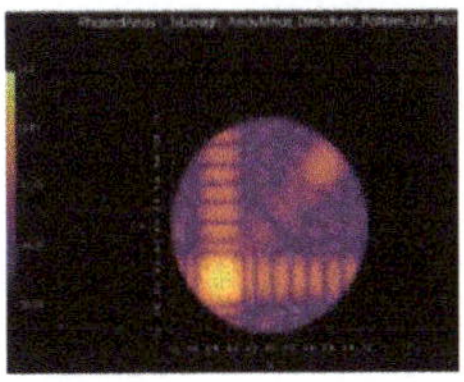

(-45°, 45°)

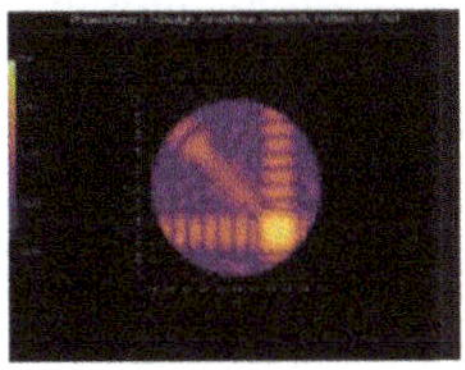

Exercises

1) How would you proceed in practice to implement the components settings (e.g., Attenuators and Phase Shifters setting) to produce the desired beam patterns?

9.4 Tutorial 3: Electronic/Warfare RADAR Performance Under Jamming Conditions

Learning objective: To enable the reader to gain intuition of the performance of an FM-CW RADAR in the presence of a jammer by examining a "toy" model.[1] A simplified simulation model in SystemVue is given amenable for

---

[1]The topic of RADAR design is a specialized one that requires a background on advanced courses in RADAR systems and signal processing. Thus, it is beyond the scope of this book. Typical books in this area are [106, 108].

the reader to vary its parameters, run simulations, and read the important quantities from the simulation results.

The toy RADAR system simulated here was inspired by the Master's Thesis "An Investigation of Jamming Techniques Through a RADAR Receiver Simulation" [105]. The scenario considered pertains to modeling a situation involving an air-to-air target and platform environment. This scenario may be representative of an aggressor RADAR platform in a high performance attack aircraft such as MIG, Mirage, F-16, etc., which has a fire control RADAR. The target, on the other hand, is a defending aircraft that possesses a self-protection electronic warfare (EW) system that can employ jamming in order to avoid fire from the aggressor. While the aggressor aircraft attempts to detect, track, and send missiles onward, the defending aircraft employs various jamming techniques to break the tracking lock or to deceive the aggressor from locking onto the defending aircraft itself. The SystemVue simulation examines the impact of the jamming-to-signal ratio (JSR) level of various jamming techniques on the measured target distance and velocity.

Prerequisites: Chapters 8.

Design task: To examine the impact of various jamming techniques employed by the target aircraft on its measured distance and velocity as determined by the attack aircraft.

9.4.1 Preliminaries: Transmitter–Receiver Simulation

The RADAR transmitter is modeled as an FM-CW RADAR possessing the following parameters, namely, operating RF frequency, PRI, PW, pulse, and sampling frequency, given in Table 9.2.

Assumptions:

- The RADAR operates at 10 GHz; thus, it is an X band airborne intercept (AI) RADAR.
- The PRI is set to 100 μs and the PW is set to 50 μs.
- The transmitter power is set to 1 kW as a typical value.

Table 9.2 RADAR parameters.

Operating RF Frequency	Pulse Repetition Interval PRI	Pulse Width PW
10 GHz	100 μs	50 μs

- As an FM-CW RADAR, a signal, whose frequency is linearly modulated periodically over a frequency bandwidth B is transmitted, and the time delay Δt and corresponding frequency shift Δf between the transmitted and received linearly frequency modulated (LFM) signals, namely, $\Delta t = \Delta f \frac{T}{B}$, is related to the range by $R = \Delta f \frac{T}{2B} c$.
- Δf is calculated during the simulation.
- We assume the attack-to-target aircraft distance is 10,000 m and assume a relative (radial) velocity at that distance of V_r=50 m/s.
- The LFM bandwidth is assumed to be B=5 MHz.
- The LFM signal is assumed to be digitized before transmission at a frequency of 20 MHz.
- The Doppler frequency, for a velocity of 50 m/s, a carrier frequency f_c=10 GHz is $f_d = 2V_r f_c/c = 3333.33 m/s$.

9.4.2 FM-CW RADAR Model and Simulations

The model for the FM-CW RADAR Transmitter plus Environment plus Jammer plus Target plus Receiver plus signal processing (to extract the target velocity and range) is shown in Figure 9.27; it is contained in the workspace **Tutorial_3_RADAR_Jammer.wsv**.

The "Jammer" building block contains four components (Figure 9.28), namely, (1) The "RADAR_EWJamming," which is a module that may produce four types of jamming signals. A red "cross" over an element icon is activated by clicking on it to select it, and then on the toolbar symbol, , and it means that doing so rendered it open or ineffective/deactivated. Thus, in Figure 9.27(a), one of the four Jammer types is activated and the other three are deactivated. This will be used in a later exercise, where the reader will study the impact of each type of jammer by selecting one type of jammer at a time. This component may be obtained from the Parts Tree under "Radar Parts." (2) The "Sampler," which is a component that digitizes an analog signal at a desired sampling rate; it is found in the "Parts Tree" under "Algorithm Design." (3) A complex-to-envelope converter that modulates a complex carrier signal of frequency f_c (represented by a real and an imaginary part) and produces an envelope signal (represented by a magnitude and a phase). This component may be obtained from the "Parts Tree" under "AlgorithmDesign/MathMatrix" category. (4) A "Gain" component, which may be found in the "Parts Tree" under "Algorithm Design/MathMatrix" category.

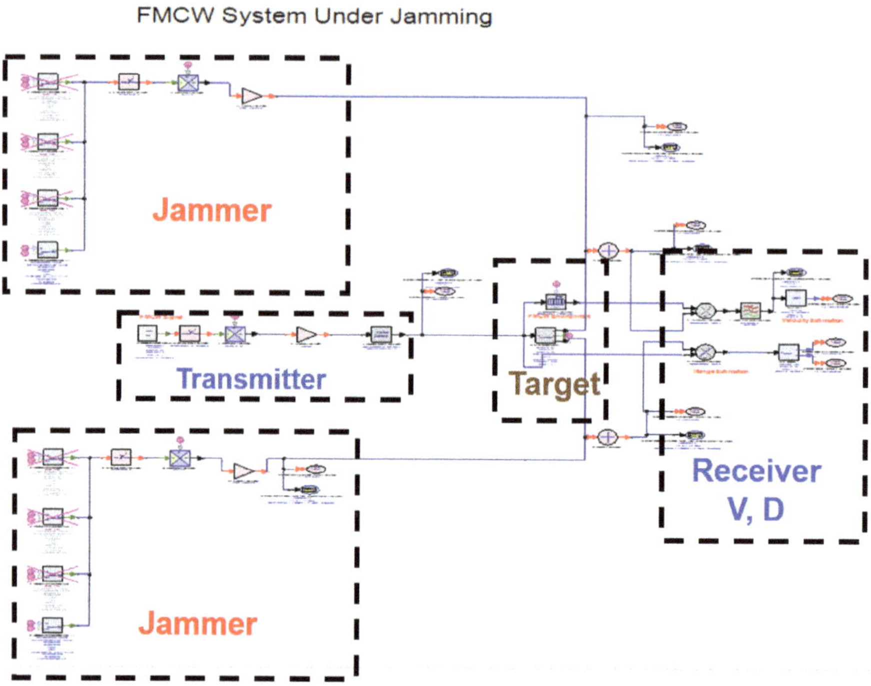

Figure 9.27 Overall system.

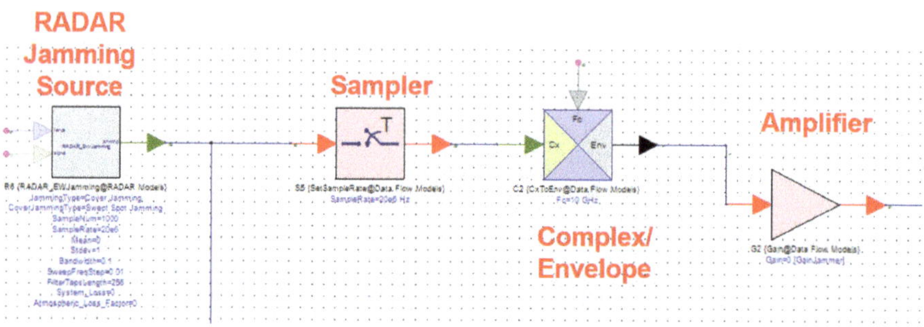

Figure 9.28 Jammer.

The transmitter (Figure 9.29), which prepares and sends the signal transmitted by the RADAR, contains four components, namely, the following. (1) an LFM source, which produces a signal whose frequency varies linearly with time. This is where the "heart" of the RADAR, namely, its pulse width, pulse repetition interval (PRI), bandwidth (B), and "Sampling

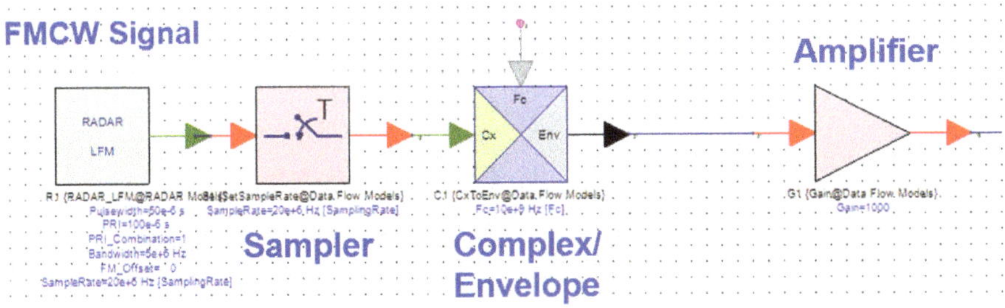

Figure 9.29 Transmitter.

Rate" are specified. This component is found in the Parts Tree under "RADAR Parts." (2) A Sampler. (3) A complex-to-envelope converter. (4) An amplifier. These last three components were described above.

The Environment plus Target (Figure 9.30), models the transmitter-to-receiver signal envelope delay and attenuation, on the one hand, and the properties of the Target such as its distance from the transmitter, its velocity, its scattering cross section, etc., on the other. The Environment is found as

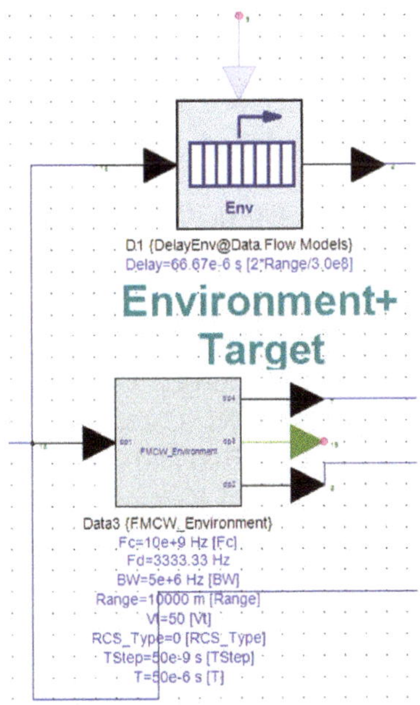

Figure 9.30 Environment plus Target.

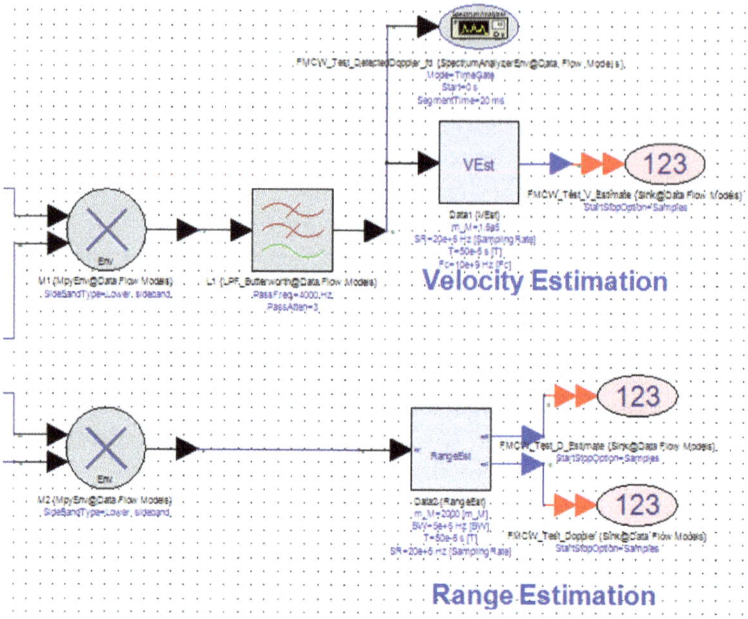

Figure 9.31 Receiver plus signal processing to extract target velocity (V) and range (D).

part of the "Algorithm Design" in the "Parts Tree." The Target is a building block developed by Keysight and included as part of the FMCW_Radar.wsv workspace template. The model is found in the "Workspace Tree" under "Models."

The Receiver (Figure 9.31), models the down-conversion of the received signal into baseband and the extraction of the target velocity and distance. It consists of two envelope multipliers (MpyEnv), which effect the signal down-conversion, a low-pass filter (LPF), a model that extracts the velocity, and one that extracts the distance of the target. The MpyEnv are found under "Algorithm Design Library > Analog RF Category>MpyEnv Part" in the "Parts Tree." The velocity extraction (VEst) and Distance extraction (Range) building blocks were also developed by Keysight and included as part of the FMCW_Radar.wsv workspace template. Their models are found in the "Workspace Tree" under "Models."

Plots of the LFM time waveform, the transmitted signal spectrum, and the jammer signal spectrum for the "multi spot jamming" case are given in Figures 9.32, 9.33, and 9.34, respectively. The gain of the amplifier in the transmitter sets its power to 1 kW.

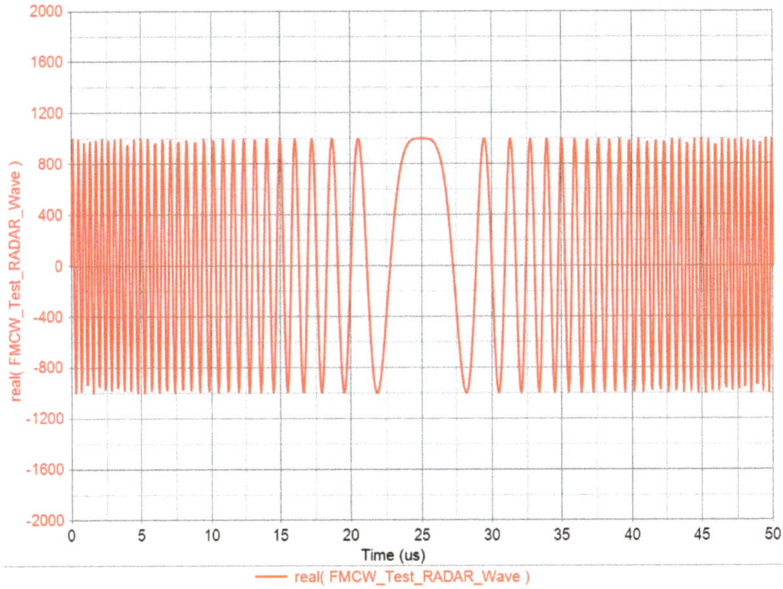

Figure 9.32 Linearly modulated frequency of RADAR signal transmitter over 50 μs.

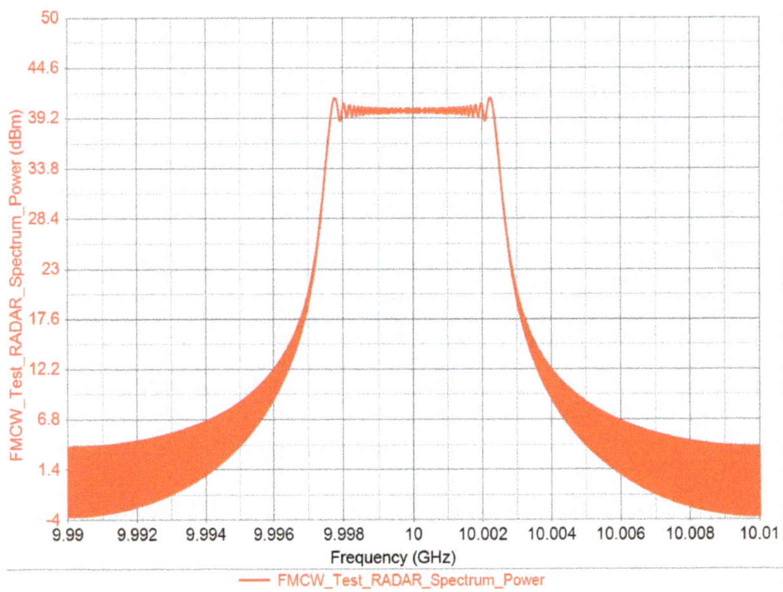

Figure 9.33 Frequency spectrum of RADAR signal transmitted.

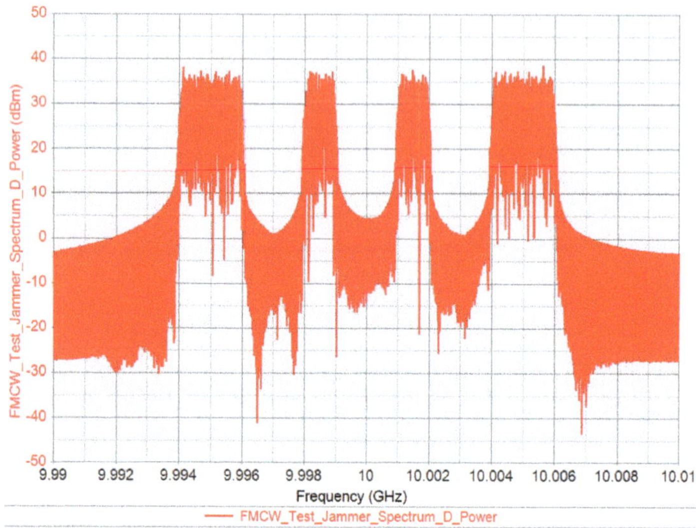

Figure 9.34 Frequency spectrum of jammer signal (Multi Spot Jamming) transmitted.

Table 9.3 Estimated target distance and velocity versus jammer power.

Multi Spot Jamming Power (W)	Distance (m)	Velocity(m/s)
0	10,019.5	51.5
1000	10,019.5	35.5
2000	10,019.5	49.2
3000	10,019.5	22.9
4000	10,019.5	5.7
5000	10,004.9	53.8
6000	10,004.9	48.1
7000	10,019.5	43.5
8000	10,019.5	6.9
9000	10,019.5	1.1
10,000	8203.1	1.1
11,000	6035.1	40.0
12,000	4423.8	18.3
13,000	5273.4	4.6
14,000	8642.6	26.3
15,000	5742.2	43.5
16,000	4614.2	4.6
17,000	615.2	4.6
18,000	6457.8	32
19,000	9199.2	12.6
20,000	7793.0	50.3

Results of running simulations to determine the impact of the jammer power on the estimated target distance and velocity as the jammer power is increased are given in Table 9.3. The gain of the amplifier in the transmitter was set so that its power was in all cases 1 kW.

9.4.3 Exercises

1) What can you say about the impact of jammer power on the estimated target distance and velocity?
2) Repeat the above runs for each of the other three cases of jamming signals, currently deactivated, in the schematic.
3) How do the four jamming signals compare in their effectiveness?
4) Explore the performance of the FM-CW RADAR by eliminating the jammers (making the parameter Gainjammer=0) and computing the distance and velocity for various parameters in Table 9.2.

Bibliography

[1] G. Miano, "The concept of field in the history of electromagnetism," ET2011-XXVII Riunione Annuale dei Ricercatori di Elettrotecnica, Bologna 16-17 giugno 2011.

[2] H. Hertz, *Electric Waves*, Authorized English Translation by D. E. JONES, B.Sc., (London and New York), Macmillan & Co., 1893.

[3] H,V. Friedburg, "Heinrich Hertz at Work in Karlsruhe," *IEEE MTT-S International Microwave Symposium Digest*, New York, NY, USA, 25-27 May 1988, pp. 267-270.

[4] G. S. Smith, "Analysis of Hertz's Experimentum Crucis (crucial experiment) on Electromagnetic Waves," *IEEE Antennas & Propagation Magazine* , October 2016, pp. 96-108.

[5] Available: [Online]: https://commons.wikimedia.org/wiki/File:Ruhmkorff_inductor_schem.svg

[6] Available: [Online]: https://en.wikipedia.org/wiki/Coherer

[7] G. Marconi, "Improvements in Transmitting Electrical impulses and Signals, and in Apparatus therefor," British patent No. 12,039 (1897). Date of Application 2 June 1896; Accepted, 2 July 1897.

[8] K.F. Braun, Patent DRP 111578 of October 14, 1898.

[9] Available: [Online]: http://ieeemilestones.ethw.org/images/6/62/Bauer_-_text.pdf

[10] Available: [Online] https://www.nobelprize.org/prizes/physics/1909/marconi/biographical/

[11] Available: [Online]: https://www.nobelprize.org/prizes/physics/1909/braun/facts/

[12] J.S. Belrose, " Fessenden and Marconi: their differing technologies and transatlantic experiments during the first decade of this century," *Proc. of the 1995 International Conference on 100 Years of Radio*, London, UK, UK, 5-7 Sept. 1995, pp. 32-43.

[13] C. E. Shannon and W. Weaver, *The Mathematical Theory of Communication*, First paperbound edition, 1963, Copyright 1949 by the Board of Trustees of the University of Illinois.

[14] W. Wiesbeck, “Radar Systems Engineering,” Lecture Script, Institut für Hochfrequenztechnik und Elektronik, Karlsruhe Institute of Technology, 16th Edition WS 2009/2010.

[15] H.J. De Los Santos, Lecture Notes, “Modern Radio Systems Engineering,” Institut für Hochfrequenztechnik und Elektronik (IHE), Karlsruher Institut für Technologie (KIT), Germany, Summer Semester, 2010-2011.

[16] M. Schwartz, *Information Transmission, Modulation, and Noise*, McGraw-Hill, 1970.

[17] N. Abramson, *Teoría de la Informacion y Codificación*, QUINTA EDICION, 1981, MADRID. PARAFINO, S.A. Version española de la obra inglesa: *INFORMATION THEORY AND CODING*. Publicado por McGRAW-HILL Book Company, Inc. Traducida por JUAN ANTONIO DE MIGUEL MENOYO, 1ngeniero de Telecomunicación.

[18] Hughes Space and Communications Co., “Local Oscillator Chains,” Receiver Design Notes.

[19] H. L. Krauss, C.W. Bostian, and F.H. Raab, *Solid State Radio Engineering*, John Wiley & Sons, Inc., 1980.

[20] Prof. Dr.-Ing. W. Wiesbeck, Skriptum zur Vorlesung, *Grundlagen der Hochfrequenztechnik*, Universität Karlsruhe (TH), Institut für Höchstfrequenztechnik und Elektronik, 2007.

[21] J. D. Kraus and K. R. Carver, *Electromagnetics*, Second Edition, McGraw-Hill, Inc., New York, 1973.

[22] W, L, Stutzman and G.A. Thiele, *Antenna Theory and Design*, Second Edition, John Wiley & Sons, Inc., New York (1998).

[23] W. L. Weeks, *Antenna Engineering*, (New York, McGraw-Hill Book Company, 1968).

[24] S. U. Rahman, Q. Cao, M. M. Ahmed, and H. Khalil, “Analysis of Linear Antenna Array for minimum Side Lobe Level, Half Power Beamwidth, and Nulls control using PSO,” *J.l of Microwaves, Optoelectronics and Electromagnetic Applications*, Vol. 16, No. 2, June 2017, pp. 577-591.

[25] T. S. Rappaport, *Wireless Communications: Principles and Practice*, Second Ed., Prentice-Hall, Inc. 2002

[26] Dipl.-Ing. Thomas Fügen , Lecture Notes, Course, “Wave Propagation and Radio Channels for Mobile Communications,” KIT, IHE, SS2010.

[27] Q. Gu, RF System Design of Transceivers for Wireless Communications, Springer, 2005.

[28] *Texas Instrument Technical Brief SWRA030*, Matt Loy, Editor, May 1999

[29] S. A. Maas, Noise in Linear and Nonlinear Circuits, Artech House (2005).

[30] J. Dabrowski, Course "Introduction to RF Electronics," Lecture Notes; Division of Electronic Devices, Department of Electrical Engineering (ISY), Linköping University, 2006.

[31] Available: [Online]: http://www.clee.freehomepage.com/teaching.html

[32] R. C. Dixon, *Spread Spectrum Systems with Commercial Applications*, Third Edition, John Wiley & Sons, Inc., New York, 1994.

[33] H. J. De Los Santos, C. Sturm, J. Pontes, *Radio Systems Engineering: A Tutorial Approach*, New York, Springer, August, 2014.

[34] B. Razavi, *RF Microelectronics*, Prentice-Hall, 1998.

[35] R. Adler, "A Study of Locking Phenomena in Oscillators," IEEE Proceedings, vol. 61, pp. 1380–1385, 1973. (Reprinted from Proceedings of the Institute of Radio Engineers, vol. 34, pp. 351–357, June 1946.)

[36] B. Razavi, "RF Transmitter Architectures and Circuits," 1999 *Proc. IEEE Custom Integrated Circuits Conf.*, pp. 197 – 204.

[37] L. A. Bronckers, A. Roc'h, and A. B. Smolders, "Wireless Receiver Architectures Towards 5G: Where Are We?" *IEEE Circuits and Systems Magazine* (Volume: 17, Issue: 3, Third Quarter 2017), pp. 6-16.

[38] J. Mitola, "The software radio architecture," *IEEE Commun. Mag.*, vol. 33, no. 5, pp. 26–38.

[39] R, M, Cerda, "Impact of ultralow phase noise oscillators on system performance,"*RF Design*, July 2006, pp. 28-34.

[40] S. Kumar, G. Gupta, K. R. Singh, "5G :Revolution of Future Communication Technology," *2015 International Conference on Green Computing and Internet of Things (ICGCIoT),* 8-10 Oct. 2015, Noida, India, pp.143-147.

[41] Number of smartphone users worldwide from 2016 to 2021, Available: [Online]: https://www.statista.com/statistics/330695/number-of-smart phone-users-worldwide/

[42] H.J. De Los Santos, *RF MEMS Circuit Design for Wireless Communications*, (Norwood, MA), Artech House, 2001.

[43] Available: [Online]: https://en.wikipedia.org/wiki/5G

[44] J. G. Andrews, S. Buzzi, W. Choi, S.V. Hanly, A. Lozano, A.C. K. Soong, and J. C. Zhang, "What Will 5G Be?", *IEEE J. on Selected Areas Comm.*, Vol. 32, NO. 6, June 2014, pp. 1065-1082.

[45] S. Onoe, "Evolution of 5G Mobile Technology Toward 2020 and Beyond," *IEEE Int. Solid State Circuits Conference (ISSCC) 2016 /* SESSION 1 / PLENARY / 1.3, Digest of Technical Papers, pp. 23-28.

[46] H.J. De Los Santos, C. Sturm, and J. Pontes, Radio Systems Engineering: A Tutorial Approach. Berlin: Springer, 2014.

[47] H.J. De Los Santos, *Understanding Nanoelectromechanical Quantum Circuits and Systems (NEMX) for the Internet of Things (IoT) Era*, River Publishers, Denmark, 2019.

[48] D. Choudhury, "5G Wireless and Millimeter Wave Technology Evolution: An Overview," *2015 IEEE MTT-S International Microwave Symposium*, 17-22 May 2015, Phoenix, AZ, USA.

[49] M. Aldababsa, M. Toka, S. Gökçeli, G.G. K. Kurt, and O. Kucur, "A Tutorial on Nonorthogonal Multiple Access for 5G and Beyond," *Hindawi Wireless Communications and Mobile Computing*, Vol. 2018, Article ID 9713450.

[50] T.L. Marzetta, "Non-cooperative cellular wireless with unlimited numbers of base station antennas," IEEE Trans. Wireless Commun., vol. 9, no. 11, Nov. 2010.

[51] G.J. Foschini, "Layered space-time architecture for wireless communication in a fading environment when using multi-element antennas," *Bell Labs Technical Journal*, Vol. 1, No. 2, Autumn 1996, pp. 41-59.

[52] G.J. Foschini and M.J. Gans, "On Limits of Wireless Communications in a Fading Environment when Using Multiple Antennas," *Wireless Personal Communications* 6: 311–335, 1998.

[53] Available: [Online]: https://en.wikipedia.org/wiki/Additive_white_Gaussian_noise

[54] A. Papoulis, *Probability, Random Variables, and Stochastic Processes*, 2nd Edition, (New York), McGraw-Hill, 1984.

[55] G. R. Cooper and C. D. McGillem, *Probabilistic Methods of Signal and System Analysis,* 3rd Edition, Oxford Press, 1999.

[56] Available: [Online]: https://en.wikipedia.org/wiki/Gaussian_integral

[57] T.M. Cover and J.A. Thomas, *Elements of Information Theory*, Second Edition, (Hoboken, New Jersey), John Wiley & Sons, Inc., 2006.

[58] A. K. Jagannatham, "Bandwidth Efficient Channel Estimation for Multiple-Input Multiple-Output (MIMO) Wireless Communication

Systems: A Study of Semi-Blind and Superimposed Schemes," Ph.D. Dissertation, University of California, San Diego, 2007.

[59] R.M. Fano, *Transmission of Information*, John Wiley and Sons, New York, 1961.

[60] M. Kumar and A. Singh, "Channel Capacity Comparison of MIMO Systems with Rician Distributions, Rayleigh Distributions and Nakagami-m," *Int. J. Eng. Res. & Tech. (IJERT)*, ISSN: 2278-0181, Vol. 2 Issue 6, June. 2013, pp. 905-909.

[61] S. M. Abuelenin, "On the similarity between Nakagami-m Fading distribution and the Gaussian ensembles of random matrix theory," Available; [Online]: https://arxiv.org/ftp/arxiv/papers/1803/1803.08688.pdf

[62] B. Noble and J.W. Daniel, *Applied Linear Algebra*, Second Edition, (Englewood Cliffs, New Jersey), Prentice-Hall, 1977.

[63] A. K. Cline and I. S. Dhillon, "Computation of the Singular Value Decomposition," *Handbook of Linear Algebra*, CRC Press, pp. 45-1—45-13, 2006.

[64] A. J. Paulraj, D. A. Gore, R. U. Nabar, and H. Bölcskei, "An Overview of MIMO Communications—A Key to Gigabit Wireless," *Proc. IEEE*, Vol. 92,No. 2, February 2004, pp. 198-218.

[65] R. Courant, *Differential and Integral Calculus*, Vol. II, (New York), Interscience Publications, Inc., 1936, pp 188-203.

[66] B. Holter, "On the Capacity of the MIMO Channel-A Tutorial Introduction," Norwegian University of Science and Technology. Available; [Online]: http://www.ux.uis.no/norsig/norsig2001/Papers/57.Capacity_of__1992001154555.pdf

[67] A. Khan and R. Vesilo, "A Tutorial on SISO and MIMO Channel Capacities," Available; [Online]: https://pdfs.semanticscholar.org/9606/2e1f897c5b75dd94375a9440d4c3702410d3.pdf

[68] E. Telatar, "Capacity of Multi-antenna Gaussian Channels," *European Transactions on Telecommunications*, Vol. 10, No. 6, November-December 1999, pp. 585-595.

[69] T. L. Marzetta, "Noncooperative cellular wireless with unlimited numbers of base station antennas," *IEEE Trans. Wireless. Commun.*, vol. 9, no. 11, Nov. 2010, pp. 3590–3600.

[70] F. Rusek *et al.*, "Scaling Up MIMO: Opportunities and Challenges with Very Large Arrays," *IEEE Sig. Proc. Mag.*, vol. 30, Jan. 2013, pp. 40–60.

[71] E. G. Larsson, O. Edfors, F. Tufvesson, and T. L. Marzetta, "Massive MIMO for Next Generation Wireless Systems," *IEEE Communications Magazine*, February 2014, pp. 186-195.

[72] Y. Karasawa, "Channel Capacity of Massive MIMO With Selected Multi-Stream Transmission in Spatially Correlated Fading Environments," *IEEE Trans. on Vehicular Tech.*, Vol. 69, No. 5, May 2020, pp. 5320-5330.

[73] S. Sun, T. S. Rappaport, R. W. Heath, Jr., A. Nix, and S. Rangan, "MIMO for Millimeter-Wave Wireless Communications: Beamforming, Spatial Multiplexing, or Both?," *IEEE Communications Magazine*, December 2014, pp. 110-121.

[74] A.J. Paulraj and T. Kailath, "Increasing capacity in wireless broadcast systems using distributed transmission/directional reception (DTDR)," US Patent # 5,345,599.

[75] J. Litva and T. K.-Y. Lo, *Digital Beamforming in Wireless Communications*, (London) Artech House, 1996.

[76] H. Bölcskei, D. Gesbert, C. B. Papadias, and A. J. van der Veen, (Editors), *Space-Time Wireless Systems: From Array Processing to MIMO Communications*, Cambridge Univ. Press, 2008.

[77] S. Huang, H. Yin, J. Wu, and V. C. M. Leung, "User Selection for Multi-user MIMO Downlink with Zero-Forcing Beamforming," IEEE Trans. on Vehicular Tech., Vol. 62 , No. 7, Sept. 2013, pp. 3084 - 3097.

[78] Available: [Online]: http://www.ece.ubc.ca/~janm/Lectures/lecture_mimo.pdf

[79] V. D. Ravva and A. McLauchlin, "The BER Analysis of MIMO System for M-PSK over Different Fading Channels using STBC Code Structure," *Int. J. of Recent Tech. and Eng. (IJRTE)* ISSN: 2277-3878, Volume-8 Issue-3, September 2019, pp. 5831-5836.

[80] S. Yang and L. Hanzo, "Fifty Years of MIMO Detection: The Road to Large-Scale MIMOs," *IEEE Comm, Surveys & Tutorials*, Vol. 17, No. 4, Fourth Quarter 2015, pp. 1941-1988.

[81] S. Mhatli, H. Mrabet and I. Dayoub, "Extensive Capacity Simulations of Massive MIMO Channels for 5G Mobile Communication System," *2019 IEEE 2nd International Conference on Computer Applications & Information Security (ICCAIS)*, 1-3 May 2019, Riyadh, Saudi Arabia, Saudi Arabia.

[82] D. Love, R. Heath Jr., W. Santipach, and M. Honig, "What is the value of limited feedback MIMO channels?" *IEEE Commun. Mag.*, vol. 42, pp. 54-59, Oct. 2004.

[83] C Jiang, M Wang, C Yang et al., "MIMO Precoding Using Rotating Codebooks," *IEEE Transactions on Vehicular Technology*, Vol. 60, No. 3, March 2011, pp 1222 - 1227.

[84] X. Su, S. Yuy, J. Zeng, Y. Kuang, N. Fang, and Z. Li, "Hierarchical Codebook Design for Massive MIMO," *2013 8th IEEE International Conference on Communications and Networking in China (CHINACOM),* pp. 178-182.

[85] A. F. Molisch, "MIMO systems with antenna selection—an overview," *IEEE Commun. Mag.*, vol. 42, pp. 68-73, Oct. 2004.

[86] S. Sanayei and A. Nosratinia, "Antenna Selection in MIMO Systems," *IEEE Communications Magazine* Âů November 2004, pp. 68-73.

[87] A. Ghrayeb, "A Survey on Antenna Selection for MIMO Communication Systems," *2006 2nd International Conference on Information & Communication Technologie*, Damascus, Syria, 24-28 April 2006.

[88] Y. Zhang, C. Ji, W. Q. Malik, D. C. O'Brien, and D. J. Edwards, "Receive Antenna Selection for MIMO Systems over Correlated Fading Channels," *IEEE Transactions on Wireless Communications, Vol. 8, No.*.9, 2009, pp. 4393–4399.

[89] G. Auer and I. Cosovic, "Pilot Design for Multi-User MIMO," *IEEE 2009 Int. Conf. on Acoustics, Speech, and Signal Processing*, April 2009, Taipei, Taiwan, pp. 3621-3624.

[90] R. Schroer, Electronic Warfare, *IEEE Aerospace and Electronic Systems Magazine*, Vol. 18 , No.: 7,. *July* 2003, pp. 49-54.

[91] F. A. Butt and M. Jalil, "An Overview of Electronic Warfare in Radar Systems," *IEEE 2013 The International Conference on Technological Advances in Electrical, Electronics and Computer Engineering (TAEECE),* Konya, Turkey, 9-11 May 2013.

[92] NAVAIR Electronic Warfare/Combat Systems, *Electronic Warfare and Radar Systems Engineering Handbook*, 01 Jun. 2012. Report No. NAWCWD TP 8347.

[93] D. Jenn, "Radar Fundamentals," Available: [Online]: https://faculty.nps.edu/jenn/Seminars/RadarFundamentals.pdf

[94] W. Wiesbeck, "Radar Systems Engineering," Lecture Script, 16th Edition, WS 2009/2010, Institut für Hochfrequenztechnik und Elektronik, Karlsruhe Institute of Technology, Germany.

[95] M. I. Skolnik, *Introduction to Radar Systems*, 2nd Edition, McGraw Hill, (1980).

[96] Available: [Online]: https://en.wikipedia.org/wiki/Range_ambiguity_resolution

[97] Available: [Online]: https://en.wikipedia.org/wiki/Range_gate

[98] Available;[Online]: https://en.wikipedia.org/wiki/Electronic_counter-countermeasure

[99] S, R. Park, I. Nam, and S. Noh, "Modeling and Simulation for the Investigation of Radar Responses to Electronic Attacks in Electronic Warfare Environments," *Hindawi, Security and Communication Networks*, Vol. 2018, Article ID 3580536.

[100] Q. J. O. Tan, R. A. Romero, "Jammer-Nulling Transmit-Adaptive Radar Against Knowledge-Based Jammers," *IEEE Access,* Vol. 7, 2019, pp. 181899-181915.

[101] Available : [Online]: http://ntuemc.tw/upload/file/2012060714010872a74.pdf

[102] L. Devlin, S. Glynn, and G. Pearson, "Power Amplifier MMICs For mmWave 5G," Available: [Online]: https://www.rfglobalnet.com/doc/power-amplifier-mmics-for-mmwave-g-0001

[103] J. -. Tsai, F. -. Lin and H. Xiao, "Low RMS phase error 28 GHz 5-bit switch type phase shifter for 5G applications," in *Electronics Letters*, vol. 54, no. 20, pp. 1184-1185.

[104] Available; [Online]: https://www.qorvo.com/products/p/CMD325#parameters

[105] A. A. KIRKPANTUR-ÇADALLI, "An Investigation of Jamming Techniques through a RADAR Receiver Simulation," A Thesis Submitted to the Graduate School of Natural and Applied Sciences of Middle East Technical University, in Partial Fulfillment of the Requirements for the degree of Master of Sciences in Electrical and Electronics Engineering, December 2007.

[106] L. M. Williams and A. S. James, *Principles of Modern Radar: Advanced Techniques*, SciTech Publishing, Edison, NJ, 2013.

[107] M. A. Richards, *Fundamentals of Radar Signal Processing*, McGraw-Hill, New York, 2005.

[108] M. A. Richards, *Principles of Modern Radar : Basic Principles,* SciTech Publishing, Raleigh, NC, 2010.

Index

A

Aerospace, 189, 216, 217
Airborne, 192, 248
Ambiguity, 99, 196, 197, 203, 204, 216
Amplifier, 12–14, 30–32, 34–38, 46, 69, 79, 83, 88, 91, 127, 130–133, 185, 219, 220, 223, 224, 237, 251, 252, 255
Antenna, 2, 4–13, 35, 37, 38, 45–56, 59–65, 70, 93, 101, 102, 127, 128, 155, 162–165, 167, 173, 176, 178–180, 183–187, 190, 192, 193, 199–201, 205, 212, 214, 218–225, 229, 234, 236, 237, 240
Array, 52–60, 65, 176–178, 180, 184, 186, 187, 218–225, 229–231, 233, 234, 236, 242
Attenuator, 219, 220, 222, 223, 242

B

Band, 95, 99, 116, 120, 122, 123, 129, 132, 133, 138, 140, 142, 147, 150, 152, 156, 162, 186, 192, 203, 211, 248
Bandpass, 34, 41, 116, 127, 131, 140
Bandwidth, 15, 22, 24, 29, 39, 69, 70, 74, 75, 77, 92, 93, 95, 104, 106, 113–116, 120, 122, 123, 136, 166, 167, 173, 176, 183, 200, 203, 207, 214, 234, 249, 250
Baseband, 11–13, 15, 17–19, 35, 70, 72, 104–111, 114, 115, 117, 118, 122, 123, 126–128, 131, 133–135, 138, 140, 141, 180, 252
Beam, 48, 49, 52, 59, 60, 150, 178, 180, 200, 201, 204, 205, 209, 214, 217, 223, 229, 232, 234–236, 240, 244, 247
Beamforming, 155, 177, 180, 182, 217, 218, 220, 225, 240
Bistatic, 190, 192, 216

C

Cancelation, 104, 140
Communication, 10–12, 25, 44, 67, 74, 125, 145, 151, 152, 155
Communications, 1, 9–12, 24, 26, 27, 29, 34, 44, 46, 59, 67, 72, 74, 100, 123, 125, 126, 135, 143, 145, 147–149, 151, 216, 217
Compression, 31–33, 36–38, 40, 65, 67, 77–79, 81, 85–87, 101, 190, 213, 214, 216, 223, 224
Conjugately matched, 93
Cross-modulation, 33, 37

D

Decomposition, 155, 164, 167, 168, 182

Detection, 1, 9, 26, 103, 107, 132, 152, 155, 176, 177, 181, 183, 187, 189, 191, 192, 199, 203, 209, 213, 214, 216
Direct-conversion, 100, 129, 130, 133, 139
Double-conversion, 137, 138
Double-sideband, 39, 97, 103, 134
down-conversion, 74, 134–140, 142, 252

E

Eigenvalues, 165, 169, 170, 182
Electromagnetic, 1, 8, 45, 189
Element, 1, 52–60, 155, 205, 218–221, 232, 234–236, 243, 249

F

Frequency-shift, 103, 109, 117
Friis, 60, 61, 65
Function, 11–13, 16–19, 21–25, 30, 33, 37, 40, 44, 45, 48, 54, 55, 57, 58, 60, 65, 69, 76, 77, 86, 88, 111, 114, 118, 126, 133, 137, 138, 140, 141, 158, 160, 163, 164, 174, 179, 189–191, 204, 213, 214, 232
Fundamental, 24, 37, 45, 59, 72, 77–82, 87, 88, 135, 143, 190, 191, 215

J

Jammer, 120, 190, 211–216, 247, 249, 250, 252, 254, 255
Jammer-nulling, 190, 215, 216
Jamming, 120, 190, 209–214, 216, 247–249, 252, 254, 255
Jamming-nulling, 216
Jamming-to-signal, 248

L

Low-pass, 105, 106, 108, 128, 139, 252

M

M-ary, 103, 111, 119, 120, 123
Matrix, 155, 163, 165, 167, 168, 171, 172, 175, 177, 180–187
Maximum ratio, 155, 179
Maximum-likelihood, 152
Maxwell's, 1, 6, 26, 27
Mixer, 13, 14, 35, 37–39, 42, 67, 73, 83, 91, 97–100, 120, 126, 127, 132, 133, 136, 138, 140, 141, 199
Mixing, 73, 80, 97, 126, 127, 131, 133, 138, 139, 142
MMICs, 243
mmWaves, 149–151, 154
Monolithic, 134, 243
MU-MIMO, 177
Multi-user, 152, 153, 177, 182

O

Offset-frequency, 126

P

Passband, 40, 70–72, 83, 86, 87, 203
PathWave, 217
Pattern, 6–9, 27, 47–49, 55, 57, 59, 178, 180, 205, 206, 209, 214, 229, 232, 234–236, 244
Phase-shift, 43, 68, 103, 110, 134
Polyphase, 133, 141
Pulse, 2, 5, 13, 44, 112, 114, 119, 189, 190, 192, 195–197, 200, 204–207, 209, 210, 213, 214, 216, 248, 250
Pulse Doppler, 189, 207, 216

Q

Q-factor, 138
QBPSK, 116

R

Radar, 9, 13–15, 26, 125, 135, 189–197, 199–203, 205–217, 247–251, 253, 255
Radar-target, 189, 216
RADARs, 14, 29, 143, 189–191, 197, 204–206, 209, 210, 215, 216
Radial, 11, 12, 15–17, 48, 202, 249
Radio, 1, 9, 10, 35, 62, 71, 93, 126, 146–148, 177, 186, 189, 191, 209, 211, 216, 217

S

Sawtooth, 207
Second-order, 34, 36, 40, 67, 133
Signal-to-noise, 37, 67, 70, 75, 90, 160, 162, 165, 177, 199
Signals, 9, 10, 24, 29–31, 35, 37, 44, 71–74, 78–80, 127–129, 133, 134, 136, 138, 139, 142, 162, 163, 173, 176–178, 180, 182, 186, 197, 207, 209–211, 214, 219, 222, 223, 249, 255
Simulation, 213, 218–220, 226, 229, 233, 234, 236, 247–249
Super-heterodyne, 73, 127, 129, 132, 137, 139–141, 208
Suppress, 127–129, 131, 133
SystemVue, 101, 217–221, 225, 232, 234, 236, 243, 247, 248

T

Target, 13, 14, 84, 148, 189–195, 197, 199–207, 209, 211, 212, 215, 216, 248, 249, 251, 252, 254, 255
Transmitted, 4, 5, 8, 10–12, 14, 25–27, 35, 44, 60, 61, 109, 120, 126–128, 162, 173, 176, 177, 192, 195, 197, 202, 205–207, 249, 250, 252–254
Transmitter, 1–3, 10–14, 24, 40, 42, 45, 46, 59–62, 64, 67, 68, 100, 161–164, 176, 177, 192, 193, 200, 202, 207, 214, 218–220, 248–253, 255
Tutorial, 217, 218, 221, 225, 236, 240, 247, 249
Tutorials, 59, 217

U

U-cut, 232, 235, 244
Up-conversion, 127, 129–131

V

Velocity, 1, 6, 9, 190, 202–204, 206, 207, 209, 212, 248, 249, 251, 252, 254, 255

W

Wave, 6–13, 22, 26, 27, 45, 46, 51, 52, 133, 189, 191, 207
Waveform, 19, 42, 68, 103, 107, 110, 112, 113, 129, 215, 252
Weaver, 139

Y

Yang, 183

Z

Zero-force, 155
Zero-forcing, 180

About the Author

Héctor J. De Los Santos received the Ph.D. degree in electrical engineering from Purdue University, West Lafayette, IN, in 1989. He founded NanoMEMS Research, LLC, Irvine, CA, a company engaged in Nanoelectromechanical Quantum Circuits and Systems (NEMX) and RF MEMS (NanoMEMS) research, consulting, and education, where he focuses on discovering fundamentally new devices, circuits and design techniques. Prior to founding NanoMEMS in 2002, he spent two years as a Principal Scientist, RF MEMS, at Coventor, Inc., Irvine, CA. From 1989 to 2000, he was with Hughes Space and Communications Company, Los Angeles, CA, where he served as Principal Investigator and the Director of the Future Enabling Technologies IR&D Program. Under this program he pursued research in RF MEMS, quantum functional devices and circuits and photonic bandgap crystal devices and circuits. He holds over 30 U.S., European, German and Japanese patents. His research interests include, theory, modeling, simulation, design and demonstration of emerging devices (electronic, plasmonic, nanophotonic, mechanical systems in the quantum regime, etc.), and wireless communications.

During the 2010-11 academic year he held a German Research Foundation (DFG) Mercator Visiting Professorship at Institute for High-Frequency Engineering and Electronics, Karlsruhe Institute of Technology/University of Karlsruhe, Germany, where his activities included teaching, and conducting research on his DFG-funded project "Nanoelectromechanical Interferometric Tuning with Non-Equilibrium Cooling for Microwave and mm-Wave Electronics". From 2001-2003 he lectured worldwide as an IEEE Distinguished Lecturer of the Microwave Theory and Techniques Society. Since 2006 he has been an IEEE Distinguished Lecturer of the Electron Devices Society. He serves as Reviewer for many Scholarly Journals and Transactions, and served as proposal reviewer for many funding agencies, including, the US National Science Foundation (NSF), the European Science Foundation (ESF), the Australian Research Council (ARC), and the Natural Sciences and Engineering Research Council of Canada (NSERC).

He has been recognized in Marqui's Who'sWho in the World and Marqui's Who'sWho in Science and Engineering. In February, 2020 he was bestowed upon the title of Honorary Professor by Amity University, Noida, Delhi-NCR, Uttar Pradesh, India. He is a member of Tau Beta Pi, Eta Kappa Nu and Sigma Xi. He is an IEEE Fellow.

For Product Safety Concerns and Information please contact our EU representative GPSR@taylorandfrancis.com
Taylor & Francis Verlag GmbH, Kaufingerstraße 24, 80331 München, Germany

www.ingramcontent.com/pod-product-compliance
Ingram Content Group UK Ltd.
Pitfield, Milton Keynes, MK11 3LW, UK
UKHW022210190926
475702UK00003B/281
* 9 7 8 8 7 7 0 2 2 3 7 5 1 *